Computation of Viscous Incompressible Flows

Scientific Computation

For further volumes:
http://www.springer.com/series/718

Dochan Kwak · Cetin C. Kiris

Computation of Viscous Incompressible Flows

 Springer

Dochan Kwak
NASA Ames Research Center
NASA Advanced Supercomputing
    Division
Mail Stop 258-5 Bldg. 258
94035-0001 Moffet Field
USA
dochan.kwak@nasa.gov

Cetin C. Kiris
NASA Ames Research Center
Applied Modeling & Simulations Branch
Mail Stop 258-2 Bldg.258
94035-0001 Moffett Field
USA
cetin.c.kiris@nasa.gov

ISSN 1434-8322
ISBN 978-94-007-0192-2   e-ISBN 978-94-007-0193-9
DOI 10.1007/978-94-007-0193-9
Springer Dordrecht Heidelberg London New York

Springer is part of Springer Science+Business Media (www.springer.com)

To my family:
Soonup Kwak
Sally, Nancy & Brian, Lawrence
and our granddaughter Subin

To my family:
Cahide Kiris
Eren

Foreword

Numerical Simulation of incompressible flows has become an essential tool for studying many important problems in science and engineering, thanks to advances both in numerical methods and computer technology. In life and earth sciences, complicated flow phenomena can be simulated today because of such sophisticated tools. For example, in biology the circulation of blood in the human heart and brain, air in the lungs, and urine in kidneys is the subject of many extensive studies. Similarly, simulations of oil well and oil field flows provide critical information to geologists. Ocean circulation and weather prediction are among the fields that have become dependent on computer simulations. Engineering applications of internal and external incompressible flows are plentiful, including laminar and turbulent flows of pipes, pumps and turbines, hydrofoils, and flow around ships and submarines.

With faster and more powerful computers available every year, scientists and engineers are running numerical simulations of highly sophisticated problems and developing efficient numerical methods. To handle complex geometry, overset grids have proven to be of practical use. Higher order upwinding schemes are used for high Reynolds number flows, and approximate (LU) factorization methods and/or relaxation schemes can be used for both structured and unstructured grids. With these advances, together with enhanced turbulence modeling (algebraic, one- and two-equation models), commercial software today is being applied to a wide spectrum of flow simulation problems.

Historically, numerical simulations of compressible and incompressible flows were based on two different mathematical formulations. For compressible flows, the density and velocity components are updated using the continuity and momentum equations, respectively, and the pressure is calculated from the energy equation together with the equation of state. On the other hand, incompressible flow calculations, where density is constant, are usually based on artificial compressibility or pressure correction methods. In the first approach, the continuity equation is augmented by an artificial, time-dependent term of the pressure, while in the second approach, a Poisson's equation for the pressure is derived by taking the divergence of the momentum equations with the constraint of mass conservation.

In this book, NASA computational fluid dynamics researchers Dochan Kwak and Cetin C. Kiris discuss and analyze these two approaches in detail. Moreover, they introduce a unified approach that is validated for both compressible and incompressible flows using standard benchmark cases.

The authors present many applications, for both laminar and turbulent flows, with an emphasis on practical applications that is clear throughout the book. Three separate chapters are devoted to simulations of liquid propellant rocket engine subsystems, turbopumps, and hemodynamics related to simulation of blood circulation in the human brain and in mechanical heart assist devices.

All calculations presented are based on finite differences or finite volumes, using structured grids. For complex geometries, overset grids are used. In order to obtain steady-state solutions in an efficient manner, several methods of convergence acceleration are included using parallel computations.

Unlike other books on incompressible flow simulations (in particular those based on finite elements), no abstract mathematics, such as functional analysis or tensors, are used in the presentation. The authors appeal to more physical approaches. Based on papers and reports written by the authors and colleagues at NASA and elsewhere over the last two decades, this collection of material is very useful for both researchers and graduate students. The book is easy to read and understand. The only mathematical prerequisites are first-level courses on linear algebra, calculus, and differential equations.

This book is a valuable contribution to the subject of incompressible flow simulations, and I am proud to have collaborated with the authors on numerous projects in this area.

Davis, CA   Mohamed Hafez
March 2010

Acknowledgements

Many colleagues contributed to the material presented in this monograph. We especially thank those who have worked with us at Ames over various periods since the 1980s; Stuart Rogers, Moshe Rosenfeld, Marcel Vinokur, Jeff Housman, Changsung Kim, Jennifer Dacles-Mariani, Seokkwan Yoon, Jong-Youb Sa, and William Chan; a number of researchers from Rocketdyne during the Space Shuttle engine redesign period, including James L.C. Chang, Steve Barson, and Gary Belie—to name just a few; researchers from NASA Marshall Space Flight Center throughout the 1980s and 1990s: Luke Schutzenhofer, Paul McConnaughey, Robert Garcia, Robert Williams, and others.

We are very grateful to Prof. Mohamed Hafez for reviewing the entire draft and writing the Foreword for this monograph. We would like to acknowledge our many years of collaboration with him in developing computational methods for incompressible flow.

There are many others not explicitly listed here—you know who you are—whom we have interacted with during the course of algorithm and applications procedure development. We truly appreciate their cooperation and encouragement throughout the course of our efforts in performing NASA mission-related tasks. We hope this monograph passes some of their ideas on numerical simulation of incompressible flow to the next generation of scientists and engineers working on real-world problems.

We would also like to thank Jill Dunbar for editing multiple drafts and Marco Librero for preparing the final manuscript.

Most notably, DK would like to thank his wife Soonup for her continuous support and encouragement throughout his CFD career over the past several decades. Without her support, completion of this monograph would not have been possible.

Moffett Field, CA   Dochan Kwak
Moffett Field, CA   Cetin C. Kiris
May 2010

Contents

Chapter 1
Introduction

The field of fluid dynamics offers a plentiful source of mathematical, experimental, and computational challenges. The computational approach for viscous incompressible flow analysis is a subset of this rich field, and has been the subject of many books and articles for many decades. This introduction gives readers a summary of the purpose and the scope of this monograph.

1.1 Flow Physics

Even though nearly all fluids are compressible in an absolute sense, incompressible flow approximation can be made when the flow speed is insignificant everywhere in the flow field compared to the speed of sound of the medium. Following this definition of incompressibility, a large number of fluid dynamic problems can be classified as incompressible and, in most cases, viscous. To name a few types of incompressible flows, there are problems related to low-speed aerodynamics, hydrodynamics such as the flow around submerged vehicles, flow through pumps, mixing of the flow in chemical reactors, coolant flow in nuclear reactors, and blood flow in the human body. When the flow is assumed to be incompressible, mathematically the flow field becomes elliptic, which introduces major challenges in computations.

Additional difficulties arise when the flow is viscous. Most notably, complications come from predicting flow physics involving turbulence and transition. In flow problems containing incompressible flow regions, physics involving multi-phase, multi-material, non-Newtonian and stress-supporting media can add complexities to the incompressible flow computation in a broad sense. Another challenge may come from resolving multi-scale dynamics such as those encountered in biomedical applications.

There are several levels of approximations in flow analysis. At a formulation level, the incompressible Navier-Stokes equations are the most commonly accepted governing equations. The "incompressibility" assumption in the governing equations is an approximation for the medium. Other than a small number of laminar flow problems, most problems of fundamental and engineering interest involves transition and turbulence. This introduces the question of how to approximate flow

D. Kwak, C.C. Kiris, *Computation of Viscous Incompressible Flows*, Scientific
Computation, DOI 10.1007/978-94-007-0193-9_1, © US Government 2011

physics to resolve the flow features involved at a reasonable level of accuracy. In most cases, a computational approach offers a viable option for flow analysis. The procedure involving computational modeling, numerical boundary conditions, and algorithms adds another source of approximation requiring assessment of computational accuracy. Therefore, it is important to define the relative contributions from mathematical formulations, physical modeling, and computational procedures to the accuracy of the analysis results.

1.2 History of Computational Approaches

Since the beginning of the computer age, the computational study of fluid dynamic problems has been of interest to researchers studying both fundamental problems and engineering applications. As computer technology progressed, a computational approach became viable to deal with increasingly complicated flow, eventually extending the computational technology to industrial problems. At the same time, flow devices became increasingly sophisticated and highly efficient, pushing the conventional operating envelope. Vast numbers of real-world problems require accurate viscous flow solutions to meet requirements for supporting engineering and science tasks—such as achieving ideal fluid dynamic performances and satisfying cost effectiveness. For example, computational analysis is indispensable, as well as economical, for developing advanced rocket-engine turbopumps and biomedical devices handling blood flow in humans. It therefore became of practical interest to have advanced computational capabilities for simulating these flow problems. Computational tools are used not only as an alternative or a complementary means to analytical or empirical approaches, but as a primary basis for preliminary engineering designs and design optimization.

The demand for advanced methods and new tools prompted a flood of research on numerical methods, flow solvers, and validation experiments. To solve fluid dynamics problems using these methods and tools requires algorithmic simplifications as well as geometric modeling. Furthermore, significant physical modeling is also required, such as turbulence modeling for high-Reynolds number flows and non-Newtonian modeling for blood flow. Accuracy of the numerical solution of these flows—especially in three dimensions—needs to be assessed in terms of errors and uncertainties involving numerical, geometric, and physical modeling. So, numerical computation of these flows, especially real-world problems in three dimensions, is often called "simulation" in the literature.

The computational fluid dynamics (CFD) for viscous, incompressible flow has been of interest for many decades to investigate fundamental fluid dynamic problems as well as engineering applications. The pioneering work by Harlow and Welch (1965) opened a new possibility of applying a computational approach to solving realistic incompressible fluid engineering problems, especially for three-dimensional problems. Their method of using pressure as a mapping parameter has been further developed into many variations ever since. Most notably, Patankar and Spalding (1972) used this approach to develop algorithms and tools especially useful for engineering applications with heat transfer. Shortly after Harlow and

Welch's Marker-and-Cell method was introduced to the incompressible flow community, Chorin (1967) proposed an artificial compressibility approach that enables the use of a wide spectrum of algorithms developed for compressible flow analysis. Some details of these two approaches are discussed in this monograph (Chapters 2, 3 and 4).

When CFD became a viable option for engineering in early 1970s, vast amounts of numerical methods and analysis were already available. Many books and reviews were published in the 1970s and 1980s, and a handful has been added since then, covering a comprehensive collection of various approaches and methods. Evolutionary advances and improvements in CFD methods have been made through the 1990s and beyond. Flow solvers developed in the 1970s and 1980s advanced to software-level CFD tools, and these became available commercially. Students and researchers in academia began using commercial codes rather than developing their own versions.

A vast number of books on CFD have been published, especially in the late 1990s. These cover mathematical formulations, numerical algorithms, and flow simulation procedures as a whole. For more comprehensive review of computational methods in general, readers are referred to these authors, to name a few: Roach (1972, 1998), Peyret and Taylor (1983), Tannehill et al. (1984, 1997), Hirsch (1988), Gunzburger and Nicolades (1993), Hafez and Oshima (1995, 1998), Gresho and Sani (1998), Lomax et al. (2001), Ferziger and Peric (2002), Hafez (2002), Chung (2002), and Drikakis and Rider (2004).

Methods for incompressible flows are usually included as special cases in these books. In addition to methods cited later in conjunction with those discussed in this monograph, various methods specific to incompressible flow computations are reported by numerous authors, for example: Ghia et al. (1977), Raithby and Schneider (1979), Leonard (1985), Guerran and Gustafsson (1986), Abdallah (1987a, b), Wesseling et al. (1992), Hafez and Soliman (1993), Chen et al. (1995), Turek (1999), Hafez (2001), Brown (2002), Glowinski et al. (2002), Gustafsson et al. (2002), Khosla and Rubin (2002), Kuwahara et al. (2002), Loner et al. (2002), Morgan et al. (2002), Nikfetrat and Hafez (2002), Satofuka et al. (2002), Tezduyar (2002), and Wendl and Agarawal (2002). While certainly not exhaustive, this list illustrates the abundance of literature addressing various aspects of the fundamentals of CFD and incompressible flows.

One might then question "why another book on computation of incompressible flow?" In the current environment, it is of crucial importance for users of existing codes to understand the algorithmic characteristics and underlying assumptions used in modeling flow physics. Users need to understand engineering issues at hand and identify what needs to be resolved through numerical simulations. Even in one code, various algorithm options and physical models are generally available, and users need to choose the most suitable procedure for the problem being solved. It is difficult to provide a universal guideline to general problem solving. However, realistic examples for illustrating the process will offer valuable information not readily available from existing numerical methods books. In this monograph, we illustrate "best practices" in solving viscous incompressible flow problems using real-world problems.

1.3 Scope of this Monograph

Back in the early 1980s, the author's major interests were roused in conjunction with industrial applications and mission support activities requiring viscous incompressible flow analysis. Specifically, our activity began with a mission task to resolve issues involving the Space Shuttle propulsion system. At that time, the available supercomputing power was not much more than that offered in later desktop computers of the early 2000s. However, CFD simulation of a complicated rocket propulsion system, when combined with engineering ideas, could make significant contributions to retrofitting a liquid-propellant rocket engine. Since then, both computing power and simulation technology have been much advanced, to the point that almost every organization has access to large-scale computers for scientific and mission computing, and CFD software has become available either through local development or through vendors. When we started with our NASA mission tasks in the early 1980s, our own applications codes had to be developed.

Our first step was to develop numerical methods and simulation tools, followed by implementation of these to the analysis of the flight engine configuration. The problem involved both complex geometry and complex flow physics. Therefore, application of these tools added another level of approximations in conjunction with geometry modeling, an engineering model of the real system including truncated boundary conditions, and flow physics modeling. These engineering tasks required deep understanding of approximations made for flow physics modeling as well as numerical methods and solvers. The tasks of generalizing and establishing guidelines for applications are difficult and often require expertise and experience to solve particular, real issues at hand. The authors have experienced these challenges over many years of developing and utilizing CFD tools, especially related to viscous incompressible flows, in order to make timely impacts on various projects related to NASA mission.

The process of utilizing CFD as an engineering tool to design or improve aerospace vehicles and to develop safe operational procedures is problem dependent and cannot be easily standardized. Simulation of incompressible flow shares many of the same fundamental algorithmic needs common to all fluid flow problems. However, there are features specific to incompressible flow. Basic CFD methods are referred to in existing literature cited throughout this monograph, so that issues relevant to *incompressible flow problem solving* can be addressed in more detail. To keep the flow of thought more self-contained, some basics specifically pertaining to incompressible flow computations are included in this monograph. The simulation details for mission computing are then presented to illustrate the entire gamut of procedures and issues. Among the vast number of application problems, limited samples—primarily from the author's experience—have been selected to illustrate different types of flow issues.

The material presented here illustrates some aspects of incompressible flow simulation not generally covered in existing CFD textbooks, and yet often encountered in practice. This monograph is, therefore, intended primarily to give a concise guide to practitioners and graduate students for applying CFD approaches to real-world problems requiring quantification of viscous incompressible flows. Although the procedural details are given with respect to particular tasks, they are very relevant to many other problems in fluid engineering. Therefore, issues in applying CFD to engineering problems are discussed extensively, assuming that readers can get CFD basics from those references cited.

The design of this monograph is as follows: Chapter 2 gives a brief review of existing methods using primitive variables. In Chapter 3, the projection method is explained in some detail, followed by a description of an artificial compressibility method in Chapter 4, which also presents a unified formulation for obtaining incompressible flow solution from compressible flow formulation. Chapter 5 contains various validation computations using fundamental problems; this is to illustrate various issues related to different flow characteristics often encountered in solving real-world problems. The remainder of the book is devoted to mission applications: liquid-propellant rocket engine subsystem in Chapter 6, turbopumps in Chapter 7, and hemodynamics and human modeling in Chapter 8.

Chapter 2
Methods for Solving Viscous Incompressible Flow Problems

In this chapter, numerical solution approaches for viscous incompressible flow are briefly compared. Detailed discussions of each approach follow in separate chapters. All discussions are from an engineering perspective and mathematical formalities are not emphasized, in keeping with our perspective for this monograph that CFD is an engineering tool for supporting mission tasks, and thus one can implement well-founded numerical algorithms and physical models to resolve engineering issues at hand. To make significant impacts on missions such as aerospace vehicle design and operation, the CFD applications procedure is just as important as tool development. Here, we present a quick summary of numerical approaches most suitable for application to tasks for supporting missions, especially space exploration missions.

2.1 Overview

The Navier-Stokes equations are generally accepted as the equations governing the flow of Newtonian fluid in a continuum regime. Mathematically, the compressible flow equations become singular at the limit where the speed of sound of the medium becomes infinity or the flow speed becomes insignificant relative to the speed of sound. This singular nature of the governing equations poses the primary difficulty in solving incompressible Navier-Stokes equations. Physically, the challenge is maintaining incompressibility during iterative processes for obtaining steady-state solutions or at each time level for computing time-accurate solutions. Several methods have been developed in the past (see references cited in Section 1.2) where the main differences among various approaches come from the way in which the incompressibility condition is satisfied computationally.

A traditional approach is to start the computational process directly from an incompressible Navier-Stokes formulation. The primary concern is then how to satisfy the continuity equation. One can use the primitive variables, namely, pressure and velocities or derived quantities like stream function-voritcy and vorticity-velocity. For general three-dimensional problems, primitive variable formulation poses the least complications in imposing physical boundary conditions.

D. Kwak, C.C. Kiris, *Computation of Viscous Incompressible Flows*, Scientific Computation, DOI 10.1007/978-94-007-0193-9_2, © US Government 2011

The summary of methods discussed in this chapter is an outgrowth of the review given by one of the authors in 1989 as a Von Karman Institute Lecture note. Some details of derivation of the equations and algorithms are given here, as well as the physical interpretation of the solution procedures. A large amount of mission support work has also been performed since the lecture notes were first assembled. The real-world application experiences show that computational approaches for obtaining engineering solutions require in-depth understanding of flow phenomena as well as the relative importance of various engineering aspects involved in the task, which involve far beyond algorithms and software issues. The mathematical foundation for several well-known approaches for solving incompressible flow equations is outlined next.

2.2 Mathematical Models

Three-dimensional incompressible flow with constant density is governed by the following Navier-Stokes equations:

$$\frac{\partial ui}{\partial xi} = 0 \tag{2.1}$$

$$\frac{\partial u_i}{\partial t} + \frac{\partial (u_i u_j)}{\partial x_j} = -\frac{\partial p}{\partial x_i} + \frac{\partial \tau_{ij}}{\partial x_j} \tag{2.2}$$

where t is the time, xi the Cartesian coordinates, ui the corresponding velocity components, p the pressure, and τ_{ij} the viscous-stress tensor. Here, all variables are non-dimensionalized by a reference velocity and length scale. The viscous stress tensor can be written as follows.

$$\tau_{ij} = 2v S_{ij} \tag{2.3a}$$

$$S_{ij} = \frac{1}{2}\left(\frac{\partial u_i}{\partial x_j} + \frac{\partial u_j}{\partial x_i}\right) \tag{2.4}$$

Here, v is the kinematic viscosity, and S_{ij} is the strain-rate tensor. Note that the bulk viscosity is reduced to "so-called" shear viscosity in the case of incompressible flow. Using conventional Reynolds decomposition, the viscous stress tensor can be written as:

$$\tau_{ij} = 2v S_{ij} - R_{ij} \tag{2.3b}$$

where, R_{ij} is the Reynolds stresses, and Equation (2.3b) represents Reynolds-averaged incompressible Navier-Stokes equations. Various levels of closure models for R_{ij} are possible. In this discussion on solution approaches, turbulence is simulated by an eddy viscosity model using a constitutive equation of the following form:

$$R_{ij} = \frac{1}{3}R_{kk}\delta_{ij} - 2\nu_T S_{ij} \tag{2.5}$$

where ν_t is the turbulent eddy viscosity. By including the normal stress, R_{kk}, in the pressure, ν in Equation (2.3b) can be replaced by $(\nu + \nu_T)$ as follows:

$$\tau_{ij} = 2\left(\nu + \nu_t\right)S_{ij} = 2\nu_T S_{ij} \tag{2.6}$$

For the remainder of this chapter, the total viscosity, ν_T, will be represented simply by ν. Therefore, in discussing solution methods and algorithms, the incompressible Navier-Stokes equations are modified to allow variable viscosity in the present formulation. The flow solvers described here include turbulent flow cases as long as turbulence is represented by an eddy viscosity model. Note that in many real cases, this assumption may not hold, causing inaccuracies in the solution. The sensitivity of the turbulence modeling on solution accuracy and convergence is problem dependent. Turbulence modeling on a higher level than eddy viscosity has been researched for many decades, and continues to be a research area. In lieu of any general recommendations, application procedures are discussed using an eddy viscosity assumption. Therefore, for flows involving non-equilibrium turbulence, the eddy viscosity formulation needs to be revisited.

One avenue worth mentioning for resolving this non-equilibrium, time-dependent turbulent flow is large eddy simulation (LES). The LES approach limits the modeling to small-scale (sub-grid scale) motion, where some degree of universality can be assumed. This approach was first used in meteorological flow simulation and later (starting in the 1970s) extended to general engineering applications. The usefulness of the LES method has been fairly limited, to date. A full account of this approach requires a separate discussion, starting with the basic formulation. There are merits to applying LES to practical problems, but they are not included here, since it is difficult to resolve the boundary layer region and requires relatively large amounts of computing time for solving realistic geometry and Reynolds numbers. Turbulence modeling issues in solving engineering problems will be discussed further, in conjunction with our work on liquid-propellant rocket propulsion system, in Chapter 6.

2.3 Formulation for General Geometry

In order to numerically solve the governing equations involving general geometry, one commonly needs to map the entire flow field using a grid system such as a Cartesian grid, body-conforming curvilinear structured grid, or an unstructured grid like a triangular or tetrahedral grid. Special techniques exist that do not require a grid system, but those grid-free approaches do not offer any general advantages over the grid-based methods, and will not be discussed here. By mapping the coordinate system to a general coordinate, it becomes easier to handle complex geometry. One advantage of body-conforming coordinates is the ability to follow the surface of an object to follow the boundary layer. For discretization of the governing equations, such as finite difference or finite volume, it is convenient to transform the governing equations into general coordinates.

Here, transformation of coordinate system will be presented first, with a discussion of ways to directly integrate the governing equations in general coordinates handled in a later chapter.

To perform calculations on 3-D arbitrarily shaped geometries, the following generalized independent variables are introduced, which transform the physical coordinates (x, y, z) into general curvilinear coordinates (ξ, η, ζ).

$$\xi = \xi(x, y, z, t)$$
$$\eta = \eta(x, y, z, t)$$
$$\zeta = \zeta(x, y, z, t)$$

The Jacobian of the transformation is defined as:

$$J = \det \frac{\partial(\xi, \eta, \zeta)}{\partial(x, y, z)} = \begin{vmatrix} \xi_x & \xi_y & \xi_z \\ \eta_x & \eta_y & \eta_z \\ \zeta_x & \zeta_y & \zeta_z \end{vmatrix} \tag{2.7}$$

where $\xi_x = \dfrac{\partial \xi}{\partial x}$, $\xi_y = \dfrac{\partial \xi}{\partial y}$, etc.

In actual coding, metric terms can be calculated as follows.

$$\begin{pmatrix} \xi_x \\ \xi_y \\ \xi_z \end{pmatrix} = \frac{1}{J'} \begin{pmatrix} y_\eta z_\zeta - y_\zeta z_\eta \\ x_\zeta z_\eta - x_\eta z_\zeta \\ x_\eta y_\zeta - x_\zeta y_\eta \end{pmatrix}, \begin{pmatrix} \eta_x \\ \eta_y \\ \eta_z \end{pmatrix} = \frac{1}{J'} \begin{pmatrix} y_\zeta z_\xi - y_\xi z_\zeta \\ x_\xi z_\zeta - x_\zeta z_\xi \\ x_\zeta y_\xi - x_\xi y_\zeta \end{pmatrix}, \begin{pmatrix} \zeta_x \\ \zeta_y \\ \zeta_z \end{pmatrix} = \frac{1}{J'} \begin{pmatrix} y_\xi z_\zeta - y_\zeta z_\xi \\ x_\eta z_\xi - x_\xi z_\eta \\ x_\xi y_\eta - x_\eta y_\xi \end{pmatrix} \tag{2.8}$$

$$J' = \det \frac{\partial(x, y, z)}{\partial(\xi, \eta, \zeta)} = \begin{vmatrix} x_\xi & x_\eta & x_\zeta \\ y_\xi & y_\eta & y_\zeta \\ z_\xi & z_\eta & z_\zeta \end{vmatrix} \tag{2.9}$$

Applying the transformation to Equations (2.1) and (2.2) yields the following governing equations in general curvilinear coordinates, (ξ, η, ζ):

$$\frac{\partial}{\partial t} \hat{u} = -\frac{\partial}{\partial \xi} \left(\hat{e} - \hat{e}_v \right) - \frac{\partial}{\partial \eta} \left(\hat{f} - \hat{f}_v \right) - \frac{\partial}{\partial \zeta} \left(\hat{g} - \hat{g}_v \right)$$

$$= -\frac{\partial}{\partial \xi_i} \left(\hat{e}_i - \hat{e}_{vi} \right) = -\hat{r} \tag{2.10}$$

$$\frac{\partial}{\partial \xi} \left(\frac{U - \xi_t}{J} \right) + \frac{\partial}{\partial \eta} \left(\frac{V - \eta_t}{J} \right) + \frac{\partial}{\partial \zeta} \left(\frac{W - \zeta_t}{J} \right)$$

$$= \frac{\partial}{\partial \xi_i} \left(\frac{U_i - (\xi_i)_t}{J} \right) = 0 \tag{2.11}$$

where $\xi_i = \xi$, η, or ζ for $i = 1, 2$ or 3

$$\hat{u} = \frac{1}{J}\begin{bmatrix} u \\ \upsilon \\ w \end{bmatrix}$$

$$\hat{e} = \frac{1}{J}\begin{bmatrix} \xi_x p + uU \\ \xi_y p + \upsilon U \\ \xi_z p + wU \end{bmatrix}, \quad \hat{f} = \frac{1}{J}\begin{bmatrix} \eta_x p + uV \\ \eta_y p + \upsilon V \\ \eta_z p + wV \end{bmatrix}, \quad \hat{g} = \frac{1}{J}\begin{bmatrix} \zeta_x p + uW \\ \zeta_y p + \upsilon W \\ \zeta_z p + wW \end{bmatrix} \qquad (2.12)$$

$$U = (\xi)_t + (\xi)_x u + (\xi)_y \upsilon + (\xi)_z w$$
$$V = (\eta)_t + (\eta)_x u + (\eta)_y \upsilon + (\eta)_z w$$
$$W = (\zeta)_t + (\zeta)_x u + (\zeta)_y \upsilon + (\zeta)_z w$$

The viscous terms are quite lengthy; for the benefit of new practitioners, the fully expanded version is given next.

$$\frac{\partial \tau_{ij}}{\partial x_j} = \frac{\partial}{\partial x_j} 2\nu S_{ij}$$

$$= \frac{\partial}{\partial x}\nu\begin{bmatrix} u_x + u_x \\ \upsilon_x + u_y \\ w_x + u_z \end{bmatrix} + \frac{\partial}{\partial y}\nu\begin{bmatrix} u_y + \upsilon_x \\ \upsilon_y + \upsilon_y \\ w_y + \upsilon_z \end{bmatrix} + \frac{\partial}{\partial z}\nu\begin{bmatrix} u_z + w_x \\ \upsilon_z + w_y \\ w_z + w_z \end{bmatrix}$$

When ν is constant, the contribution of the second terms in the bracket sum up to be zero for incompressible flow, since the velocity field is divergence free. However, in general, ν varies in space and time, such as in the case of eddy viscosity formulation, and so these terms must be kept. Then, the viscous terms in transformed coordinates are as follows.

$$\hat{e}_\upsilon = \frac{\nu}{J}\left\{ \nabla\xi \cdot \left(\nabla\xi\frac{\partial}{\partial\xi} + \nabla\eta\frac{\partial}{\partial\eta} + \nabla\zeta\frac{\partial}{\partial\zeta}\right)\begin{bmatrix} u \\ \upsilon \\ w \end{bmatrix} + \left(\xi_x\frac{\partial u}{\partial\xi_i} + \xi_y\frac{\partial \upsilon}{\partial\xi_i} + \xi_z\frac{\partial w}{\partial\xi_i}\right)\begin{bmatrix} \frac{\partial}{\partial x}\xi_i \\ \frac{\partial}{\partial y}\xi_i \\ \frac{\partial}{\partial z}\xi_i \end{bmatrix}\right\}$$

$$\hat{f}_\upsilon = \frac{\nu}{J}\left\{ \nabla\eta \cdot \left(\nabla\xi\frac{\partial}{\partial\xi} + \nabla\eta\frac{\partial}{\partial\eta} + \nabla\zeta\frac{\partial}{\partial\zeta}\right)\begin{bmatrix} u \\ \upsilon \\ w \end{bmatrix} + \left(\eta_x\frac{\partial u}{\partial\xi_i} + \eta_y\frac{\partial \upsilon}{\partial\xi_i} + \eta_z\frac{\partial w}{\partial\xi_i}\right)\begin{bmatrix} \frac{\partial}{\partial x}\xi_i \\ \frac{\partial}{\partial y}\xi_i \\ \frac{\partial}{\partial z}\xi_i \end{bmatrix}\right\}$$

$$\hat{g}_\upsilon = \frac{\nu}{J}\left\{ \nabla\zeta \cdot \left(\nabla\xi\frac{\partial}{\partial\xi} + \nabla\eta\frac{\partial}{\partial\eta} + \nabla\zeta\frac{\partial}{\partial\zeta}\right)\begin{bmatrix} u \\ \upsilon \\ w \end{bmatrix} + \left(\zeta_x\frac{\partial u}{\partial\xi_i} + \zeta_y\frac{\partial \upsilon}{\partial\xi_i} + \zeta_z\frac{\partial w}{\partial\xi_i}\right)\begin{bmatrix} \frac{\partial}{\partial x}\xi_i \\ \frac{\partial}{\partial y}\xi_i \\ \frac{\partial}{\partial z}\xi_i \end{bmatrix}\right\}$$

$$(2.13a)$$

The above fully expanded terms can be simplified if an orthogonal coordinate system is used.

$$\nabla \xi_i \cdot \nabla \xi_j = 0; \quad \text{for } i \neq j$$

For constant v (that is, constant density laminar flow), the contribution by the second group of terms in the parentheses of Equation (2.13a) is zero. Therefore, for flows with constant v in orthogonal coordinates, the full viscous term in (2.13a) can be simplified as shown below.

$$\hat{e}_v = \left(\frac{v}{J}\right)\left(\xi_x^2 + \xi_y^2 + \xi_z^2\right)\begin{bmatrix} u_\xi \\ v_\xi \\ w_\xi \end{bmatrix}$$

$$\hat{f}_v = \left(\frac{v}{J}\right)\left(\eta_x^2 + \eta_y^2 + \eta_z^2\right)\begin{bmatrix} u_\eta \\ v_\eta \\ w_\eta \end{bmatrix} \quad (2.13b)$$

$$\hat{g}_v = \left(\frac{v}{J}\right)\left(\zeta_x^2 + \zeta_y^2 + \zeta_z^2\right)\begin{bmatrix} u_\zeta \\ v_\zeta \\ w_\zeta \end{bmatrix}$$

Satisfying the mass conservation equation, where the pressure term is decoupled from the momentum equations, is the primary issue in solving the above set of equations. Physically, incompressible flow is characterized by elliptic behavior of the pressure waves, the speed of which, in a truly incompressible medium is infinite. The pressure field is wanted as a part of the solution. However, the pressure condition has to be imposed on the boundary for numerical computation that poses the difficulty in designing numerical boundary conditions for incompressible flow computations.

Instead of using primitive variables, one can utilize other formulations using derived quantities, such as vorticity, to eliminate the pressure from the boundary conditions; however, this is at the expense of introducing boundary conditions for the derived variables. In realistic 3-D problems, these derived quantities are difficult to define or impractical to use. The primitive variable formulation, namely, using pressure and velocities as dependent variables, then becomes more convenient and flexible in 3-D applications. Keep in mind that in this formulation, the pressure solver has to be designed to satisfy mass conservation while achieving computational efficiency in obtaining the pressure-field solution. Various techniques have been developed in the past, none of which have proven to be universally better than another.

2.4 Overview of Solution Approaches

Depending on the flow features to quantify, the solution method of choice can vary. For flows involving thin viscous layers, it is advantageous to have large time steps, possibly using an implicit method. For time accurate solutions, the physical time step required to resolve unsteady motion could be very small, in which case, explicit schemes can be used effectively. For spatial differencing, the usual finite difference and finite volume schemes can be implemented along with central or upwind differencing. Stable central differencing schemes need dissipation terms, while upwind schemes include dissipation automatically in the differencing. The addition of dissipation effectively lowers the Reynolds number of the flow. In addition, dissipation is affected by grid effect and turbulence.

Grid topology and the "goodness" of grid can affect the solution accuracy in significant ways—not only from a numerical dissipation point of view but also in designing boundary conditions. Requirements on grid density and distribution are also realistic factors affecting the order of differencing schemes. All these factors should be considered in developing a flow solver and implementation guidelines. Solution procedures discussed below represent different approaches but are not unique combinations of these methods. A quick overview of the different approaches is discussed below to give readers a general outline of methods commonly used to date. A more detailed discussion on production-oriented methods will be discussed in separately in later chapters.

2.4.1 Pressure-Based Method

The basic idea of this method is to solve the pressure field such that a divergence-free velocity field is maintained at every time step. This approach was first started with the marker-and-cell (MAC) method discussed below, followed by proliferation in simplified and generalized forms. Several variations of this method have been developed as computer speed and the numerical methods have advanced.

2.4.1.1 MAC Method

The MAC Method was probably the first primitive variable method for incompressible flow using a derived Poisson equation for pressure to satisfy mass conservation. The first paper written on this approach was published by Harlow and Welch from Los Alamos National Laboratory in 1965. They called this the marker-and-cell (MAC) method, in which the pressure is used as a mapping parameter to satisfy the continuity equation. This laid the foundation for subsequent variations of the pressure-based method by many researchers. The MAC method can be viewed as a special case of the pressure projection method or—from the operator point of view—one variation of the fractional step method. A pressure projection method in general coordinates, using a fractional step approach, will be discussed in detail in Chapter 3.

In a fractional-step procedure, the time-dependant governing equations can be solved in several steps, which can be convenient for time-dependent computations of the incompressible Navier-Stokes equations (see Chorin, 1968; Yanenko, 1971; Marchuk, 1975). In this procedure, the time evolution of the flow field can be approximated through several steps. Operator splitting can be accomplished in several ways by treating the momentum equations as a combination of convection, pressure, and viscous terms. The common application of this method to incompressible Navier-Stokes equations is done in two steps.

The first step is to solve for an auxiliary velocity field using the momentum equation, in which the pressure-gradient term can be computed from the pressure in the previous time step (for example, Dwyer et al., 1986) or can be excluded entirely in this step. In the second step, the pressure is computed, which can map the auxiliary velocity onto a divergence-free velocity field.

The concept of this approach is illustrated by the following simplified example:

Step 1: Using the following form of the momentum equations (2.2):

$$\frac{\partial u_i}{\partial t} = -\frac{\partial p}{\partial x_i} - \frac{\partial(u_i u_j)}{\partial x_j} + \frac{\partial \tau_{ij}}{\partial x_j} = -\frac{\partial p}{\partial x_i} + R_i \qquad (2.2a)$$

Calculate the auxiliary or intermediate velocity, $\hat{u}_i$, for example by:

$$\frac{\hat{u}_i - u^n}{\Delta t} = -\frac{\delta p^n}{\delta x_i} + R_i^n \qquad (2.14)$$

Step 2: Solve for the pressure correction.

$$\frac{u^{n+1} - \hat{u}_i}{\Delta t} = -\frac{\delta\left(p^{n+1} - p^n\right)}{\delta x_i} \qquad (2.15)$$

In the second step the pressure correction is computed. To minimize the pressure correction in the next time step, the pressure gradient term at the previous time step can be included. The velocity field has to satisfy the following continuity equation.

$$\nabla \cdot u^{n+1} = 0 \qquad (2.16)$$

The above Equation (2.16), can be satisfied by combining Equations (2.14) and (2.15), resulting in the following Poisson equation for pressure.

$$\nabla^2\left(p^{n+1} - p^n\right) = \frac{1}{\Delta t}\frac{\delta \hat{u}_i}{\delta x_i} \qquad (2.17)$$

Once the pressure correction is computed, new velocities can be obtained.

$$\frac{u^{n+1} - u^n}{\Delta t} = -\nabla p^{n+1} + R_i \qquad (2.18)$$

Many researchers have used essentially the same procedure shown above. One particular aspect of the fractional-step method requiring special care is that of intermediate boundary conditions. Orszag et al. (1986) discussed this extensively. As

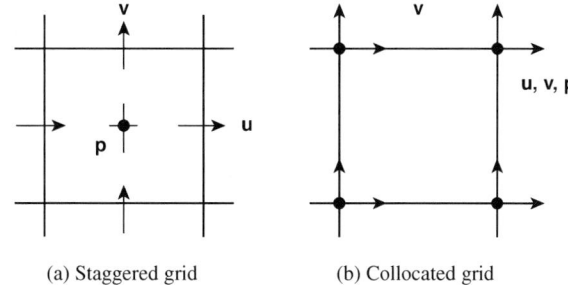

Fig. 2.1 Staggered vs. collocated variable arrangements on a Cartesian grid. **a** Staggered grid. **b** Collocated grid

(a) Staggered grid   (b) Collocated grid

will be explained in Chapter 3, Rosenfeld et al. (1988) and Kiris and Kwak (2001) devised a generalized scheme where physical boundary conditions can be used at intermediate steps.

The original MAC method is based on a staggered-grid arrangement on 2-D Cartesian coordinates (Fig. 2.1). The staggered grid arrangement conserves mass, momentum, and kinetic energy in a natural way and avoids odd-even point decoupling of the pressure encountered in a regular collocated grid (Gresho and Sani, 1987). A differencing method using the staggered-grid arrangement is essentially a finite-volume discretization while the regular grid produces a finite difference form of discretization. As will be discussed in more detail in Chapter 3, the differences between these two approaches in Cartesian coordinates become unclear when using generalized curvilinear coordinates. A more complete presentation of these grid arrangements can be found in Ferziger and Peric (2002), Abdallah (1987a, b), and Roach (1998).

The major drawback of the MAC method is the large amount of computing time required for solving the Poisson equation for pressure. When the physical problem requires a very small time step, the penalty paid for an iterative solution procedure for the pressure may be tolerable. But the method as a whole is slow and the pressure boundary condition is difficult to specify. Although the original method used an explicit Euler solver, various time advancing schemes can be implemented here. Since its introduction, numerous variations of the MAC method have been devised and successful computations have been made. Many more examples can be found in the literature, for example, Ferziger (1987), and Orszag and Israelli (1974).

2.4.1.2 Pressure Field Solution for MAC-Type Method

One important aspect of the numerical solution of the Poisson equation for pressure is tied to the spatial differencing of the second derivatives. To satisfy the mass conservation in grid space, the difference form of the second derivative in the Poisson equation has to be constructed consistent with the discretized momentum equation (Kwak, 1989). To explain this intuitive comment in a more convincing way (primarily for those not experienced with incompressible flow) the following comparison of various discretization approaches is given.

Let's compare the different methods for discretizing the Poisson equation for pressure.

The equation for pressure is obtained by taking the divergence of the momentum equation, as shown below:

$$\nabla^2 p = \frac{\partial h_i}{\partial x_i} - \frac{\partial}{\partial t}\frac{\partial u_i}{\partial x_i} = g \qquad (2.19)$$

where

$$h_i = -\frac{\partial u_i u_j}{\partial x_j} + \frac{\partial \tau_{ij}}{\partial x_j}$$

Three methods for solving the pressure field are compared, below.

Method 1:

First, an exact form of the Laplacian operator is used in solving Equation (2.19). The Fourier transform of Equation (2.19) is:

$$-k^2 \hat{p} = \hat{g}' \qquad (2.20a)$$

where

$\hat{p} =$ Fourier tranform of p
$k^2 = k_x^2 + k_y^2 + k_z^2$
$k_x^2, k_y^2, k_z^2 =$ wave numbers in the x-, y-, and z-directions
$g' =$ finite difference approximation to g
$\hat{g}' =$ Fourier transform of g'

The wave number, k_i, in discrete Fourier expansion is defined as below.

$k_i = \dfrac{2\pi}{N\Delta}n=$ wave number in the x_i-direction
$n = -\dfrac{N}{2},\ldots 0, 1, \ldots \left(\dfrac{N}{2} - 1\right)$
$\Delta =$ mesh space
$N =$ number of mesh points in the x_i-direction

By inverse transformation, p can be obtained.

Method 2:

A second approach is to use the difference form of the second derivatives in Equation (2.19), as below.

$$\left(\frac{\delta^2}{\delta x^2} + \frac{\delta^2}{\delta y^2} + \frac{\delta^2}{\delta z^2}\right)p = g'$$

The Fourier transform of the above equation is:

$$- \tilde{k}_i \tilde{k}_i \hat{p} = \hat{g}'$$
(2.20b)

where $\tilde{k}_i$ is the Fourier transform of the difference form of the second derivative.

Method 3:

The finite difference form of the governing equations is:

$$\frac{\delta u_i}{\delta x_i} = Du_i = 0$$
(2.1')

$$\frac{\delta u_i}{\delta t} - h'_i = \frac{\delta p}{\delta x_i} = Gp$$
(2.2')

where h'_i is the finite difference form of h_i. By applying the divergence operator, D, to the above equations, the following difference form of the Poisson equation is obtained.

$$DGp = \frac{\delta Du_i}{\delta t} - Dh'_i = g'_i$$

Then, taking the Fourier transform of the above equation:

$$- k'_i k'_i \hat{p} = \hat{g}'$$
(2.20c)

Since the governing equations are solved in difference form, the above three methods are compared in solving Equations (2.1') and (2.2'). The Fourier transform of Equation (2.21) is as below.

$$\frac{\delta \hat{u}_i}{\delta t} - \hat{h}'_i = -k'_i \hat{p}$$
(2.21)

To satisfy the continuity equation in grid space, the following equation has to be satisfied for a flow that has $Du_i = 0$ at the beginning.

$$\frac{\delta D}{\delta t} = 0$$

In Fourier space, this is equivalent to $k'_i u_i = 0$ at the first and next time steps. Substituting $\hat{p}$ from the above three methods into Equation (2.21), the following results are obtained.

For Method 1:

$$\frac{\delta}{\delta t} \left(k'_i \hat{u}_i \right) = \left(\hat{h}'_i - \frac{k'_i k'_j}{k^2} \hat{h}'_j \right) k'_i \neq 0$$

For Method 2:

$$\frac{\delta}{\delta t}\left(k'_i \hat{u}_i\right) = \left(\hat{h}'_i - \frac{k' k'_j}{\tilde{k}^2}\hat{h}'_j\right) k'_i \neq 0$$

For Method 3:

$$\frac{\delta}{\delta t}\left(k'_i \hat{u}_i\right) = \left(\hat{h}'_i - \frac{k' k'_j}{k'^2}\hat{h}'_j\right) k'_i = 0$$

The source of error introduced by different choices of the ∇^2 operator in discretized space can be seen by observing the magnitude of k^2, $\tilde{k}^2$ and k'^2.

For the purpose of comparison, modified wave numbers, $\tilde{k}^2$ and k'^2 are given below, where five-point fourth-order central differencing is applied.

$$(\tilde{k}_i)^2 = \frac{1}{6\Delta^2}\left[15 - 16\cos(\Delta k_i) + \cos(2\Delta k_i)\right]$$

$$(k'_i)^2 = \frac{1}{72\Delta^2}\left[65 - 16\cos(\Delta k_i) - 64\cos(2\Delta k_i) - \cos(4\Delta k_i)\right]$$

(2.23)

The magnitude of these quantities is compared in Fig. 2.2. Method 3 satisfies the continuity equation at the next time step in grid space, and should be used for the pressure field solution. Therefore, in the solution method using a Poisson equation for pressure, the divergence gradient (DG) operator plays an important role in satisfying the mass conservation in grid space. As will be explained later in Chapter 4, this requirement can be relaxed in an artificial compressibility approach.

$$k_1^2 = \text{exact}, \quad \tilde{k}_1^2 = \text{fourth order } \frac{\delta^2}{\delta x_1^2}, \quad k_1'^2 = \text{fourth order DG operator} \left(\frac{\delta}{\delta x_1}\frac{\delta}{\delta x_1}\right)$$

2.4.1.3 Simplified Pressure Iteration (SIMPLE-Type) Method

The major drawback of the MAC method is that a Poisson equation must be solved for pressure. A direct solver is practical only for simple 2-D cases. However, for 3-D problems, an iterative procedure is the best available choice. The strict requirement of obtaining correct pressure for a divergence-free velocity field in each step significantly slows down the overall computational efficiency. Since, for a steady-state solution, the correct pressure field is needed only when the solution is converged, then the iteration procedure for the pressure can be simplified such that it requires only a few iterations at each time step. The best-known method using this approach is the Semi-Implicit Method for Pressure-Linked Equations (SIMPLE) developed by Caretto et al. (1972); see also Patankar et al. (1972, 1979) or Patankar (1980).

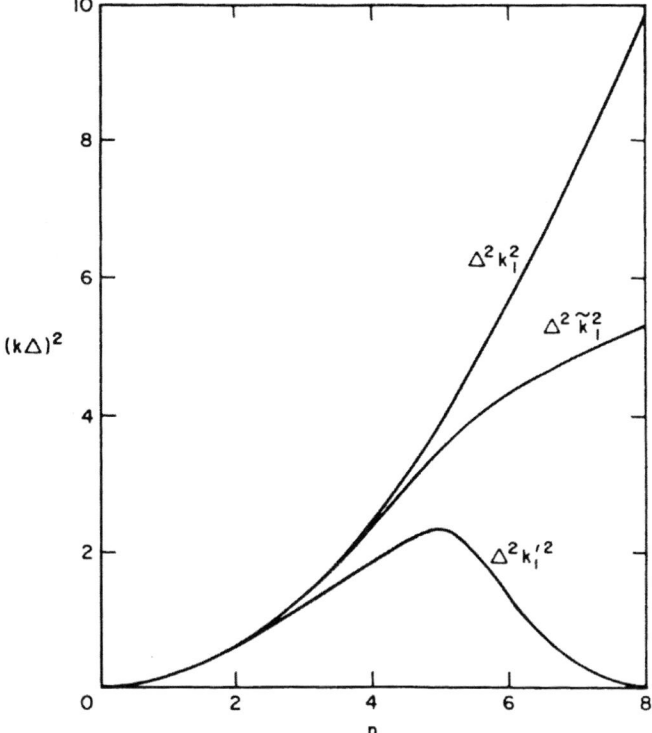

Fig. 2.2 Comparison of ∇^2 operators for 16 equally spaced mesh points (see Equation 2.23)

This particular method is presented here only for historical reasons and has not been used for computing any of the examples included in this monograph.

A brief explanation of the SIMPLE method key features begins with a guessed pressure p^*, which is usually assumed to be p^n at the beginning of the cycle. Then, the momentum equation is solved to obtain an intermediate velocity u_i^* as below:

$$u_i^* - u_i^n = \Delta t \left[fcn(u^n, v^n, u^*, v^*) - \frac{\delta p^*}{\delta x_i} \right] \tag{2.24}$$

The corrected pressure is obtained by setting:

$$p = p^* + p' \tag{2.25}$$

The velocity correction is introduced in a similar manner:

$$u_i = u_i^* + u_i' \tag{2.26}$$

Now the relation between the pressure correction, p', and the velocity correction, u_i', is obtained from a simplified momentum equation. First, the equation for p'

and u_i' is obtained from the linearized momentum equations. Then, by dropping all terms involving neighboring velocities, the following form of the pressure correction equation is obtained:

$$u_i' = \varpi \frac{\delta p'}{\delta x_i} \tag{2.27}$$

where ϖ is a function of the particular differencing scheme chosen. Substituting Equations (2.26) and (2.27) into the continuity equation, a pressure correction equation is obtained. This is equivalent to taking the divergence of Equation (2.27). This procedure in essence results in a simplified Poisson equation for pressure, which can be solved iteratively, for example, line by line.

The unique feature of this method is the simple way of estimating the velocity correction u_i'. This feature simplifies the computation but introduces empiricism into the method. Despite its empiricism, the method has been used successfully for many computations, especially when the computing resources were rather limited. Further details of SIMPLE, SIMPLER, and other variations such as the pressure-implicit with splitting of operators (PISO) algorithm, can be found in the literature, for example, see Patankar (1980) and Issa (1985). As computing power rapidly increases, this type of simplification has become unnecessary and, moreover, unwanted for more accurate prediction of physics, especially involving three-dimensional applications. Subsequently, a rather general formulation called the "pressure-based" approach has been used more, in practice.

2.4.2 Artificial Compressibility Method

Large advances in the state of the art in CFD have been made in conjunction with the field of aerodynamics. Therefore, it is of significant interest to be able to use some of the compressible flow algorithms. To do this, the "artificial compressibility" approach of Chorin (1967), who first introduced the term, can be used.

Later, the method was fully extended to general three dimensions in the form presented here by Kwak et al. (1984) and Chang and Kwak (1984). To reflect the physical nature of the pressure projection in this method more accurately, a new term "pseudo-compressibility" was then introduced. The two terms, artificial compressibility and pseudo-compressibility have been used interchangeably ever since.

In this formulation, the continuity equation is modified by adding a time-derivative of the pressure term, resulting in:

$$\frac{1}{\beta} \frac{\partial p}{\partial t} + \frac{\partial u_i}{\partial x_i} = 0 \tag{2.28}$$

where β is an artificial compressibility or a pseudo compressibility parameter.

Together with the unsteady momentum equations, Equation (2.28) forms a hyperbolic-parabolic type of time-dependent system of equations. Thus, implicit schemes developed for compressible flows can be implemented. Note that the no longer represents a true physical time in this formulation.

Physically, this means that waves of finite speed are introduced into the incompressible flow field as a medium to distribute the pressure. For a truly incompressible flow, the wave speed is infinite, whereas the speed of propagation of the pseudo waves introduced by this formulation depends on the magnitude of the artificial compressibility parameter. In a true incompressible flow, the pressure field is affected instantaneously by a disturbance in the flow, but with artificial compressibility, there will be a time lag between the flow disturbance and its effect on the pressure field. Ideally, the chosen value of the artificial compressibility can be as high as the particular choice of algorithm allows so that incompressibility is recovered quickly. This must be done without lessening the accuracy and the stability property of the numerical method implemented.

On the other hand, if the artificial compressibility is chosen such that these waves travel too slowly, then the variation of the pressure field accompanying these waves is very slow. This will interfere with the proper development of the viscous effects, such as the boundary layer for wall-bounded flows. In wall-bounded viscous flows, the behavior of the boundary layer is very sensitive to the stream wise pressure gradient, especially when the boundary layer is separated. If separation is present, a pressure wave traveling with finite speed will cause a change in the local pressure gradient, which will affect the location of the flow separation. This change in separated flow will feed back to the pressure field, possibly preventing convergence to a steady state. Especially for internal flow, the viscous effect is important for the entire flow field, and the interaction between the pseudo pressure-waves and the viscous flow field becomes very important.

Artificial compressibility relaxes the strict requirement of satisfying mass conservation in each step. To utilize this convenient feature, it is essential to understand the nature of artificial compressibility from both the physical and mathematical points of view. Chang and Kwak (1984) reported physical characteristics of artificial compressibility, and suggested some guidelines for choosing the artificial compressibility parameter. An extensive mathematical account of the artificial compressibility approach is presented by Temam (1979).

Various applications that evolved from this concept were reported for obtaining steady-state solutions (for example, Steger and Kutler, 1977; Kwak et al., 1984; Chang et al., 1984, 1985a, b, 1988a, b; Choi and Merkle, 1985). To obtain time-dependent solutions using this method, an iterative procedure can be applied in each physical time step such that the continuity equation is satisfied. Merkle and Athavale (1987) and Rogers and Kwak (1988, 1989, 1991) reported on this pseudo-time iteration approach. Further details are given in Chapter 4.

2.4.3 Methods Based on Derived Variables

So far, we have outlined various strategies where incompressibility was satisfied using pressure as a mapping parameter. To avoid solving for pressure directly, other approaches have been developed that introduce other variables, allowing elimination of pressure from the formulation. The most commonly used variables are stream function and vorticity.

Here, two approaches from those categories are introduced briefly.

2.4.3.1 Stream Function-Vorticity

For two-dimensional flow, introducing a stream function, ψ, as below, the incompressibility condition is identically satisfied.

$$u = \frac{\partial \psi}{\partial y}, \quad v = -\frac{\partial \psi}{\partial x}$$

The vorticity, ω, is defined in the conventional way as:

$$\omega = \nabla \times \mathbf{u}$$

then the stream function satisfies:

$$\nabla^2 \psi = -\omega \tag{2.29}$$

From the momentum equations, the following vorticity transport equation can be derived.

$$\frac{\partial \omega_i}{\partial t} + \frac{\partial \omega_i u_i}{\partial x_j} = \omega_j \frac{\partial u_i}{\partial x_j} + \nu \nabla^2 \omega \tag{2.30}$$

This, essentially, is the stream function-vorticity formulation—an approach used since the early days of CFD mainly to solve two-dimensional fluid dynamic problems of fundamental interest. To extend this approach to three dimensions involves velocity vector potential that adds much complexity to the formulation. Thus, this approach has not been the method of our choice for three-dimensional real-world applications. More extensive coverage of this and related approaches is found in the book by Quartapelle (1993).

2.4.3.2 Vorticity-Velocity Method

Among various methods using derived variables, the vorticity-velocity method has a good potential for 3-D applications.

A vorticity-velocity method was proposed, for example, by Fasel (1972), to study the boundary layer stability problems in two dimensions. Other authors have used this approach to solving incompressible flow problems (for example, Dennis et al., 1979; Gatski et al., 1982; Osswald et al., 1987; Hafez et al., 1988). However, a three-dimensional extension of this method has been limited to simple geometries, to date.

The basic equations can be summarized below. Instead of the momentum equation, the vorticity transport Equation (2.30), is used. Taking the curl of the above definition of vorticity and using the continuity equation, the following Poisson equation for velocity is obtained:

$$\nabla^2 \mathbf{u} = -\nabla \times \omega \tag{2.31}$$

Equations (2.29) and (2.30) can be solved for velocity and vorticity. Here, the pressure term and the continuity equation are removed at the expense of introducing three vorticity equations. This requires vorticity boundary conditions on a solid surface. As in the pressure projection approach, the computational efficiency of this method depends on the Poisson solver. In general 3-D applications, overall performance will determine the competitiveness of this approach compared to the artificial compressibility or pressure projection approaches.

Chapter 3
Pressure Projection Method in Generalized Coordinates

In Chapter 2, a general idea of the pressure projection method is introduced. This method is described in detail for developing general three-dimensional simulation capability. While an artificial compressibility approach modifies the nature of governing equations, the pressure projection method is formulated time accurately, and so is used both in time-dependant problems and for obtaining steady-state solutions. In light of successful computations in Cartesian coordinates using its numerous variants, Rosenfeld et al. (1991a, b, 1992) developed a staggered grid-based fractional step method in general curvilinear coordinates. Later, Kiris and Kwak (2001) developed a more robust implicit procedure for "not- so-nice" grids using the same finite-volume framework. Among many variations in projection methods, the details presented in this chapter are extracted from these activities by the authors and their colleagues at NASA Ames Research Center.

3.1 Overview

As was pointed out in the previous chapter, in the pressure projection method, pressure is used as a mapping parameter to satisfy the continuity equation. The usual computational procedure involves choosing the pressure field at the current time step such that continuity is satisfied at the next time step. The time step can be advanced in multiple steps (fractional step method), which is computationally convenient. However, the governing equations are not coupled as in the artificial compressibility approach. In the usual fractional step approach, an auxiliary velocity field is first obtained by solving momentum equations. Then, a Poisson equation for pressure is formed by taking the divergence of the momentum equations and using a divergence-free velocity field constraint. The numerical solution of the Poisson equation for pressure with the Neuman-type boundary conditions exists *only* if the compatibility condition is satisfied. In three-dimensional curvilinear coordinates, efficiently solving the resulting algebraic equations from Poisson and momentum equations is one of the important features of the pressure projection approach.

Spatial discretization, especially for pressure field solutions, needs special attention in developing flow solvers in order to satisfy incompressibility conditions in

D. Kwak, C.C. Kiris, *Computation of Viscous Incompressible Flows*, Scientific Computation, DOI 10.1007/978-94-007-0193-9_3, © US Government 2011

the grid space. In Cartesian coordinates, a staggered grid orientation is convenient for satisfying the continuity equation. But in general curvilinear coordinates, the formulation is not straightforward, and geometric quantities have to be discretized consistently with the flow differencing.

For the Poisson solver, a multi-grid acceleration procedure can be incorporated, which is consistent with the elliptic nature of incompressible flow equations. In this respect, the governing equations can be discretized by a finite-volume scheme on a staggered grid. As for the choice of dependent variables, the volume fluxes across each face of the computational cells can be used, thereby improving the stability of the algorithm by treating both convective and viscous fluxes implicitly. In this way, the discretized mass conservation equation can be satisfied consistently with the flux-balancing scheme explained later.

In solving the momentum equations, a significant part of the viscous terms is solved implicitly to minimize the time-step limitation resulting from use of a viscous grid. A staggered grid has favorable properties in Cartesian coordinates, such as coupling odd-even points. However, it is debatable whether a staggered grid has clear advantages over a regular grid in generalized curvilinear coordinates since flux computations in generalized coordinates require various interpolations. In the procedure discussed below, a staggered arrangement is chosen to simplify the pressure boundary condition in devising a Poisson solver.

The pressure projection method presented here is formulated time-accurately. Consequently, the flow solver developed using this particular approach has been used in time-dependant problems as well as for obtaining steady-state solutions. The method developed by Rosenfeld et al. (1991a) and Kiris and Kwak (2001) is presented in some detail to give readers an in-depth understanding involved in this approach.

3.2 Formulation in Integral Form

The equations governing the flow of isothermal, constant density, incompressible, viscous fluids in a time-dependent control volume can be written for the conservation of mass with the face area vector $\mathbf{S}(t)$ and volume $V(t)$. In order to use a finite-volume discretization, it will be convenient to write the governing equations in the following integral form:

$$\frac{\partial V}{\partial t} + \oint_S d\mathbf{S} \cdot (\mathbf{u} - \mathbf{v}) = 0 \tag{3.1}$$

and for the conservation of momentum:

$$\frac{\partial}{\partial t} \int_V \mathbf{u} dV = \oint_S d\mathbf{S} \cdot \overline{T} \tag{3.2}$$

where t is the time, $\mathbf{u}$ is the velocity vector, $\mathbf{v}$ is the surface element velocity resulting from the motion of the grid, and $d\mathbf{S}$ is a surface area vector. The tensor $\overline{T}$ is given by:

$$\overline{T} = -(\mathbf{u} - \mathbf{v})\mathbf{u} - p\mathbf{I} + v\left(\nabla\mathbf{u} + (\nabla\mathbf{u})^{\mathrm{T}}\right) \tag{3.3}$$

where $\nabla\mathbf{u}$ is the gradient of $\mathbf{u}$ while $(\cdot)^T$ is the transpose operator.

The only differences between the fixed and the moving-grid equations are the terms that include the surface element velocity $\mathbf{v}$ and the time dependence of the cell geometry (volume and face area). The volume conservation of each time-varying cell requires the following:

$$\frac{\partial V}{\partial t} - \oint_S d\mathbf{S} \cdot \mathbf{v} = 0 \tag{3.4}$$

where the term $d\mathbf{S} \cdot \mathbf{v}$ represents the volume swespt out by the face $\mathbf{S}$ over the time increment . Thus, the mass conservation equation has exactly the same form as for fixed grids.

$$\oint_S d\mathbf{S} \cdot \mathbf{u} = 0 \tag{3.5}$$

The usual practice is to transform Equations (3.2) and (3.5) into a differential form. Here, the integral formulation is maintained for convenience in deriving the finite-volume scheme for arbitrarily moving geometries.

3.3 Discretization

In the present integral formulation, spatial derivative terms are discretized in another way from the differential formulation described in Chapter 4. In this section, details of the discretization method for governing equations are discussed.

3.3.1 Geometric Quantities

A general non-orthogonal coordinate system (ξ, η, ζ) is defined by the following equation.

$$\mathbf{r} = \mathbf{r}(\xi, \eta, \zeta, t) \tag{3.6}$$

The center of each primary cell is designated by the indices (i, j, k), as shown in Fig. 3.1.

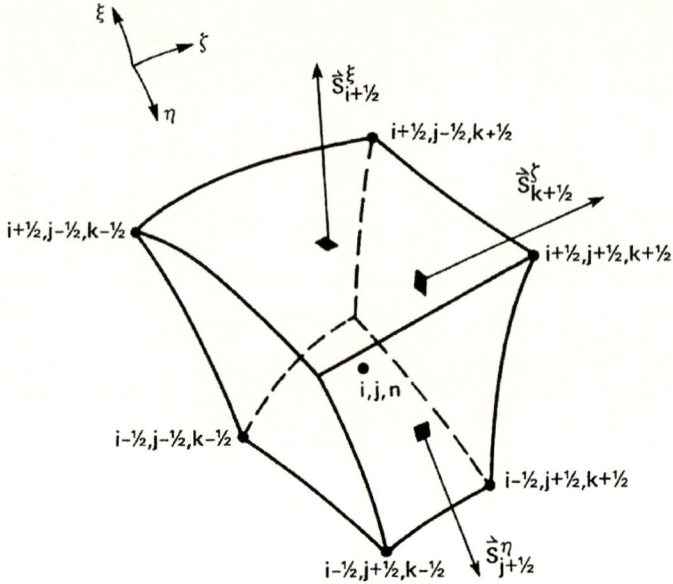

Fig. 3.1 Staggered grid in generalized coordinates: primary cell

The surface area of the face l of a primary cell is given by the vector quantity:

$$\mathbf{S}^l = \frac{\partial \mathbf{r}}{\partial(l+1)} \times \frac{\partial \mathbf{r}}{\partial(l+2)} \tag{3.7}$$

where the computational coordinates $l = \xi, \eta,$ or ζ are in cyclic order and $\times$ is the cross product operator. The vector $\mathbf{S}^l$ has the magnitude of the face area and a direction normal to it. The equivalent differential formulation is the contravariant base vector ∇l scaled by the inverse of the Jacobian $1/J$, i.e., $\mathbf{S}^l = (1/J)\nabla l$.

An accurate discretization should satisfy certain geometric identities, as pointed out by Vinokur (1986). The condition that a computational cell is closed in integral form is shown below.

$$\oint_S d\mathbf{S} = 0 \tag{3.8a}$$

This condition should be satisfied exactly in discrete form, as well:

$$\sum_l \mathbf{S}^l = 0 \tag{3.8b}$$

where the summation is over all the faces of the computational cell. Equation (3.8b) can be satisfied if $\mathbf{S}^l$ is approximated from Equation (3.7) by a proper approximation of $d\mathbf{r}/d\mathbf{l}$.

For example, S^ξ is computed at the point $\left(i + \frac{1}{2}, j, k\right)$ by using the second order approximation.

$$
\left(\frac{\partial \mathbf{r}}{\partial \eta}\right)_{i+\frac{1}{2}} = \frac{1}{2}\left(\mathbf{r}_{j+\frac{1}{2},k-\frac{1}{2}} - \mathbf{r}_{j-\frac{1}{2},k-\frac{1}{2}} + \mathbf{r}_{j+\frac{1}{2},k+\frac{1}{2}} - \mathbf{r}_{j-\frac{1}{2},k+\frac{1}{2}}\right)_{i+\frac{1}{2}}
$$

$$
\left(\frac{\partial \mathbf{r}}{\partial \varsigma}\right)_{i+\frac{1}{2}} = \frac{1}{2}\left(\mathbf{r}_{j-\frac{1}{2},k+\frac{1}{2}} - \mathbf{r}_{j-\frac{1}{2},k-\frac{1}{2}} + \mathbf{r}_{j+\frac{1}{2},k+\frac{1}{2}} - \mathbf{r}_{j+\frac{1}{2},k-\frac{1}{2}}\right)_{i+\frac{1}{2}}
$$

The volumes of all discrete computational cells will sum up to the total volume at a given time. The definition of a cell volume is not unique. However, the volume of each computational cell can be computed by dividing the cell into three pyramids having in common the main diagonal and one vertex of the cell, resulting in the following:

$$
V = \frac{1}{3}\left(\mathbf{S}^\xi_{i-\frac{1}{2}} + \mathbf{S}^\eta_{j-\frac{1}{2}} + \mathbf{S}^\varsigma_{k-\frac{1}{2}}\right) \cdot \left(\mathbf{r}_{i+\frac{1}{2},j+\frac{1}{2},k+\frac{1}{2}} - \mathbf{r}_{i-\frac{1}{2},j-\frac{1}{2},k-\frac{1}{2}}\right) \tag{3.9}
$$

For time-varying or moving grids, the volume conservation Equation (3.4) must be satisfied discretely. This can be done by interpreting the term $d\mathbf{S} \cdot \mathbf{v}$ in Equation (3.4) as the rate of the volume swept by the face $d\mathbf{S}$. For example, the volume swept by the face $\mathbf{S}^\xi_{i+1/2}$ can be computed by a formula similar to Equation (3.9), as shown below:

$$
\left(\delta V^\xi_{i-\frac{1}{2}}\right)^{n+\frac{1}{2}} = \frac{1}{3}\left(\left(\mathbf{S}^\xi_{i-\frac{1}{2}}\right)^n + \delta\mathbf{S}^\eta_{j-\frac{1}{2}} + \delta\mathbf{S}^\varsigma_{k-\frac{1}{2}}\right) \cdot \left(\mathbf{r}^{n+1}_{i-\frac{1}{2},j+\frac{1}{2},k+\frac{1}{2}} - \mathbf{r}^{n}_{i-\frac{1}{2},j-\frac{1}{2},k-\frac{1}{2}}\right)
$$
$$\tag{3.10}$$

where the time level is given by n. The quantities $\delta\mathbf{S}^\eta_{j-1/2}$ and $\delta\mathbf{S}^\varsigma_{k-1/2}$ are the areas swept by the motion of the face $\mathbf{S}^\xi_{i-1/2}$.

The area $\delta\mathbf{S}^\eta_{j-1/2}$ is computed from:

$$
\delta\mathbf{S}^\eta_{j-\frac{1}{2}} = \frac{1}{2}\left(\left(\mathbf{r}^{n+1}_{k-\frac{1}{2}} - \mathbf{r}^{n}_{k+\frac{1}{2}}\right) \times \left(\mathbf{r}^{n+1}_{k+\frac{1}{2}} - \mathbf{r}^{n}_{k-\frac{1}{2}}\right)\right)_{i-\frac{1}{2},j-\frac{1}{2}} \tag{3.11}
$$

and $\delta\mathbf{S}^\varsigma_{k-1/2}$ can be computed in a similar way. The volume of the cell at the time level $(n + 1)$ is computed from Equation (3.4), as shown below.

$$
V^{n+1} = V^n + \sum_l \left(\delta V^l\right)^{n+\frac{1}{2}} \tag{3.12}
$$

Here, the summation is over all the faces of the computational cell. Figure 3.2 illustrates the moving grid notations.

Fig. 3.2 Notations for a
time-varying grid

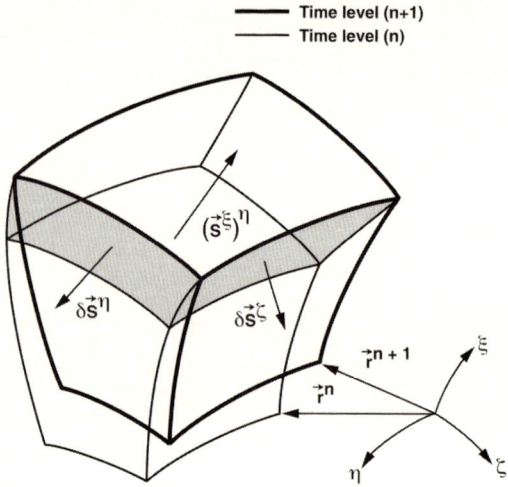

Note that $\delta V^l / \Delta t$ is the volume flux resulting from the motion of the coordinate system and has a meaning similar to the volume-flux U^l as defined in the next section on the mass conservation equation. An accurate computation of the volume of each cell at the time level $(n+1)$ is important for an accurate representation of the momentum equation.

In our finite volume formulation, no coordinate derivatives appear directly in the discrete equations, as in the case of finite-difference formulas. Instead, quantities with clear geometric meaning, such as the volume and the face areas of the computational cells, are used. The discrete approximation of these quantities satisfies the geometric conservation laws. A principal difference between the finite-volume and the finite-difference approaches to moving grids is in the interpretation of the quantity $\delta V^l / \Delta t$. In the finite-volume method, it is treated as a geometric quantity that expresses the rate of displacement of a cell face, whereas in the finite-difference method, the grid velocity is combined with the fluid velocity to define a "relative flow velocity" (see Vinokur, 1986).

3.3.2 Mass Conservation Equation

The discretization of the mass conservation Equation (3.5) over the faces of the primary computational cells yields:

$$\left(S^\xi \cdot u\right)_{i+\frac{1}{2}} - \left(S^\xi \cdot u\right)_{i-\frac{1}{2}} + \left(S^\eta \cdot u\right)_{j+\frac{1}{2}} - \left(S^\eta \cdot u\right)_{j-\frac{1}{2}} + \left(S^\zeta \cdot u\right)_{k+\frac{1}{2}} - \left(S^\zeta \cdot u\right)_{k-\frac{1}{2}} = 0$$

$$(3.13)$$

Note that the default indices (i, j, k) and the time-level $(n+1)$ are omitted for simplicity. Each term on the left-hand side of Equation (3.13) approximates the volume-flux over a face of the primary cell. A simple, discretized mass conservation equation can be obtained by using the following variables as the unknowns instead of the Cartesian velocity components.

$$U^\xi = \mathbf{S}^\xi \cdot \mathbf{u}$$
$$U^\eta = \mathbf{S}^\eta \cdot \mathbf{u} \qquad\qquad (3.14)$$
$$U^\varsigma = \mathbf{S}^\varsigma \cdot \mathbf{u}$$

The quantities U^ξ, U^η, and U^ς are the volume fluxes over the ξ, η, and ς faces of a primary cell, respectively. In tensor algebra nomenclature, these are the contravariant components of the velocity vector scaled by the inverse of the Jacobian $(1/J)$, which is equivalent to volume of the computational cell. With this choice of the dependent variables, the continuity equation takes a form identical to the Cartesian case as shown below.

$$U^\xi_{i+\frac{1}{2}} - U^\xi_{i-\frac{1}{2}} + U^\eta_{j+\frac{1}{2}} - U^\eta_{j-\frac{1}{2}} + U^\varsigma_{k+\frac{1}{2}} - U^\varsigma_{k-\frac{1}{2}} = D_{iv}U^l = 0 \qquad (3.15)$$

This is crucial to satisfying the discrete mass conservation equation. Therefore, the simple form of Equation (3.15) suggests that the volume fluxes can be chosen as the dependent variables for fractional step methods. Treating the mass fluxes as dependent variables in a finite volume formulation is equivalent to using contravariant velocity components, scaled by the inverse of the transformation Jacobian, in a finite-difference formulation. The choice of mass fluxes as dependent variables complicates the discretization of the momentum equations. Here, we have chosen volume fluxes and the pressure fluxes are chosen as the dependent variables.

3.3.3 Momentum Conservation Equation

In order to replace $\mathbf{u}$ by the new dependent variables U^l, the corresponding area vectors are dotted with the momentum equations. Then, the integral momentum equation is evaluated on each cell surfaces for the unknown U^l.

Each cell has the dimensions of $\Delta\xi \times \Delta\eta \times \Delta\zeta$ with the centers of each cell surface located at $(i + 1/2, j, k)$, $(i, j + 1/2, k)$, and $(i, j, k + 1/2)$ for the U^ξ, U^η, and U^ς momentum equations, respectively. In Fig. 3.3, the computational cell for U^ξ-momentum equation is shown with the cell volume, $V_{j+1/2,j,k}$. The staggered grid orientation eliminates "checkerboard-like" pressure oscillations and provides more compact stencils.

The derivation of the U^ξ-momentum equation is outlined next. The U^η- and U^ς-momentum equations can be obtained similarly by using cyclic permutation.

Spatial discretization of the momentum conservation law Equation (3.2) for a computational cell with volume V yields:

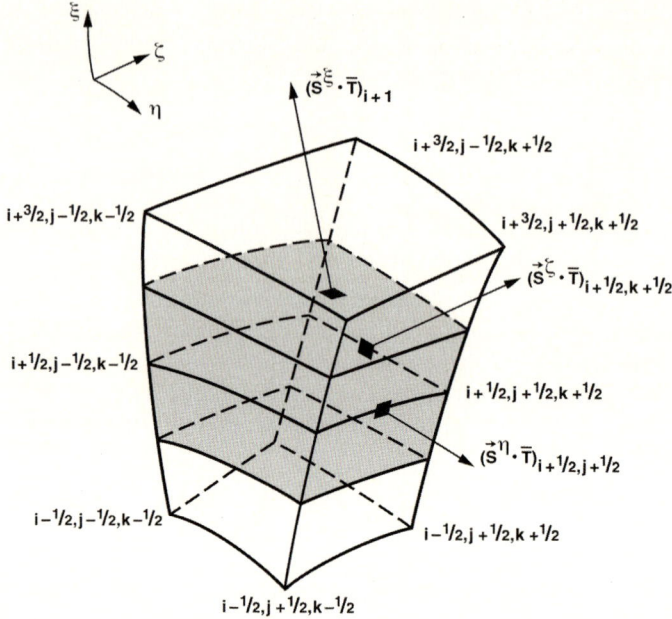

Fig. 3.3 Computational cell for ξ-direction momentum equation

$$\frac{\partial}{\partial t}(Vu) = \sum_l \mathbf{S}^l \cdot \overline{\mathbf{T}} \tag{3.16}$$

The dot product of the above equation and the surface vector $\mathbf{S}^\xi_{j+1/2}$ results in:

$$\frac{\partial}{\partial t}\left(VU^\xi\right) = \mathbf{S}^\xi_{j+\frac{1}{2}} \cdot \sum_l \mathbf{S}^l \cdot \overline{\mathbf{T}} \tag{3.17}$$

where the summation is over all faces of a computational cell. Note that:

$$\mathbf{u} = \mathbf{S}_\xi U^\xi + \mathbf{S}_\eta U^\eta + \mathbf{S}_\zeta U^\zeta \tag{3.18}$$

and

$$U^l = \mathbf{S}^l \cdot u = \mathbf{S}^l \cdot \mathbf{S}_m U^m \tag{3.19}$$

The invariance of the velocity vector requires:

$$\mathbf{S}^l \cdot \mathbf{S}_m = \delta_{lm} \tag{3.20}$$

where δ_{lm} is the Kronecker delta, and $\mathbf{S}^l$ is the inverse base to $\mathbf{S}_m$ and has the differential analogue $\mathbf{S}_m = J\partial\mathbf{r}/\partial m$. In terms of tensor algebra, $\mathbf{S}_m$ is the covariant base vector scaled by the Jacobian J while $\mathbf{S}^l$ is the contravariant base vector scaled by $1/J$. A uniform velocity field can be numerically preserved if the base $\mathbf{S}_m$ is computed at each point from the relation:

$$\mathbf{S}_m = \frac{\mathbf{S}^{m+1} \times \mathbf{S}^{m+2}}{\mathbf{S}^m \cdot \left(\mathbf{S}^{m+1} \times \mathbf{S}^{m+2}\right)} \quad (3.21)$$

which satisfies (3.20) identically. The variable m is the cyclic permutation of (ξ, η, ζ).

In constructing momentum equations, the product $\mathbf{S}^l \cdot \overline{\mathbf{T}}$ should be computed for each face of each momentum equation; see Equation (3.17). For example, the ξ face-center for the U^ξ momentum cell is located at (i, j, k). The flux over this face is computed from below.

$$\left(\mathbf{S}^\xi \cdot \overline{\mathbf{T}}\right)_{i,j,k} = \left(-\left(U^\xi - \frac{\delta V^\xi}{\Delta t}\right) U^l \mathbf{S}_l - \mathbf{S}^\xi P + \mathbf{S}^\xi \cdot v\left(\nabla \mathbf{u} + (\nabla \mathbf{u})^T\right)\right)_{i,j,k} \quad (3.22)$$

The conservative form of the velocity vector gradient is as below.

$$\nabla \mathbf{u} = \frac{\oint_s d\mathbf{S}\mathbf{u}}{V} \quad (3.23)$$

Applying Equation (3.26) for the computation of $\nabla \mathbf{u}_{i,j,k}$ yields:

$$\nabla \mathbf{u}_{i,j,k} = \frac{1}{V}\left(\mathbf{S}^\xi_{i+\frac{1}{2}}\mathbf{u}_{i+\frac{1}{2}} - \mathbf{S}^\xi_{i-\frac{1}{2}}\mathbf{u}_{i-\frac{1}{2}} + \mathbf{S}^\eta_{j+\frac{1}{2}}\mathbf{u}_{j+\frac{1}{2}} - \mathbf{S}^\eta_{j-\frac{1}{2}}\mathbf{u}_{j-\frac{1}{2}} + \mathbf{S}^\varsigma_{k+\frac{1}{2}}\mathbf{u}_{k+\frac{1}{2}} - \mathbf{S}^\varsigma_{k-\frac{1}{2}}\mathbf{u}_{k-\frac{1}{2}}\right) \quad (3.24)$$

where only those indices different from (i, j, k) are given.

The η or ζ face-centers are located at $(i+1/2, j-1/2, k)$ and $(i+1/2, j, k-1/2)$, respectively. The fluxes over these faces are computed in a similar way. The convection and diffusion fluxes in Equation (3.22) can be approximated in various ways. In the present formulation, all of the unknowns at the point (i, j, k) are computed by simple averaging, and therefore, the scheme is equivalent to the results computed by second-order central differencing.

The difference between a fixed-grid and a moving-grid case is in the computation of the convection term, which should include the motion of the grid. For example, the convection flux of the ξ-momentum equation on the ξ-face center (i, j, k) is given by:

$$\left(-\left(U^\xi - \frac{\delta V^\xi}{\Delta t}\right) U^l \mathbf{S}_l\right)_{i,j,k}$$

The difference equations are second-order accurate in time if δV^ξ would not lag in time by $\Delta t/2$ over the volume-flux terms U^l. The resulting discrete equations are conservative in any moving coordinate system and are second-order accurate, spatially. For high Reynolds number flows, fourth-order dissipation can be added to eliminate high-frequency components of the solution (Rosenfeld et al., 1991). The dissipation terms are interpreted in terms of fluxes, and therefore the conservation properties of the equations are maintained.

3.4 Solution Procedure

In Chapter 2, the basic idea of the pressure projection method was illustrated using a MAC approach. Similar procedures can be devised, in combination with other algorithms, in advancing advection terms. From an operator point of view, the MAC method is a form of the fractional step approach. In this section, other variations we have implemented in solution procedures are discussed.

3.4.1 Fractional-Step Procedure

The time-advancing scheme can be designed using numerous combinations of existing algorithms. For example, for computing intermediate velocity, Kiris and Kwak (2001) used a second-order Runge-Kutta scheme. The procedure employing pressure and volume fluxes as dependent variables combined with the Adams-Bashforth method was used by Rosenfeld et al. (1988) and Rosenfeld and Kwak (1989). Here we explain a simple example where the momentum equations are discretized in time using a three-point-backward difference formula as below.

Rewriting the momentum equation (2.2) for simplicity of explanation as below:

$$\frac{\partial u_i}{\partial t} = -\frac{\partial p}{\partial x_i} - \frac{\partial (u_i u_j)}{\partial x_j} + \frac{\partial \tau_{ij}}{\partial x_j} = -\frac{\partial p}{\partial x_i} + R_i \qquad (2.2a)$$

then:

$$\frac{1}{\Delta t}(3u_i^* - 4u_i^n + u_i^{n-1}) = -\frac{\partial p^n}{\partial x_i} + R(u_i^*) \qquad (3.25)$$

where u_i^* denotes the auxiliary velocity field. The $R(u_i^*)$ term in the momentum equations includes the convective and viscous terms. Note that the time derivatives can be differenced using the backward Euler formula for steady-state calculations. The velocity field that satisfies the incompressibility condition is obtained by using the following correction step:

$$\frac{1}{\Delta t}(u^{n+1} - u^*) = -\nabla p' \qquad (3.26)$$

where $p' = p^{n+1} - p^n$. At the $n + 1$ time level, the velocity field must satisfy the incompressibility condition that is to satisfy the following continuity equation:

$$\nabla \cdot u^{n+1} = 0 \tag{3.27}$$

This is done by using a Poisson equation for pressure.

$$\nabla^2 p' = \frac{1}{\Delta t} \nabla \cdot u^* \tag{3.28}$$

The Poisson equation for pressure is obtained by taking the divergence of Equation (3.26) and using Equation (3.27).

In Equation (3.25), both convective and viscous terms are treated implicitly. In order to maintain second-order temporal accuracy, the linearization error in the implicit solution of Equation (3.25) needs to be reduced. This is achieved by using sub-iterations. In most cases, three sub-iterations are sufficient to reduce the linearization error. Here, the purpose of this sub-iteration procedure is quite different from that in the artificial compressibility method for time-accurate computations. The artificial compressibility formulation requires the solution of a steady-state problem at each physical time step. Therefore, the number of sub-iterations for time accuracy in an artificial compressibility approach can be an order of magnitude higher than the number of sub-iterations for the present formulation.

3.4.2 Solution of Momentum Equations Using an Upwind Scheme

The convective and viscous terms in Equation (3.22) can be approximated in various ways. Rosenfeld et al. (1991) implemented an approximate factorization method. In order to relax the inherent Courant-Friedrichs-Lewy (CFL) number restriction in three dimensions and to be able to use the "not so smoothly" varying grids, Kiris and Kwak (2001) implemented a line relaxation scheme where both convective and viscous terms are treated implicitly. In this example, the convective flux terms in Equation (3.22) are computed using a high-order upwind-biased stencil. This alleviates the need for specifying smoothing terms required for central differencing. The numerical flux, for the convective terms is given by:

$$\tilde{f}_{i+\frac{1}{2}} = \frac{1}{2}\left[f(u_{i+1}) + f(u_i) - \phi_{i+\frac{1}{2}} \right] \tag{3.29}$$

where $\phi_{i+\frac{1}{2}}$ is a dissipation term. If $\phi_{i+\frac{1}{2}} = 0$, this represents a second-order central difference scheme. A first-order upwind scheme is given by:

$$\phi_{i+\frac{1}{2}} = \left[\Delta f^{+}_{i+\frac{1}{2}} - \Delta f^{-}_{i+\frac{1}{2}} \right] \tag{3.30}$$

and a third-order upwind is defined as below.

$$\phi_{i+\frac{1}{2}} = -\frac{1}{3}\left[\Delta f^{+}_{i-\frac{1}{2}} - \Delta f^{+}_{i+\frac{1}{2}} + \Delta f^{-}_{i+\frac{1}{2}} - \Delta f^{-}_{i+\frac{3}{2}} \right] \tag{3.31}$$

A fifth-order accurate, upwind-biased stencil that requires seven points, as derived by Rai (1987), is as follows:

$$\phi_{i+\frac{1}{2}} = -\frac{1}{30}\left[-2\Delta f^+_{i-\frac{3}{2}} + 11\Delta f^+_{i-\frac{1}{2}} - 6\Delta f^+_{i+\frac{1}{2}} - 3\Delta f^+_{i+\frac{3}{2}} + 2\Delta f^-_{i+\frac{5}{2}} - 11\Delta f^-_{i+\frac{3}{2}} + 6\Delta f^-_{i+\frac{1}{2}} + 3\Delta f^-_{i-\frac{1}{2}}\right]$$

(3.32)

where $\Delta f^\pm$ is the flux difference across positive and negative traveling waves. The flux difference is computed by:

$$\Delta f^\pm_{i+\frac{1}{2}} = a^\pm\left(\bar{\mathbf{u}}\right)\Delta\mathbf{u}_{i+\frac{1}{2}}$$

(3.33)

where the Δ operator is given by:

$$\Delta\mathbf{u}_{i+\frac{1}{2}} = \mathbf{u}_{i+1} - \mathbf{u}_i$$

(3.34)

The plus (minus) Jacobian is computed by:

$$a^\pm = \frac{1}{2}\left(a \pm |a|\right)$$

(3.35)

The Roe properties (1981), which are necessary for a conservative scheme, are satisfied if the following averaging procedure is employed.

$$\bar{\mathbf{u}} = \frac{1}{2}\left(\mathbf{u}_{i+1} + \mathbf{u}_i\right)$$

(3.36)

An implicit delta form approximation applied to the momentum equations, after linearization in time, results in the following hepta-diagonal scalar matrix equation:

$$\bar{b}\delta q_{i-1} + \bar{a}\delta q + \bar{c}\delta q_{i+1} + \bar{d}\delta q_{j-1} + \bar{e}\delta q_{j+1} + \bar{f}\delta q_{k-1} + \bar{g}\delta q_{k+1} = RHS$$ (3.37)

where $\delta q = U^{n+1} - U^n$ and $\bar{a}, \bar{b}, \bar{c}, \bar{d}, \bar{e}, \bar{f}, \bar{g}$ are diagonals.

The Gauss-Seidel line relaxation scheme by MacCormack (1985) can be employed to solve the matrix equations. The right-hand-side (RHS) term in Equation (3.45) can be stored for the entire domain during a relaxation procedure. The line relaxation procedure is composed of three stages, each involving a scalar tri-diagonal inversion in one direction. In the first stage, δq is solved line-by-line in one direction at a time. Before the tri-diagonal equation can be solved, off-diagonal terms are multiplied by the current value of δq and are shifted over to the RHS of the equation. The same procedure is repeated in the second and third stages by inverting the tri-diagonal matrix in each remaining direction, and treating the off-diagonal terms for the other two directions in Gauss-Seidel fashion. One forward and one backward sweep in each computational direction are sufficient for most problems. However, the number of sweeps can be increased as needed.

3.4.3 Pressure Poisson Solver

Since the early days of computation, Poisson solvers have been studied extensively, and many variations of iterative approach are available. Efficiency of the pressure projection method is largely dependant on the computational efficiency of the pressure Poisson equation. In the particular arrangement of the variable definition discussed in this chapter, the pressure is defined at the center of each cell. Computationally, the pressure boundary condition is not needed in this arrangement. For a general, non-orthogonal coordinate system, the 19-point discrete equations pose some challenges in solving the Poisson equation. Here, the method employed by Rosenfeld et al. (1991) is briefly introduced.

In this method, a four-color ZEBRA scheme is used. The three-dimensional ZEBRA scheme is an iterative scheme that solves implicitly all equations along one coordinate line—say along ξ—as in the successive line over-relaxation (SLOR) method. For the ZEBRA scheme, the order in which the lines are processed is not the usual order by rows or columns. Rather, a "colored" order is devised such that the implicit solution of a line is decoupled from the solution of the other lines. For the non-orthogonal grid discussed here, instead of using the usual two-color ordering (red-black scheme), the points in the (ξ, η) plane are grouped into four different color labels (see Fig. 3.4). First, all the "black" lines are swept, then the "red," "blue," and "green" lines, respectively. The implicit solution of one coordinate line is decoupled from the same color lines. For example, when solving for a "black" line, all the neighboring lines are of different color. The convergence properties of this approach are similar to the SLOR method. In typical cases, more than 80% of

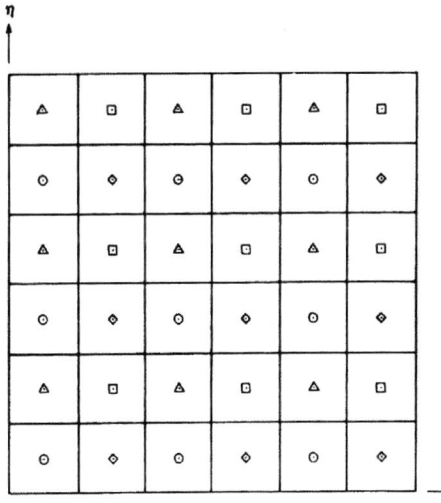

Fig. 3.4 Four-color labeling of the points in the (ξ, η) plane for the Poisson solver

the computational time was required for the Poisson solver. Therefore, a multi-grid acceleration algorithm was implemented by Rosenfeld and Kwak (1993) to substantially improve the computational efficiency. The four-color scheme has good smoothing properties and is a good choice for a multi-grid procedure.

When used in conjunction with the line relaxation scheme, the Poisson solver does not need to be solved in each sweep. Kiris and Kwak (1996) solved the Poisson solver after two sweeps in each direction of the Gauss-Seidel line relaxation procedure. This saves some computational time. In general, however, the pressure projection method discussed in this chapter is expensive and requires small time steps. For unsteady flow where flow physics requires small time steps, this approach can be comparable to others. However, for steady-state computations, this approach is generally more expensive than the artificial compressibility approach. In the next chapter, we describe in detail an artificial compressibility method as another option for solving incompressible flow problems. The two methods will be compared more extensively using benchmark problems, in Chapter 5.

3.5 Validation of the Solution Procedure

The fractional step procedure presented below, combined with the generalized coordinate systems chosen, involves complicated algebraic manipulation. Therefore, it is in the readers' best interest to verify the procedure using a simple idealized problem. This validation is designed to study the effects of grid quality. The test case selected by Kiris and Kwak (1996) is a laminar Couette flow with the grid intentionally generated in a saw-tooth shape to introduce metric discontinuity and non-orthogonality (Fig. 3.5). Even with this "not-so-smooth" mesh point distribution, the numerical procedure should be able to produce a linear u-velocity profile that is the exact solution for the laminar Couette flow.

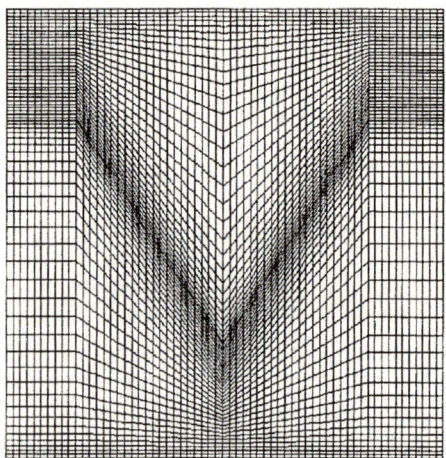

Fig. 3.5 Grid (63×63) for Couette flow

Fig. 3.6 U-velocity contours
for the Couette flow

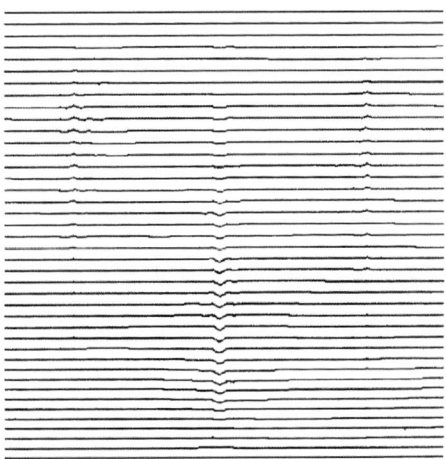

The flow is started with free-stream velocity everywhere except at the stationary wall. The stationary wall has a no-slip condition and the upper wall is moving at a constant velocity. Periodic conditions are imposed on inflow and outflow boundaries. A Courant-Friedrich-Levy (CFL) number of 100 is used for this computation, where the CFL number is defined as:

$$CFL = \max \left(\left| U^{\xi} \right| + \left| U^{\eta} \right| + \left| U^{\zeta} \right| \right)^{*} dt / V$$

Figure 3.6 shows axial (U) velocity contours at steady state. The velocity contours show very small kinks where metric discontinuities are present in the mesh.

In Fig. 3.7, the U-velocity profile at x/L = 0.5 station is compared with the exact solution of the Couette flow. This test case shows that our approach presented here introduces minimal grid effects where a sudden change in the slope of grid lines

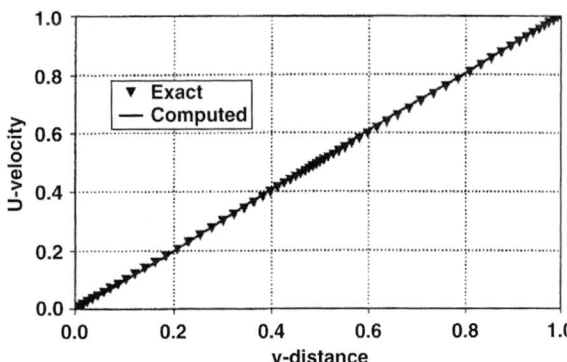

Fig. 3.7 U-velocity profile
for Couette flow

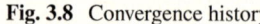 **Fig. 3.8** Convergence history

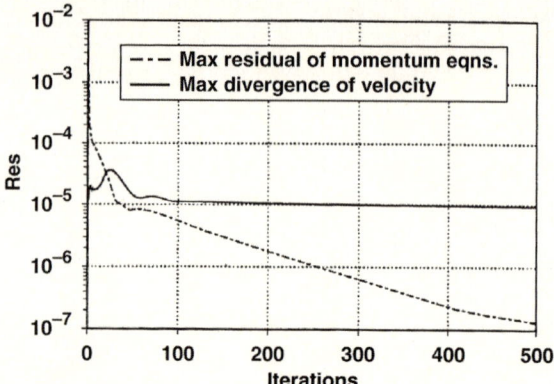

occurs. The small errors caused by discontinuity in grid slope have an insignificant effect on the solution.

The convergence history is plotted in Fig. 3.8. The solid line shows the maximum residual of the momentum equations, and the dashed line represents the maximum divergence of the velocity. The maximum divergence of velocity flattens at about 10–5 because the iteration procedure in the Poisson equation solver is terminated after achieving 10−5 accuracy. With further iterations in the Poisson solver, the error in the divergence of velocity can be lowered to machine accuracy. Since the purpose of this test case is to obtain steady-state solutions, this accuracy limit is adequate. Moreover, the convergence rate is good for a large CFL number. In time-accurate calculations, the Poisson solver may need to be converged more tightly. Further validations where time accuracy and viscous effects are important are presented in Chapter 5.

Chapter 4
Artificial Compressibility Method

The artificial compressibility method is quite different from the pressure projection approach in both the nature of the formulation and the subsequent numerical algorithm. In an artificial compressibility method, a fictitious time derivative of pressure is added to the continuity equation so that the set of equations modified from the incompressible Navier-Stokes equations can be solved implicitly by marching in pseudo time. When a steady-state solution is reached, the original equations are recovered. To obtain time accuracy, an iterative technique can be employed at each time level, which is equivalent to solving the governing equations for steady state at each time level. Using a large, artificial compressibility parameter to spread artificial waves quickly throughout the computational domain, and allowing some residual level of the mass conservation equation, the computing time requirement for time accurate solutions may be controlled within approximately one order-of-magnitude higher than the steady-state computations. In the artificial compressibility approach, the mass conservation does not have to be strictly enforced at each time step, and this gives robustness during iteration.

In this chapter, the physical characteristics of the artificial compressibility method are examined, followed by a more detailed discussion on the solution procedures developed utilizing this approach. As you will see, the addition of pressure term in the continuity equation introduces mathematical pressure waves into the incompressible flow field. This addition is more accurately termed "pseudo-waves" or "pseudo-compressibility." However, for historical reason, the term "artificial compressibility" has been used more frequently. We use both terms interchangeably throughout.

Some details are presented next. At the end of this chapter, a more generalized form stemming from the artificial compressibility idea is included to unify the compressible and incompressible flow regimes, and some test cases are presented.

4.1 Artificial Compressibility Formulation and Physical Characteristics

The artificial compressibility method, as shown by Equation (2.28), results in a system of hyperbolic-parabolic equations of motion. Physically, this means that waves

D. Kwak, C.C. Kiris, *Computation of Viscous Incompressible Flows*, Scientific Computation, DOI 10.1007/978-94-007-0193-9_4, © US Government 2011

of finite speed are introduced into the incompressible flow field as a medium to distribute the pressure. For a truly incompressible flow, the wave speed is infinite, whereas the speed of propagation of these pseudo-waves depends on the magnitude of the artificial compressibility. Ideally, the value of the artificial compressibility is to be chosen as high as the particular choice of algorithm allows, so that the incompressibility is recovered quickly. This has to be done without lessening the accuracy and the stability property of the numerical method implemented. On the other hand, if the artificial compressibility is chosen such that these waves travel too slowly, then the variation of the pressure field accompanying these waves is very slow. This will interfere with the realistic development of the viscous boundary layer, especially when the flow separates. For internal flow, the viscous effect is important for the entire flow field, and the interaction between the pseudo pressure-waves and the viscous flow field also becomes very important.

Artificial compressibility relaxes the strict requirement of satisfying mass conservation in each time step in an iterative process. However, to utilize this convenient feature, it is essential to understand the nature of the artificial compressibility both physically and mathematically. A few key questions need to be answered with respect to this "perturbed" system resulting from the addition of an artificial pressure term. These are:

- What are the characteristics of the pseudo-waves introduced by the addition of an artificial pressure term?
- How does this pseudo-wave interact with vorticity due to viscosity?
- When converged, do the modified governing equations become incompressible flow equations, e.g., incompressible Navier-Stokes equations?
- What are factors affecting the rate of convergence?

These questions can be answered from a mathematical viewpoint with respect to a system of equations perturbed from the incompressible Navier-Stokes formulation. However, since our primary focus is to develop a CFD capability for solving real-world problems, these questioned will be examined from a flow physics point of view.

During the early development of the computational procedure, Chang and Kwak (1984) reported details of the physical characteristics of this method, of which some key features are presented below. From a mathematical point of view, this approach can be viewed as a special case of a preconditioned compressible Navier-Stokes solution procedure. This will be discussed at the end of the chapter in the context of a unified formulation with compressible flow.

For simplicity of analysis, the following 1-D form of the governing equations is used here:

$$\frac{\partial p}{\partial t} + \beta \frac{\partial u}{\partial x} = 0 \tag{4.1a}$$

$$\frac{\partial u}{\partial t} + \frac{\partial u^2}{\partial x} = -\frac{\partial p}{\partial x} + \tau_w \tag{4.1b}$$

where τ_w is the normalized shear stress term, and is equal to $(1/\mathrm{Re})(\partial^2 u/\partial x^2)$ in the 1-D case. The shear stress term in the 1-D formulation contributes to stream-wise viscous diffusion, which is small. To investigate the interaction between the pseudo-wave and vorticity from the wall, the wall shear stress needs to be considered; this aspect will be discussed later in this section.

4.1.1   Characteristics of Pseudo Waves

In the above governing equations, the normal stress term in the stream-wise direction contributes to the diffusion of the waves. To study the wave propagation phenomena, the shear stress term in the above equations is neglected for simplicity, and then linearization around a steady-state velocity component, $\bar{u}$, produces the following system of equations.

$$\begin{pmatrix} \dfrac{\partial p}{\partial t} \\[2mm] \dfrac{\partial u}{\partial t} \end{pmatrix} = \begin{pmatrix} 0 & -\beta \\ -1 & -2\bar{u} \end{pmatrix} \begin{pmatrix} \dfrac{\partial p}{\partial x} \\[2mm] \dfrac{\partial u}{\partial x} \end{pmatrix} \tag{4.2}$$

From this, the following characteristic equation can be obtained:

$$\lambda^2 + 2\bar{u}\lambda - \beta = 0$$

The corresponding eigenvalues are: $\lambda = -\bar{u} \pm \sqrt{\bar{u}^2 + \beta}$
then we can write the 1-D equations without the viscous stress term as:

$$\left[\frac{\partial u}{\partial t} + \frac{1}{(\bar{u} \pm c)} \frac{\partial p}{\partial t} \right] + (\bar{u} \pm c) \left[\frac{\partial u}{\partial x} + \frac{1}{(\bar{u} \pm c)} \frac{\partial p}{\partial x} \right] = 0 \tag{4.3}$$

and the pseudo speed of sound, c, is given as below:

$$c = \sqrt{\bar{u}^2 + \beta}$$

Relative to this sound speed, the pseudo Mach number, M, can be expressed as below:

$$M = \frac{\bar{u}}{\sqrt{\bar{u}^2 + \beta}} < 1 \tag{4.4}$$

It is clear that M is always less than 1 for all $\beta > 0$; therefore, the artificial compressibility formulation does not introduce artificial shock waves into the system and the flow remains subsonic with respect to the pseudo-sound speed.

Equation (4.3) suggests that quantities similar to $(u + \dfrac{p}{\bar{u} + c})$ propagate with $(\bar{u} + c)$ and $(u + \dfrac{p}{\bar{u} - c})$ propagates with $(\bar{u} - c)$. This is heuristically comparable to invariants propagating along characteristic lines in compressible flow. Next, the nature of the pseudo-wave propagation will be examined further.

4.1.2 Wave-Vorticity Interaction

To understand the main features of interaction between the pseudo-wave propagation and spreading of the vorticity, such as in boundary layer development, it is of interest to study one-dimensional linear waves. First, the momentum equation is locally linearized and then, by cross-differencing equations (4.1a) and (4.1b), the following equations are obtained:

$$
\frac{\partial^2 p}{\partial t^2} + 2\bar{u}\frac{\partial^2 p}{\partial t \partial x} - \beta\frac{\partial^2 p}{\partial x^2} = \beta\frac{\partial \tau_w}{\partial x}
$$

$$
\frac{\partial^2 u}{\partial t^2} + 2\bar{u}\frac{\partial^2 u}{\partial t \partial x} - \beta\frac{\partial^2 p}{\partial x^2} = -\frac{\partial \tau_w}{\partial t}
$$

(4.5)

These equations may be expressed as

$$
\left[\frac{\partial}{\partial t} + (\bar{u} + c)\frac{\partial}{\partial x}\right]\left[\frac{\partial}{\partial t} + (\bar{u} - c)\frac{\partial}{\partial x}\right]\begin{pmatrix} p \\ u \end{pmatrix} = \begin{pmatrix} \beta\dfrac{\partial \tau_w}{\partial x} \\ -\dfrac{\partial \tau_w}{\partial t} \end{pmatrix}
$$

(4.6)

If the shear stress term on the right-hand side of Equation (4.6) were absent, characteristic equations for the linear waves would take a simple form expressed by:

$$
\left[\frac{\partial}{\partial t} + (\bar{u} + c)\frac{\partial}{\partial x}\right]\begin{pmatrix} p^+ \\ u^+ \end{pmatrix} = 0
$$

(4.7a)

$$
\left[\frac{\partial}{\partial t} + (\bar{u} - c)\frac{\partial}{\partial x}\right]\begin{pmatrix} p^- \\ u^- \end{pmatrix} = 0
$$

(4.7b)

Waves denoted by the "+" sign propagate downstream with a speed $|\bar{u} + c|$, and waves denoted by a "−" sign travel against the stream with a speed $|\bar{u} - c|$. The quantities with "+" or "−" signs are not defined rigorously here, but may be considered as quantities propagating down- or upstream similar to invariants in compressible flow formulation. For compressible flow, these quantities are functions of density, while in the artificial compressibility formulation these are related to the pressure.

The presence of the shear stress term, however, complicates the wave systems because the shear stress depends on the velocity field. The coupling between the pseudo pressure-waves and the vorticity spreading depends on their respective time scales. An order-of-magnitude analysis can be performed here to obtain a general guideline for determining the magnitude of artificial compressibility. To investigate the interaction between the vorticity spreading from the wall and upstream propagating waves, we will consider a channel with width x_{ref} and length L (normalized by x_{ref}). This can be studied using the following characteristic equation.

$$\frac{\partial u^-}{\partial t} + (u - c)\frac{\partial u^-}{\partial x} = -\frac{\partial \tau_w}{\partial t} \tag{4.8a}$$

The vorticity development during the iteration process in an artificial compressibility approach resembles the boundary layer being developed from a suddenly started flow. Therefore, the rate of growth of vorticity thickness, δ, can be approximated as below.

$$\frac{\partial \delta^2}{\partial t} \cong 4\tilde{v} = \frac{4}{\text{Re}}, \quad \text{where } \text{Re} = \frac{u x_{ref}}{v}$$

Now we will consider the wave with the lowest wave number with the length scale of L. Defining the following non-dimensional quantities:

$$\tilde{x} = \frac{x}{L}, \tilde{t} = \frac{(u - c)t}{L}, \quad t_v = \frac{4}{\text{Re}}t \quad \text{such that } \frac{d\delta^2}{dt_v} \cong 1$$

Equation (4.8) can be written as below.

$$\frac{\partial u^-}{\partial \tilde{t}} + \frac{\partial u^-}{\partial \tilde{x}} = -\left[\frac{4L}{(u - c)\text{Re}}\right]\frac{\partial}{\partial t_v}\tau_w(u, \delta(t_v)) \tag{4.8b}$$

In this equation, the variation of the wave with respect to $\tilde{x}$ and $\tilde{t}$, as well as the shear stress term with respect to t_v, are of order 1. Therefore, the interaction between the waves and the vorticity can be decoupled if:

$$\left[\frac{4L}{(u - c)\text{Re}}\right] << 1 \tag{4.9}$$

This relation can be from the following physical reasoning: suppose the distance from a point in the flow field (such as a point on a flat plate or channel) to the downstream boundary is x_L; then the time required for the upstream propagating wave to reach that point from the downstream boundary, τ_L, can be estimated by the following relation.

$$\tau_L = \frac{L}{c - u}$$

The time scale, τ_δ, for the viscous effect to spread for a distance, δ, can be estimated as follows.

$$\tau_\delta \cong \frac{\text{Re}}{4}\left(\frac{\delta}{x_{ref}}\right)^2$$

The pseudo wave introduced by the artificial compressibility formulation must distribute the pressure such that the viscous boundary layer adjusts to its new pressure environment properly, and must avoid slow fluctuations of the separation region when it is present; so it is required that:

$$\tau_\delta >> \tau_L \tag{4.10}$$

If we set δ/x_{ef} to be order 1, we will have the wave-vorticity decoupling relation, Equation (4.9).

Now, substituting the pseudo-sound speed into Equation (4.9) and letting the dimensionless flow speed to unity, i.e. $u = 1$, we obtain the following criterion for β for obtaining a converged solution for laminar flows.

$$\beta >> \left(1 + \frac{4L}{\text{Re}}\right)^2 - 1 \tag{4.11}$$

This gives an estimate of the lower bound of the artificial compressibility parameter, β. Here, δ is the length scale for the viscous effect to cover, and can be on the same order of the viscous reference length, x_{ref}. For example, in a channel flow, x_{ref} can be the channel height. This guideline is based on physical interpretation of the artificial compressibility formulation. The physical phenomena described above will be illustrated by numerical experiments later.

4.1.3 Rate of Convergence

The rate at which the solution converges to the incompressible solution depends on β as well as on the stream-wise length, L, which represents convective flow geometry such as the channel length. The time required for the pseudo-waves to travel downstream and back upstream for a total distance of $2L$ is obtained as below.

$$\tau_{2L} = \frac{L}{c + u} + \frac{L}{c - u} \tag{4.12}$$

To converge the solution to a steady state, pseudo-waves have to travel the length of the entire flow field at least one complete sweep so that the pseudo pressure-wave can interact with the viscous layer spreading from the wall into the flow field. Therefore, the physical time required for an iterative process has to be greater than τ_{2L}. Substituting the definition of the pseudo-sound speed, Equation (4.3) into

Equation (4.12), and taking the reference velocity, $u = 1$, then the number of computational time steps required to achieve a steady-state solution is:

$$N > \frac{\sqrt{1+\beta}}{\beta}\frac{2L}{\Delta\tau}$$

(4.13)

where $\Delta\tau$ is the dimensionless computational time step.

4.1.4 Limit of Incompressibility

It is also of interest to see whether the solution obtained using the present artificial compressibility method indeed converges to an incompressible solution in the limit. To look into this, the pressure and velocities are expressed into steady-state part, $\bar{p}$ and $\bar{u}_i$, and transient part due to artificial compressibility, p' and u'_i as below:

$$p(x,t) = \bar{p}(x) + p'(x,t)$$
$$u(x,t) = \bar{u}(x) + u'(x,t)$$

(4.14)

Substituting these into the artificial compressibility Equations (4.1a) and (4.1b), the following equations for fluctuating components are obtained.

$$\frac{\partial p'}{\partial t} + \beta\frac{\partial u'}{\partial x} = 0$$

(4.15)

$$\frac{\partial u'}{\partial t} + 2u\frac{\partial u'}{\partial x} = -\frac{\partial p'}{\partial x} + \frac{1}{Re}\frac{\partial^2 u'}{\partial x^2}$$

(4.16)

Moreover, by cross-differencing these equations, the coupling of velocity and pressure can be eliminated. Solving the decoupled equations, solutions of the following form can be obtained:

$$\binom{p'^{+}}{u'^{+}} \approx f\{\alpha[x-(u+c)t]\}\exp\left[-\frac{\alpha^2}{2Re}(1+M^2)\right] \to 0$$

(4.17)

$$\binom{p'^{-}}{u'^{-}} \approx g\{\alpha[x-(u-c)t]\}\exp\left[-\frac{\alpha^2}{2Re}(1-M^2)\right] \to 0$$

(4.18)

where M is the pseudo Mach number based on $\bar{u}$, and α is the wave number. Since M is always less than 1 for all $\beta > 0$, the pseudo-waves vanish as time progresses, resulting in an incompressible steady-state solution. The rate of decay of the transient waves depends on their wave numbers. Heuristically speaking, for a given problem with fixed boundary conditions, any compression waves generated in the course of computation will always be accompanied by the generation of expansion waves. When these two families of waves meet, they will either cancel one another out or break themselves into several waves. The broken waves have a shorter wave-length that corresponds to larger wave numbers, and therefore will decay faster.

4.2 Steady-State Formulation

Now, let's derive the generalized equations for obtaining steady-state solutions using an artificial compressibility approach. Artificial compressibility is introduced by adding a time derivative term for pressure to the continuity equation, resulting in:

$$\frac{1}{\beta}\frac{\partial p}{\partial \tau} + \frac{\partial}{\partial \xi_i}\left(\frac{U_i - (\xi_i)_t}{J}\right) = 0 \tag{4.19}$$

In the steady-state formulation the equations are to be marched in a time-like fashion until the divergence of velocity in Equation (4.19) converges to zero. The time variable for this process no longer represents physical time, so in the momentum equations t is replaced with τ, which can be thought of as a pseudo-time or iteration parameter. Combining Equation (4.19) with the momentum equations gives the following system of equations:

$$\frac{\partial}{\partial \tau}\hat{D} = -\frac{\partial}{\partial \xi}(\hat{E} - \hat{E}_v) - \frac{\partial}{\partial \eta}(\hat{F} - \hat{F}_v) - \frac{\partial}{\partial \zeta}(\hat{G} - \hat{G}_v) = -\hat{R} \tag{4.20}$$

where $\hat{R}$ is the right-hand side of the momentum equation and can be defined as the residual for the steady-state computations, where:

$$\hat{D} = \frac{D}{J} = \frac{1}{J}\begin{bmatrix} p \\ u \\ v \\ w \end{bmatrix}$$

$$\hat{E} = \begin{bmatrix} \beta(U - \xi_t)/J \\ \hat{e} \end{bmatrix} = \frac{1}{J}\begin{bmatrix} \beta(U - \xi_t) \\ \xi_x p + uU \\ \xi_y p + vU \\ \xi_z p + wU \end{bmatrix}$$

$$\hat{F} = \begin{bmatrix} \beta(V - \eta_t)/J \\ \hat{f} \end{bmatrix} = \frac{1}{J}\begin{bmatrix} \beta(V - \eta_t) \\ \eta_x p + uV \\ \eta_y p + vV \\ \eta_z p + wV \end{bmatrix} \tag{4.21a}$$

$$\hat{G} = \begin{bmatrix} \beta(W - \zeta_t)/J \\ \hat{g} \end{bmatrix} = \frac{1}{J}\begin{bmatrix} \beta(W - \zeta_t) \\ \zeta_x p + uW \\ \zeta_y p + vW \\ \zeta_z p + wW \end{bmatrix}$$

$$\hat{E}_v = \begin{bmatrix} 0 \\ \hat{e}_v \end{bmatrix}$$

$$\hat{F}_v = \begin{bmatrix} 0 \\ \hat{f}_v \end{bmatrix}$$

$$\hat{G}_v = \begin{bmatrix} 0 \\ \hat{g}_v \end{bmatrix}$$

and for flow with constant ν in orthogonal coordinates:

$$\hat{E}_v = \left(\frac{\nu}{J}\right)\left(\xi_x^2 + \xi_y^2 + \xi_z^2\right)I_m\frac{\partial D}{\partial \xi} = \gamma_1 D$$

$$\hat{F}_v = \left(\frac{\nu}{J}\right)\left(\eta_x^2 + \eta_y^2 + \eta_z^2\right)I_m\frac{\partial D}{\partial \eta} = \gamma_2 D$$

$$\hat{G}_v = \left(\frac{\nu}{J}\right)\left(\zeta_x^2 + \zeta_y^2 + \zeta_z^2\right)I_m\frac{\partial D}{\partial \zeta} = \gamma_3 D \qquad (4.21b)$$

$$I_m = \begin{bmatrix} 0 & 0 & 0 & 0 \\ 0 & 1 & 0 & 0 \\ 0 & 0 & 1 & 0 \\ 0 & 0 & 0 & 1 \end{bmatrix}$$

This set of equations is to be solved for obtaining a steady-state solution in generalized coordinates.

4.3 Steady-State Algorithm

This section focuses on iterative schemes. Even though the algorithm is explained in steady-state formulation, the iterative schemes can be used for a time-accurate solution procedure, as discussed in Section 4.4.

4.3.1 Difference Equations

For spatial differencing, there are several different ways of defining variables in a grid system. For example, a standard cell-node oriented grid or a staggered grid arrangement can be chosen. In Cartesian coordinates, a staggered grid arrangement has some favorable properties such as natural coupling of variables at odd-even points. In generalized coordinates, these advantages become obscured because of the interpolation required. However, a fully conservative differencing scheme can be devised that maintains the convenience of a staggered arrangement such as in a Poisson solver (Rosenfeld et al., 1988). Using any grid system, spatial differencing can be done either in finite-difference (Steger and Kutler, 1977) or finite-volume form. The finite-volume scheme usually produces better results near geometric singularities. Since most of the results presented later in this monograph were obtained using finite-difference schemes, algorithms based on a finite-difference approach are discussed in this chapter. The numerical procedure using central differencing will be discussed in this section and an upwind differencing scheme will be discussed in a later section.

If the pseudo-time derivative is replaced by a trapezoidal rule finite-differencing scheme, the time difference term results in:

$$\hat{D}^{n+1} = \hat{D}^n + \frac{\Delta\tau}{2}\left[\left(\frac{\partial\hat{D}}{\partial\tau}\right)^n + \left(\frac{\partial\hat{D}}{\partial\tau}\right)^{n+1}\right] + O\left(\Delta\tau^3\right) \qquad (4.22)$$

where the superscript n refers to the nth pseudo-time iteration level. By substituting Equation (4.20) into Equation (4.22), one obtains the following.

$$D^{n+1} + \frac{\Delta\tau}{2}J\left[\delta_\xi(\hat{E}-\hat{E}_v)^{n+1} + \delta_\eta(\hat{F}-\hat{F}_v)^{n+1} + \delta_\zeta(\hat{G}-\hat{G}_v)^{n+1}\right]$$
$$= D^n - \frac{\Delta\tau}{2}J\left[\delta_\xi(\hat{E}-\hat{E}_v)^n + \delta_\eta(\hat{F}-\hat{F}_v)^n + \delta_\zeta(\hat{G}-\hat{G}_v)^n\right] \qquad (4.23)$$

The objective is to solve for D^{n+1}, and this is nonlinear in nature since $\hat{E}^{n+1} = \hat{E}(D^{n+1})$ is a nonlinear function of D^{n+1} as are $\hat{F}^{n+1}$ and $\hat{G}^{n+1}$. The following linearization procedure is applied to solve for these quantities. A local Taylor expansion about u^n yields:

$$\hat{E}^{n+1} = \hat{E}^n + \hat{A}^n(D^{n+1}-D^n) + O(\Delta\tau^2)$$
$$\hat{F}^{n+1} = \hat{F}^n + \hat{B}^n(D^{n+1}-D^n) + O(\Delta\tau^2) \qquad (4.24)$$
$$\hat{G}^{n+1} = \hat{G}^n + \hat{C}^n(D^{n+1}-D^n) + O(\Delta\tau^2)$$

where $\hat{A}$, $\hat{B}$ and $\hat{C}$ are the Jacobian matrices defined as below.

$$\hat{A} = \frac{\partial\hat{E}}{\partial D}, \quad \hat{B} = \frac{\partial\hat{F}}{\partial D}, \quad \hat{C} = \frac{\partial\hat{G}}{\partial D} \qquad (4.25)$$

The Jacobian matrices can all be represented by the following equation:

$$\hat{A}_i = \frac{1}{J}\begin{bmatrix} 0 & L_1\beta & L_2\beta & L_3\beta \\ L_1 & Q+L_1u & L_2u & L_3u \\ L_2 & L_1v & Q+L_2v & L_3v \\ L_3 & L_1w & L_2w & Q+L_3w \end{bmatrix} \qquad (4.26)$$

where $\hat{A}_i = \hat{A}$, $\hat{B}$ or $\hat{C}$ for $i = 1, 2$, or 3, respectively.

$$Q = L_0 + L_1u + L_2v + L_3w$$
$$L_0 = (\xi_i)_t, \ L_1 = (\xi_i)_x, \ L_2 = (\xi_i)_y, \ L_3 = (\xi_i)_w$$
$$\xi_i = (\xi, \eta, \text{or } \zeta) \text{ for } (\hat{A}, \hat{B}, \text{ or } \hat{C})$$

Substituting Equation (4.24) into Equation (4.23) results in the governing equation in delta form:

$$\left\{ I + \frac{h}{2} J \left[\delta_\xi (\hat{A}^n - \Gamma_1) + \delta_\eta (\hat{B}^n - \Gamma_2) + \delta_\zeta (\hat{C}^n - \Gamma_3) \right] \right\} (D^{n+1} - D^n)$$

$$= -\Delta \tau J \left[\delta_\xi (\hat{E} - \hat{E}_v)^n + \delta_\eta (\hat{F} - \hat{F}_v)^n + \delta_\zeta (\hat{G} - \hat{G}_v)^n \right] \tag{4.27}$$

where

$$\Gamma_i = \left(\frac{\nu}{J} \right) \nabla \xi_i \cdot \left(\nabla \xi \frac{\partial}{\partial \xi} + \nabla \eta \frac{\partial}{\partial \eta} + \nabla \zeta \frac{\partial}{\partial \zeta} \right) I_m$$

$$h = \Delta \tau \quad \text{for trapezoidal differencing}$$

$$h = 2 \Delta \tau \quad \text{for Euler implicit scheme}$$

At this point it should be noted that the notation of the form $\left[\delta_\xi (A - \Gamma) \right] D$ refers to:

$$\frac{\partial}{\partial \xi} (AD) - \frac{\partial}{\partial \xi} (\Gamma D) \quad \text{and not} \quad \frac{\partial A}{\partial \xi} D - \frac{\partial \Gamma}{\partial \xi} D$$

4.3.2 Approximate Factorization Scheme

The solution of Equation (4.27) would involve a formidable matrix inversion problem. With the use of an alternating direction implicit (ADI) type scheme, the problem could be reduced to the inversion of three matrices of small bandwidth, for which there exist some efficient solution algorithms. The particular ADI form used here is known as approximate factorization (AF) (Beam and Warming, 1978); a similar scheme was developed independently by Briley and McDonald in 1977. However, it is difficult to apply the AF scheme to Equation (4.27) in its full matrix form. Noting that at the steady state, the left-hand side of Equation (4.27) approaches zero, a simplified expression for the viscous term as shown in Equation (4.21b) is used on the left-hand side. To maintain the accuracy of the solution, the entire viscous term is used on the right-hand side. Using these terms, the governing equation becomes:

$$L_\xi L_\eta L_\zeta (D^{n+1} - D^n) = RHS \tag{4.28a}$$

where

$$L_\xi = \left[I + \frac{\Delta \tau}{2} J^{n+1} \delta_\xi (\hat{A}^n - \gamma_1) \right]$$

$$L_\eta = \left[I + \frac{\Delta \tau}{2} J^{n+1} \delta_\eta (\hat{B}^n - \gamma_2) \right] \tag{4.28b}$$

$$L_\zeta = \left[I + \frac{\Delta \tau}{2} J^{n+1} \delta_\zeta (\hat{C}^n - \gamma_3) \right]$$

and *RHS* is the right-hand side of Equation (4.27). When second-order central differencing is used, the solution to this problem becomes the inversion of three block tridiagonal matrices. Then the inversion problem is reduced to the three inversions as below.

$$
\begin{aligned}
(L_\eta)\Delta\bar{D} &= RHS \\
(L_\xi)\Delta\tilde{D} &= \Delta\bar{D} \\
(L_\zeta)\Delta D^{n+1} &= \Delta\tilde{D}
\end{aligned}
\tag{4.29}
$$

These inversions are carried out for all interior points, and the boundary conditions can be implemented explicitly. Later on, we'll discuss how to implement the boundary conditions implicitly.

A guideline for estimating the lower bound of β was given by Equation (4.10), which was derived from physical reasoning. To make the pressure wave travel fast, it is advantageous to choose β as large as possible. There is, however, a bound of β that comes from the particular algorithm chosen here; namely, the error introduced by the approximate factorization. In implementing the AF scheme leading to Equation (4.28), the following second-order cross product terms are introduced into the following equation.

$$
\left(\frac{\Delta\tau}{2}J^{n+1}\right)^2\left[\delta_\xi(\hat{A}^n - \Gamma_1)\delta_\eta(\hat{B}^n - \Gamma_2) + \ldots\ldots\right]
$$

This term must be kept smaller than the original terms in the equation. Including only the terms that contain β, this restriction can be expressed as:

$$
\frac{\Delta\tau}{2}J^{n+1}\delta_{\xi_i}\hat{A}_i^n\delta_{\xi_j}\hat{A}_j^n\left\langle\delta_{\xi_i}\hat{A}_i^n, \ i\neq j\right.
$$

or

$$
\frac{\Delta\tau}{2}J^{n+1}\delta_{\xi_j}\hat{A}_j^n < 1
$$

Recalling the expression for $\hat{A}_i^n$ given by Equation (4.26), the terms that have β in them give the following.

$$
\frac{\Delta\tau}{2}\beta\delta_{\xi_j}\left(\frac{\partial\xi_j}{\partial x_i}\right) < 1
$$

The term to the right of β in this inequality is essentially the change in $1/\Delta x_i$ in either the ξ, η, or ζ direction. An estimate of the order of magnitude of this term is given by:

$$
O\left[\delta_{\xi_j}\left(\frac{\partial\xi_j}{\partial x_i}\right)\right] \approx 2
$$

which puts the restriction on β

$$O(\beta \Delta \tau) < 1 \tag{4.30}$$

For most problems, the restrictions for β given by Equations (4.10) and (4.30) are satisfied with a value for β in the range from 1 to 10. As will be shown later, this restriction on the upper value of β can be relaxed if the factorization error involving β is removed, for example, by implementing a line relaxation scheme.

4.3.2.1 Diagonal Algorithm

To gain computational efficiency, the Jacobian matrices can be diagonalized. In a diagonal algorithm, a similarity transform can be implemented to uncouple the governing set of equations. The equations can then be answered by solving scalar tridiagonal matrices instead of block tridiagonal matrices. A similarity transform, which symmetrizes and diagonalizes the matrices of the compressible gas dynamic equations, has been used by Warming et al. (1975) and Turkel (1973). This method was extended by Pulliam and Chaussee (1981) to produce a diagonal algorithm for the Euler equations. This method can be applied to the compressible Navier-Stokes equations to obtain a considerable savings in computing time (Flores, 1985).

Here, similarity transforms for the matrices used in the artificial compressibility method are presented, which results in a substantial reduction in computer time (Rogers et al., 1987).

Similarity transformations exist that diagonalize the Jacobian matrices:

$$\hat{A}_i = T_i \hat{\Lambda}_i T_i^{-1} \tag{4.31}$$

where $\hat{\Lambda}_i$ is a diagonal matrix whose elements are the eigenvalues of the Jacobian matrices and which is given by:

$$\hat{\Lambda}_i = \begin{bmatrix} Q & 0 & 0 & 0 \\ 0 & Q & 0 & 0 \\ 0 & 0 & Q - L_0/2 + c & 0 \\ 0 & 0 & 0 & Q - L_0/2 - c \end{bmatrix} \tag{4.32}$$

and where c is the pseudo-speed of sound, which is given by:

$$c = \sqrt{(Q + L_0/2)^2 + \beta(L_1^2 + L_2^2 + L_3^2)} \tag{4.33}$$

The T_i matrix is composed of the eigenvectors of the Jacobian matrix. For more details on the derivation of T_i, its inverse, and eigenvectors, see Rogers et al. (1987).

The implementation of the diagonal scheme involves replacing the Jacobian matrices in the implicit operators with the product of the similarity transform matrices and the diagonal matrix, as given in Equation (4.31). The identity matrix in the implicit operators is replaced by the product of the similarity transform matrix and its inverse. Modification is made to the implicit viscous terms by replacing the I_m matrix with an identity matrix so that the transformation matrices may also be factored out of these terms. This implicitly adds an additional viscous dissipation term to the pressure. Adding higher order smoothing terms for stability, the transformation matrices are now factored out of the implicit operators to give:

$$L_\xi = T_\xi \left[I + \frac{\Delta\tau}{2} J\delta_\xi \left(\hat{\Lambda}_\xi - \hat{\gamma}_1 \right) + \varepsilon_i \nabla_\xi \Delta_\xi \right] T_\xi^{-1}$$

$$L_\eta = T_\eta \left[I + \frac{\Delta\tau}{2} J\delta_\eta \left(\hat{\Lambda}_\eta - \hat{\gamma}_2 \right) + \varepsilon_i \nabla_\eta \Delta_\eta \right] T_\eta^{-1} \qquad (4.34)$$

$$L_\zeta = T_\zeta \left[I + \frac{\Delta\tau}{2} J\delta_\zeta \left(\hat{\Lambda}_\zeta - \hat{\gamma}_3 \right) + \varepsilon_i \nabla_\zeta \Delta_\zeta \right] T_\zeta^{-1}$$

where the implicit viscous terms are now given by:

$$\hat{\gamma}_i = \frac{\nu}{J} \nabla\xi_i \cdot \nabla\xi_i I\delta\xi_i \qquad (4.35)$$

and ∇ and Δ represent forward and backward spatial-differencing operators, respectively. The higher order smoothing terms are discussed further in Section 4.3.5.

Since the transformation matrices are dependent on the metric quantities, factoring them outside the difference operators introduces an error. No modification has been made to the right-hand side of the equation, and therefore, these linearization errors will not affect the steady-state solution. Only the convergence path toward the solution is affected using this diagonal algorithm.

The implementation of this algorithm over the block algorithm will result in a substantial reduction in computational time per iteration because of the decrease in the number of operations performed. Additionally, considerably less memory is required to store the elements on the left-hand side. This space can be used in data and memory management, depending on the computer architecture. For example, when vector machines were the workhorse systems, this additional memory was used to further vectorize the existing code. Since the solution of a tri-diagonal block or scalar matrix is recursive, it is not vectorizable for loops that use the current sweep direction as the inner do-loop index. However, if a large number of these matrices are passed into the inversion routines at once, then vectorization can take place in the "non-sweep" direction. Computer architectures continuously evolve and coding strategy can also change to accelerate memory speed. Whatever the computer architecture may be, lower memory requirements from the algorithm can be utilized in gaining overall computational efficiency.

4.3.3 LU-SGS Scheme

In 1987, Yoon and Jameson developed an implicit lower-upper symmetric-Gauss-Seidel (LU-SGS) scheme for the compressible Euler and Navier-Stokes equations. A similar scheme was devised for the artificial compressibility formulation (Yoon and Kwak, 1989). The LU-SGS scheme is not only unconditionally stable but also completely vectorizable in three dimensions if a vector computer is used. Spatial differencing is equivalent to either central or upwind schemes, depending on the numerical dissipation model that augments the finite volume method (Yoon and Kwak, 1988). This scheme is briefly described below.

Starting from an un-factored implicit scheme similar to Equation (4.27):

$$\left\{ I + \frac{h}{2} J \left[\delta_\xi \hat{A} + \delta_\eta \hat{B} + \delta_\zeta \hat{C} \right] \right\} (D^{n+1} - D^n)$$
$$= -\Delta t \left[\delta_\xi (\hat{E} - \hat{E}_v) + \delta_\eta (\hat{F} - \hat{F}_v) + \delta_\zeta (\hat{G} - \hat{G}_v) \right] \tag{4.27'}$$

The LU-SGS implicit factorization scheme can be derived as:

$$L_l L_d^{-1} L_u (D^{n+1} - D^n) = RHS \tag{4.36a}$$

where

$$L_l = I + \frac{h}{2} (\delta_\xi^- \hat{A}^+ \delta_\eta^- \hat{B}^+ + \delta_\zeta^- \hat{C}^+ - \hat{A}^- - \hat{B}^- - \hat{C}^-)$$

$$L_d = I + \frac{h}{2} (\hat{A}^+ - \hat{A}^- + \hat{B}^+ - \hat{B}^- + \hat{C}^+ - \hat{C}^-) \tag{4.36b}$$

$$L_u = I + \frac{h}{2} (\delta_\xi^+ \hat{A}^- + \delta_\eta^+ \hat{B}^- + \delta_\zeta^+ \hat{C}^- + \hat{A}^+ + \hat{B}^+ + \hat{C}^+)$$

and where δ_ξ^- and δ_ξ^+ are the backward- and forward-difference operators respectively.

This particular scheme has been coded using finite volume discretization and second-order central differencing for the viscous fluxes, which require numerical dissipation terms for stability (Yoon and Kwak, 1989). By choosing different numerical dissipation models and Jacobian matrices, a variety of other schemes can be developed. For example, when there is no source term, the Jacobian matrices of the flux vectors can be constructed to yield diagonally dominant approximate Jacobian matrices. This will eliminate the need for block inversion and enables scalar inversion. This makes the cost per iteration much lower than block inversion and can accelerate convergence, and thus can be a useful alternative for obtaining steady-state solutions. However, it is essential to make sure that the accuracy of the solution does not suffer due to the approximation of the flux Jacobian. In general, approximations at the algorithm level need to be minimized to enhance prediction capability for analyzing complex flow physics. Further details of this method are found in the references cited.

4.3.4 Line Relaxation Scheme

The line-relaxation implicit scheme is formed through an iterative solution process rather than through factorization of the left-hand-side matrix. The discrete form of the matrix on the left-hand side of Equation (4.27) is a banded matrix composed of seven diagonals, where each entry of a diagonal consists of a 4×4 block. The discrete version of Equation (4.27) is written as:

$$[T, 0, \ldots, 0, U, 0, \ldots, 0, X, Y, Z, 0, \ldots, 0, V, 0, \ldots, 0, W]\Delta D = RHS \qquad (4.37)$$

where T, U, V, W, X, Y and Z are the diagonals, with the Y vector being the main (center) diagonal. This matrix equation is approximately solved using an iterative approach. One of the three computational directions is chosen to be the implicit direction, and sweeping through the domain proceeds in the other two directions. Using, for example, the ξ family, a tridiagonal matrix is formed by keeping the X, Y and Z diagonals on the left-hand side, and multiplying the remaining diagonals by the latest known iteration of the ΔD solution vector, and shifting them to the right-hand side. A forward sweep is composed of solving a block-tridiagonal system of the form:

$$[X, Y, Z]\Delta D^{l+1} = RHS - [T, 0, \ldots, 0, U]\Delta D^{l+1}$$
$$- [V, 0, \ldots, 0, W]\Delta D^{l}$$

and a backward sweep is similarly composed of solving the following:

$$[X, Y, Z]\Delta D^{l+1} = RHS - [T, 0, \ldots, 0, U]\Delta D^{l}$$
$$- [V, 0, \ldots, 0, W]\Delta D^{l+1}$$

where the l superscript denotes the sweep iteration number.

The process is initialized by setting $\Delta D^0 = 0$. The algorithm is implemented so that any or all of the three computational directions can be chosen for the sweep direction. The optimum direction and number of sweeps is very much problem dependent. Our experience with this algorithm has revealed that for most problems it is best to use the wall-normal direction as the implicit direction, and that on the order of 10 sweeps should be used (Rogers et al., 1991a, b).

4.3.5 Numerical Dissipation or Smoothing

In applying the above factored schemes, it has been found that the stability of the scheme is dependent on the use of some higher-order smoothing terms. These terms help to damp out the higher-order oscillations and odd- and even-point decoupling in the solutions, which are caused by the use of central differencing. The smoothing term can be related to an upwind finite-difference approximation. The idea of splitting the upwinding scheme into the central differencing scheme plus dissipation was successfully implemented by Kreiss (1964) and others (for example, Jameson et al., 1981). Pulliam (1986) discussed an implicit dissipation model extensively. Later, the dissipation models were unified in the framework of a finite volume total variation diminishing (TVD) method for high-speed flow. Since for incompressible flow we are not dealing with shocks, numerical dissipation is added primarily for stability. However, we generally followed the strategy used for compressible flow formulation, and these models were then extended into the artificial compressibility formulation (Yoon and Kwak, 1989). Here, we focus on only those specifics relevant to the constant coefficient model.

By including these smoothing terms, Equations (4.28a) and (4.28b) become:

$$L_\xi L_\eta L_\zeta (D^{n+1} - D^n) = \text{RHS of (4.27)} - \varepsilon_e [(\nabla_\xi \Delta_\xi)^2 + (\nabla_\eta \Delta_\eta)^2 + (\nabla_\zeta \Delta_\zeta)^2] D^n$$

$$\text{(4.28c)}$$

where

$$L_\xi = \left[I + \frac{\Delta\tau}{2} J^{n+1} \delta_\xi \left(\hat{A}_1^n - \gamma_1 \right) + \varepsilon_i \nabla_\xi \Delta_\xi \right]$$

$$L_\eta = \left[I + \frac{\Delta\tau}{2} J^{n+1} \delta_\eta \left(\hat{A}_2^n - \gamma_2 \right) + \varepsilon_i \nabla_\eta \Delta_\eta \right] \qquad \text{(4.28d)}$$

$$L_\zeta = \left[I + \frac{\Delta\tau}{2} J^{n+1} \delta_\zeta \left(\hat{A}_3^n - \gamma_3 \right) + \varepsilon_i \nabla_\zeta \Delta_\zeta \right]$$

Here, ∇ and Δ represent forward and backward spatial-differencing operators, respectively. To preserve the tridiagonal nature of the system, only second-order smoothing can be used on the left-hand side of the equation, whereas fourth-order smoothing is used on the right-hand side. When the diagonal algorithm (described in Section 4.3.2) is used, however, it is feasible to increase the bandwidth of the system to a pentadiagonal. This makes it possible to use fourth-order smoothing on the left-hand side of the equation, as well. The AF algorithm will be stable if ε_i and ε_e satisfy a certain relation (see Pulliam, 1986; Jameson and Yoon, 1986) as discussed below.

To study the nature of the numerical smoothing, a 1-D form of the dissipation terms is represented as below.

$$\left[1 - \varepsilon_i \nabla_\xi \Delta_\xi \right] (p^{n+1} - p^n) = -\varepsilon_e (\nabla_\xi \Delta_\xi)^2 p^n \qquad \text{(4.38)}$$

Suppose p is represented by the discrete Fourier expansion:

$$p = \sum_n \hat{p}(k) e^{ik\xi} \qquad \text{(4.39)}$$

where

$\hat{p} = $ Fourier transform of p

$k = \dfrac{2\pi}{N\Delta\xi} n = $ wave number

$n = -N/2, \ldots 0, 1, \ldots (N/2 - 1)$

$N = $ number of mesh points

Then Equation (4.37) can be written as:

$$[1 - \varepsilon_i k'](\hat{p}^{n+1} - \hat{p}^n) = -\varepsilon_e (k')^2 \hat{p}^n \qquad \text{(4.40)}$$

where

$$k' = -2 + 2\cos(k)$$
$$(k')^2 = 6 - 8\cos(k) + 2\cos(2k)$$

From this, the amplification factor can be defined.

$$\sigma = \frac{\hat{p}^{n+1}}{\hat{p}^n} = \frac{[1 - \varepsilon_i k' - \varepsilon_e (k')^2]}{[1 - \varepsilon_i k']} \tag{4.41}$$

To damp out the numerical fluctuations as time advances, the absolute value of the amplification factor σ has to be less than one for all possible frequencies.

$$|\sigma| < 1$$

Noting that k' is always negative, this requirement leads to the following relation.

$$\varepsilon_e \leq 2(1 - \varepsilon_i k') \tag{4.42a}$$

It can be shown that the above inequality is always satisfied if:

$$2\varepsilon_e \leq \varepsilon_i \tag{4.42b}$$

The exact relation between these two coefficients can be determined only by a nonlinear stability analysis. In the cases presented in this monograph, ε_i is taken to be three times larger than ε_e. From the expression given in Equation (4.40), it is clear that if ε_i is too large, the rate of damping will be diminished, so it may not be advantageous to take a very large value for ε_i over ε_e. The choice of ε_e depends on the Reynolds number and the grid spacing. However, as discussed later, large values of ε_e adversely affect the accuracy of the continuity equation, which is why the magnitude of ε_e is usually taken to be small. If grid sizes are fine enough to resolve the changes in the flow field, then ε_e can be as small as 10^{-3}.

In computing incompressible flow problems, two major sources of inaccuracy are associated with the numerical dissipation terms, namely: (1) the numerical dissipation terms effectively change the Reynolds number of the flow, and (2) the explicit smoothing terms added to the continuity equation do not conserve mass. In particular, the explicit smoothing on the pressure can affect whether or not the computational procedure converges to an incompressible flow solution. Chang and Kwak (1984) showed that the pseudo-pressure waves decay exponentially with time, and vanish as the solution converges. Thus, the change in pressure with time approaches zero. When there is no explicit smoothing added to the equation, the divergence of the velocity field identically approaches zero. However, when explicit smoothing is included, as the change in pressure approaches zero, the divergence of the velocity approaches the following:

$$\frac{\delta u_i}{\delta x_i} \rightarrow \frac{\varepsilon_{e1}}{\beta \Delta \tau} [(\nabla_\xi \Delta_\xi)^2 + (\nabla_\eta \Delta_\eta)^2 + (\nabla_\zeta \Delta_\zeta)^2] p \tag{4.43}$$

where $\delta/\delta x_i$ is a difference form of the divergence operator and ε_{e1} is the explicit smoothing parameter for the pressure. If ε_e is scaled by h, for example, $\varepsilon_e = \Delta\tau\varepsilon_e$, Equation (4.41) becomes independent of the time step. Depending on the magnitude of β and the local pressure gradient, this term can deteriorate the conservation of mass. However, note that in the numerical computations, the central differencing scheme is modified to include numerical dissipation terms resulting in what is essentially an upwinding scheme for momentum equations.

4.3.6 Boundary Conditions

Once the flow solver is developed, a boundary condition procedure has to be devised to be compatible with the solution algorithm. Boundary conditions play an important part in determining the overall accuracy, stability property, and convergence speed of the solution process. Different types of boundaries are encountered in numerical simulation, including solid surface, inflow and outflow, and far-field boundaries, discussed below.

4.3.6.1 Solid Surface

At a solid surface boundary, the usual no-slip condition is applied. To design a pressure boundary condition on the boundary in generalized coordinates, the grid curvature on the boundary needs to be considered. In general, however, the grid points adjacent to the surface are sufficiently close to the boundary so that constant pressure normal to the surface in the viscous boundary layer can be assumed. For a $\zeta = $ constant surface, this can be expressed as:

$$\left(\frac{\partial p}{\partial \zeta}\right)_{L=1} = 0 \tag{4.44}$$

This approximate boundary condition is good for high Reynolds numbers, and can be implemented either explicitly or implicitly. The implicit implementation, however, will enhance the stability of the code. This can be done during the $\zeta -$ sweep by including the following in the matrix to be inverted:

$$I\Delta D_{j,k,1} + \hat{c}\Delta D_{j,k,2} = \hat{f} \tag{4.45}$$

where

$$\hat{c} = \begin{bmatrix} -1 & 0 & 0 & 0 \\ 0 & 0 & 0 & 0 \\ 0 & 0 & 0 & 0 \\ 0 & 0 & 0 & 0 \end{bmatrix},$$

$$\hat{f} = \begin{bmatrix} p_{L=2} - p_{L=1} \\ 0 \\ 0 \\ 0 \end{bmatrix}$$

4.3.6.2 Inflow, Outflow and Far-Field Conditions

The inflow and outflow boundary conditions for an internal flow problem and the far-field boundary conditions for an external flow problem are handled in much the same way. The incoming flows for both problems are set to an appropriate constant as dictated by the problem. For example, at the inlet to a pipe, the pressure can be set to a constant and the velocity profile can be specified as uniform. Downstream conditions, however, are the most difficult to provide. Simple upwind extrapolation is not well posed. The best convergence rate is obtained if global mass is conserved. So to ensure the best results, the velocities and pressure are first updated using a second-order upwind extrapolation. For an exit at $L = LMAX$ this is written as:

$$Q^{\tilde{n}}_{l\,\mathrm{max}} = \frac{Q^{n+1}_{l\,\mathrm{max}-1}\left(\dfrac{\Delta z_2}{\Delta z_1}\right) - Q^{n+1}_{l\,\mathrm{max}-2}}{\dfrac{\Delta z_2}{\Delta z_1} - 1} \tag{4.46}$$

where

$$\Delta z_1 = z_{l\,\mathrm{max}} - z_{l\,\mathrm{max}-1}$$
$$\Delta z_2 = z_{l\,\mathrm{max}} - z_{l\,\mathrm{max}-2}$$

Then, these extrapolated velocities are integrated over the exit boundary to obtain the outlet mass flux.

$$m\acute{Y}_{out} = \int\limits_{exit} \bar{V}^{\tilde{n}} \cdot d\hat{a} \tag{4.47}$$

Next, the velocity components are weighted by the mass flux ratio to conserve global mass:

$$\bar{V}^{n+1} = \frac{n\acute{Y}_{in}}{n\acute{Y}_{out}} \bar{V}^{\tilde{n}} \tag{4.48}$$

If nothing further is done to update the boundary pressure, this can lead to discontinuities in the pressure because momentum is not being conserved. A method of weighting the pressure by a momentum correction was presented by Chang et al. (1985a), where the pressure condition is obtained by the mass weighted velocities:

$$p^{n+1} = p^{\tilde{n}} - \frac{1}{\zeta_z}[(wW)^{n+1} - (wW)^{\tilde{n}}] + \frac{\nu}{\zeta_z}(\nabla\zeta \cdot \nabla\zeta)\left[\left(\frac{\partial w}{\partial \zeta}\right)^{n+1} - \left(\frac{\partial w}{\partial \zeta}\right)^{\tilde{n}}\right] \tag{4.49}$$

where W is the contravariant velocity. In obtaining this formula, we assume that the streamlines near the exit plane are nearly straight. Any appreciable deviation will cause a discontinuity in the pressure and may lead to an instability. To avoid this, we used a momentum-weighted pressure, obtained by integrating the momentum-corrected pressure p^{n+1} and the extrapolated pressure $p^{\tilde{n}}$ across the exit, as below.

$$I_p^{n+1} = \int\limits_{exit} p^{n+1} d\hat{a}$$

$$I_p^{\tilde{n}} = \int\limits_{exit} p^{\tilde{n}} d\hat{a}$$

The final outlet pressure is then taken to be:

$$p^{n+1} = \left(\frac{I_p^{n+1}}{I_p^{\tilde{n}}}\right) p^{\tilde{n}} \tag{4.50}$$

Under these downstream boundary conditions, global conservation of mass and momentum are ensured. Many practical applications have been solved using the above procedure. However, the nonreflecting-type boundary conditions, according to Rudy and Strikwerda (1980), may enhance the convergence speed.

4.4 Time-Accurate Procedure

Time-dependent calculations of incompressible flows are especially time consuming due to the elliptic nature of the governing equations. Physically, this means that any local change in the flow has to be felt by the entire flow field instantaneously. Numerically, this means that in each time step, the pressure field has to go through one complete steady-state iteration cycle; for example, by Poisson-solver-type pressure iteration or the pseudo-compressibility iteration method.

In transient flow, the physical time step has to be small—consequently the change in the flow field may be small. In this situation, the number of iterations at each time step for getting a divergence-free flow field may not be as high as regular steady-state computations. However, time-accurate computations are in general extremely time consuming. Therefore, it is particularly useful to develop computationally efficient methods by implementing a fast algorithm and by utilizing computer characteristics such as parallel processing. In this section, a way of obtaining time-accurate solutions using an artificial compressibility approach is reviewed (see Rogers and Kwak, 1988, 1989).

Using a second-order, three-point, backward-difference formula, the time derivatives in the momentum equations are differenced:

$$\frac{3\hat{u}^{n+1} - 4\hat{u}^n + \hat{u}^{n-1}}{2\Delta t} = \hat{r}^{n+1} \tag{4.51}$$

where the superscript n denotes the quantities at time $t = n\Delta t$ and $\hat{r}$ is the right-hand side given in Equation (2.10). To solve Equation (4.51) for a divergence-free velocity at the $n+1$ time level, a pseudo-time level is introduced and is denoted by the superscript m. The equations are iteratively solved such that $\hat{u}^{n+1,m+1}$ approaches the new velocity $\hat{u}^{n+1}$ as the divergence of $\hat{u}^{n+1,m+1}$ approaches zero. To drive the divergence of this velocity to zero, the following artificial compressibility relation is introduced:

$$\frac{p^{n+1,m+1} - p^{n+1,m}}{\Delta \tau} = -\beta \nabla \cdot \hat{u}^{n+1,m+1} \tag{4.52}$$

where τ denotes pseudo-time and β is an artificial compressibility parameter. Combining Equation (4.52) with the momentum equations gives:

$$I_{tr}\left(\hat{D}^{n+1,m+1} - \hat{D}^{n+1,m}\right) = -\hat{R}^{n+1,m+1} - \frac{I_m}{\Delta t}\left(1.5\hat{D}^{n+1,m} - 2\hat{D}^n + 0.5\hat{D}^{n-1}\right) \tag{4.53}$$

where $\hat{D}$ is the same vector defined in Equation (4.21), $\hat{R}$ is the same residual vector defined in Equation (4.20), I_{tr} is a diagonal matrix, and I_m is a modified identity matrix given by:

$$I_{tr} = diag\left[\frac{1}{\Delta \tau}, \frac{1.5}{\Delta t}, \frac{1.5}{\Delta t}, \frac{1.5}{\Delta t}\right]$$

$$I_m = diag[0, 1, 1, 1]$$

Finally, the residual term at the $m+1$ pseudo-time level is linearized, giving the following equation in delta form.

$$\left[\frac{I_{tr}}{J} + \left(\frac{\partial \hat{R}}{\partial D}\right)^{n+1,m}\right](D^{n+1,m+1} - D^{n+1,m})$$

$$= -\hat{R}^{n+1,m} - \frac{I_m}{\Delta t}(1.5\hat{D}^{n+1,m} - 2\hat{D}^n + 0.5\hat{D}^{n-1}) \tag{4.54}$$

As can be seen, this equation is very similar to the steady-state formulation given by Equation (4.27), which can be rewritten for the Euler implicit case as below.

$$\left[\frac{I}{J\Delta \tau} + \left(\frac{\partial \hat{R}}{\partial D}\right)^{n}\right](D^{n+1} - D^n) = \hat{R}^n \tag{4.55}$$

Both systems of equations will require discretization of the same residual vector $\hat{R}$. The derivatives of the viscous fluxes in this vector are approximated using second-order central differences. The convective flux terms can be discretized using central differences, as was done in Section 4.3. This will require numerical dissipation terms for stability. Since, in the artificial compressible formulation, the governing equations are changed into the hyperbolic-parabolic type, some of the upwind differencing schemes developed for the compressible Euler and Navier-Stokes equations by numerous authors (e.g. Roe, 1981; Chakravarthy and Osher, 1985; Steger and Warming, 1981; Harten et al., 1983) can be utilized.

In this section, the method of Roe (1981) was adopted in differencing the convective terms. Here, the upwind method is biased by the signs of the eigenvalues of the local flux Jacobian. This is accomplished by casting the governing equations in their characteristic form and then forming the differencing stencil such that it accounts for the direction of wave propagation. In this formulation, the set of numerical equations is solved using a nonfactored line relaxation scheme, similar to that employed by MacCormack (1985). This implicit scheme, described in the next section, makes use of large amounts of processor memory for efficient coding. However, the particulars of coding may vary depending on ever-evolving computer architectures.

4.5   Time-Accurate Algorithm Using Upwind Differencing

Earlier, in Section 4.3.5, higher order smoothing was explained in conjunction with the central differencing scheme. Upwind differencing is essentially a combined form of differencing where numerical dissipation is embedded in the central differencing. Although the current method is explained in conjunction with time-accurate calculations, the same method can be used for steady-state problems as well.

4.5.1   Upwind Differencing Scheme

The upwind scheme for the convective flux derivatives is derived from the 1-D theory, and is then applied to each of the coordinate directions separately. Flux-difference splitting is used here to structure the differencing stencil, based on the sign of the eigenvalues of the convective flux Jacobian. The scheme presented here was originally derived by Roe (1981) as an approximate Riemann solver for the compressible gas dynamics equations.

The derivative of the convective flux in the ξ-direction is approximated by:

$$\frac{\partial \hat{E}}{\partial \xi} \approx \frac{\left[\tilde{E}_{i+1/2} - \tilde{E}_{i-1/2} \right]}{\Delta \xi} \tag{4.56}$$

where $\tilde{E}_{i+1/2}$ is a numerical flux and the subscript i is the discrete spatial index for the ξ-direction.

The numerical flux is given by:

$$\tilde{E}_{i+1/2} = \frac{1}{2} \left[\hat{E}(D_{i+1}) + \hat{E}(D_i) - \phi_{i+1/2} \right] \tag{4.57}$$

where the $\phi_{i+1/2}$ is a dissipation term. For $\phi_{i+1/2} = 0$ this represents a second-order central difference scheme. A first-order upwind scheme is given by:

$$\phi_{i+1/2} = \left(\Delta E_{i+1/2}^+ - \Delta E_{i+1/2}^- \right) \tag{4.58}$$

where $\Delta E^{\pm}$ is the flux difference across positive or negative traveling waves. The flux difference is computed as:

$$\Delta E^{\pm} = A^{\pm}(\bar{D})\Delta D_{i+1/2} \tag{4.59}$$

where the Δ operator is given by:

$$\Delta D_{i+1/2} = D_{i+1/2} - D_i \tag{4.60}$$

The plus (minus) Jacobian matrix has only positive or negative eigenvalues and is computed from:

$$\begin{aligned} A^{\pm} &= X_1 \Lambda_1^{\pm} X_1^{-1} \\ \Lambda_1^{\pm} &= \frac{1}{2}(\Lambda_1 \pm |\Lambda_1|) \end{aligned} \tag{4.61}$$

where the subscript 1 denotes matrices corresponding to the ξ-direction flux. The matrices X_1 and X_1^{-1} are the right and left eigenvectors of the Jacobian matrix of the flux vector and Λ_1 is a diagonal matrix consisting of its eigenvalues. All matrices appearing in the upwind dissipation term must be evaluated at a half-point (denoted by i+1/2). To do this, a special averaging of the dependent variables at neighboring points must be performed. The following averaging procedure is employed.

$$\bar{D} = \frac{1}{2}(D_{i+1} + D_i) \tag{4.62}$$

A scheme of arbitrary order may be derived using these flux differences. Implementation of higher-order-accurate approximations in an explicit scheme does not require significantly more computational time if the flux differences $\Delta E^{\pm}$ are all computed simultaneously for a single line. A third-order upwind flux is then defined by:

$$\phi_{i+1/2} = -\frac{1}{3}\left(\Delta E_{i-1/2}^{+} - \Delta E_{i+1/2}^{+} + \Delta E_{i+1/2}^{-} - \Delta E_{i+3/2}^{-}\right) \tag{4.63}$$

The primary problem with using schemes of greater than third-order accuracy occurs at the boundaries. Special treatment is needed, requiring a reduction of order—or a much more complicated scheme. Therefore, when going to a higher-order-accurate scheme, compactness is advantageous. Such a scheme was derived by Rai (1987), using a fifth-order-accurate, upwind-biased stencil. A fifth-order, fully upwind difference would require 11 points, but this upwind-biased scheme requires only seven points, and is given below.

$$\begin{aligned} rl\phi_{i+1/2} = -\frac{1}{30}\Big[&-2\Delta E_{i-3/2}^{+} + 11\Delta E_{i-1/2}^{+} - 6\Delta E_{i+1/2}^{+} - 3\Delta E_{i+3/2}^{+} \\ &+2\Delta E_{i+5/2}^{-} - 11\Delta E_{i+3/2}^{-} + 6\Delta E_{i+1/2}^{-} + 3\Delta E_{i-1/2}^{-}\Big] \end{aligned} \tag{4.64}$$

Next to the boundary, near-second-order accuracy can be maintained by the third-and fifth-order schemes, by using the following:

$$\phi_{i+1/2} = \varepsilon \left(\Delta E_{i+1/2}^+ - \Delta E_{i+1/2}^- \right) \tag{4.65}$$

For $\varepsilon = 0$, this flux leads to a second-order central difference. For $\varepsilon = 1$, this is the same as the first-order dissipation term given by Equation (4.58). By including a nonzero ε, dissipation is added to the second-order, central-difference scheme to help suppress any oscillations. A value of $\varepsilon = 0.01$ is commonly used for many applications.

To form the delta fluxes used in this scheme, the eigensystem of the convective flux Jacobian is needed. For the current formulation, a generalized flux vector is given by Equation (4.21), and the Jacobian matrix $A_i = \partial \hat{E}_i / \partial D$ of the flux vector is given by Equation (4.26). The normalized metrics are redefined as:

$$k_x = \frac{1}{J} \frac{\partial \xi_i}{\partial x}, i = 1, 2, 3$$

$$k_y = \frac{1}{J} \frac{\partial \xi_i}{\partial y}, i = 1, 2, 3$$

$$k_z = \frac{1}{J} \frac{\partial \xi_i}{\partial z}, i = 1, 2, 3$$

$$k_t = \frac{1}{J} \frac{\partial \xi_i}{\partial t}, i = 1, 2, 3$$

As explained in conjunction with the diagonal algorithm in the previous section, a similarity transform for the Jacobian matrix is introduced here:

$$\hat{A}_i = X_i \Lambda_i X_i^{-1}$$

where Λ_i is defined by Equation (4.39). The matrix of the right eigenvectors is given by:

$$X_i = \begin{vmatrix} 0 & 0 & \beta(c - k_t/2) & -\beta(c + k_t/2) \\ x_k & x_{kk} & u\lambda_3 + \beta k_x & u\lambda_4 + \beta k_x \\ y_k & y_{kk} & v\lambda_3 + \beta k_y & v\lambda_4 + \beta k_y \\ z_k & z_{kk} & w\lambda_3 + \beta k_z & w\lambda_4 + \beta k_z \end{vmatrix} \tag{4.66}$$

where

$$x_k = \frac{\partial x}{\partial \xi_{i+1}}, \quad y_k = \frac{\partial y}{\partial \xi_{i+1}}, \quad z_k = \frac{\partial z}{\partial \xi_{i+1}}$$

$$x_{kk} = \frac{\partial x}{\partial \xi_{i+2}}, \quad y_{kk} = \frac{\partial y}{\partial \xi_{i+2}}, \quad z_{kk} = \frac{\partial z}{\partial \xi_{i+2}} \tag{4.67}$$

$\xi_{i+1} = \eta, \zeta, \quad \text{or } \xi \text{ for } i = 1, 2, \text{ and } 3 \text{ respectively}$

$\xi_{i+2} = \zeta, \xi, \quad \text{or } \eta \text{ for } i = 1, 2, \text{ and } 3 \text{ respectively}$

and its inverse can be similarly obtained. Note that this transformation is nonsingular in any combination of metrics.

4.5.2 Implicit Scheme

Here, we describe the way in which Equations (4.54) and (4.55) are numerically represented and solved. The first consideration is the formation of the Jacobian matrix of the residual vector $\hat{R}$ required for the implicit side of the equation. Applying the difference formula given in Equation (4.56) to the convective flux vectors and applying a second-order, central-difference formula to the viscous terms, the residual at a discrete point $\left(x_{ijk}, y_{ijk}, z_{ijk}\right)$ is given by:

$$
\hat{R}_{ijk} = \frac{\tilde{E}_{i+1/2,j,k} - \tilde{E}_{i-1/2,j,k}}{\Delta \xi} + \frac{\tilde{F}_{i,j+1/2,k} - \tilde{F}_{i,j-1/2,k}}{\Delta \eta} + \frac{\tilde{G}_{i,j,k+1/2} - \tilde{G}_{i,j,k-1/2}}{\Delta \zeta}
$$

$$
- \frac{\left(\hat{E}_v\right)_{i+1,j,k} - \left(\hat{E}_v\right)_{i-1,j,k}}{2\Delta \xi} + \frac{\left(\hat{F}_v\right)_{i,j+1,k} - \left(\hat{F}_v\right)_{i,j-1,k}}{2\Delta \eta} + \frac{\left(\hat{G}_v\right)_{i,j,k+1} - \left(\hat{G}_v\right)_{i,j,k-1}}{2\Delta \zeta}
$$

$$(4.68)$$

The generalized coordinates are chosen so that $\Delta \xi$, $\Delta \eta$, and $\Delta \zeta$ are equal to one. To limit the bandwidth of the implicit system of equations, the Jacobian of the residual vector will be formed by considering only first-order contributions to the upwind numerical fluxes, and the second-order differencing of the viscous terms. So, the only portion of the residual vector that is actually linearized is the following:

$$
\hat{R}_{ijk} = \frac{1}{2} \left(\hat{E}_{i+1,j,k} - \hat{E}_{i-1,j,k} + \dots \right.
$$

$$
- \Delta E^+_{i+1/2,j,k} + \Delta E^-_{i+1/2,j,k} + \Delta E^+_{i-1/2,j,k} - \Delta E^-_{i-1/2,j,k} - \cdots \qquad (4.69)
$$

$$
\left. - \left(\hat{E}_v\right)_{i+1/2,j,k} + \left(\hat{E}_v\right)_{i-1/2,j,k} - \cdots \right.
$$

The exact Jacobian of this residual vector will result in a banded matrix of the following form.

$$
\frac{\partial \hat{R}}{\partial D} = B \left[\frac{\partial \hat{R}_{ijk}}{\partial D_{i,j,k-1}}, 0, \dots, 0, \frac{\partial \hat{R}_{ijk}}{\partial D_{i,j-1,k}}, 0, \dots, 0, \frac{\partial \hat{R}_{ijk}}{\partial D_{i-1,j,k}}, \frac{\partial \hat{R}_{ijk}}{\partial D_{i,j,k}}, \frac{\partial \hat{R}_{ijk}}{\partial D_{i+1,j,k}}, \right.
$$

$$
\left. 0, \dots, 0, \frac{\partial \hat{R}_{ijk}}{\partial D_{ij+1,k}}, 0, \dots, \frac{\partial \hat{R}_{ijk}}{\partial D_{i,j,k+1}} \right]
$$

$$(4.70)$$

These exact Jacobians can be very costly to form; therefore, approximate Jacobians of the flux differences as derived and analyzed by both Yee (1986) and Barth (1987) are used. These are given as follows:

$$\frac{\partial \hat{R}_{ijk}}{\partial D_{i,j,k-1}} \approx \frac{1}{2}\left(-\hat{C}_{i,j,k-1} - \hat{C}^+_{i,j,k-1/2} + \hat{C}^-_{i,j,k-1/2}\right) - (\bar{\gamma}_3)_{i,j,k-1/2}$$

$$\frac{\partial \hat{R}_{ijk}}{\partial D_{i,j,k}} \approx \frac{1}{2}\Big(A^+_{i+1/2,j,k} + A^+_{i-1/2,j,k} - A^-_{i+1/2,j,k} - A^-_{i-1/2,j,k}$$
$$+ B^+_{i,j+1/2,k} + B^+_{i,j-1/2,k} - B^-_{i,j+1/2,k} - B^-_{i,j-1/2,k}$$
$$+ C^+_{i,j,k+1/2} + C^+_{i,j,k-1/2} - C^-_{i,j,k+1/2} - C^-_{i,j,k-1/2}\Big)$$
$$+ (\bar{\gamma}_1)_{i+1/2,j,k} + (\bar{\gamma}_2)_{i,j+1/2,k} + (\bar{\gamma}_3)_{i,j,k+1/2}$$
$$+ (\bar{\gamma}_1)_{i-1/2,j,k} + (\bar{\gamma}_2)_{i,j-1/2,k} + (\bar{\gamma}_3)_{i,j,k-1/2}$$

$$\frac{\partial \hat{R}_{ijk}}{\partial D_{i+1,j,k}} \approx \frac{1}{2}\left(\hat{A}_{i+1,j,k} - \hat{A}^+_{i+1/2,j,k} + \hat{A}^-_{i+1/2,j,k}\right) - (\bar{\gamma}_1)_{i+1/2,j,k}$$

(4.71)

where $\hat{A} = \hat{A}_1$, $\hat{B} = \hat{A}_2$ and $\hat{C} = \hat{A}_3$ as given by Equation (4.26), and where only the orthogonal mesh terms are retained for the implicit viscous terms. This set of matrix equations can be solved using the line-relaxation method presented in Section 4.3.4.

4.5.3   Boundary Conditions for Upwind Scheme

Implicit boundary conditions at all of the boundaries enable the use of large time steps. At a viscous no-slip surface, the velocity is specified to be zero, and the pressure at the boundary is obtained by specifying that the pressure gradient normal to the wall be zero, as discussed for the steady-state solution procedure. The boundary conditions used for the inflow and outflow regions are based on the method of characteristics. The formulation of these boundary conditions is similar to that given by Merkle and Tsai (1986), but the implementation we use here is slightly different.

Here, we derive boundary conditions in two-dimensional space, first for a constant ξ boundary, with similar results for a constant η or a constant ζ boundary in a three-dimensional case. The finite-speed waves that arise with the use of artificial compressibility are governed by the following:

$$\frac{\partial \hat{D}}{\partial \tau} = -\frac{\partial \hat{E}}{\partial \xi} = -\frac{\partial \hat{E}}{\partial D}\frac{\partial D}{\partial \xi} = -\hat{A}\frac{\partial D}{\partial \xi} = -X\Lambda X^{-1}\frac{\partial D}{\partial \xi}$$

Multiplying by X^- gives:

$$X^{-1}\frac{\partial \hat{D}}{\partial \tau} = \Lambda X^{-1}\frac{\partial D}{\partial \xi}$$

(4.72)

If the X^{-1} matrix is moved inside the spatial- and time-derivative, for example for the far field, the result is a system of independent scalar equations, each having the form of a wave equation. The sign of the eigenvalues in the Λ matrix determines the traveling direction of each wave. For a positive or negative eigenvalue, a corresponding wave propagates information in the positive or negative ξ-direction. The number of positive or negative eigenvalues determines the number of characteristic equations propagating information from the interior of the computational domain to the boundary. Thus, at the boundary, the characteristics equations that bring information from the interior will be chosen as part of the boundary conditions. The rest of the information should come from outside the computational domain, which leaves some variables to be specified.

Either one or two characteristics will be traveling toward the boundary from the interior because there is always at least one positive eigenvalue and one negative eigenvalue. To select the proper waves, Equation (4.72) is multiplied by a diagonal selection of matrix L, which has an entry of one in the position of the eigenvalue we wish to select, and zeros elsewhere. Therefore:

$$LX^{-1}\frac{\partial \hat{D}}{\partial \tau} = -L\Lambda X^{-1}\frac{\partial D}{\partial \xi} \tag{4.73}$$

Replacing the time derivative with an implicit Euler time step gives:

$$\left(\frac{LX^{-1}}{J\Delta \tau} + L\Lambda X^{-1}\frac{\partial}{\partial \xi}\right)\left(D^{n+1} - D^n\right) = -L\Lambda X^{-1}\frac{\partial D^n}{\partial \xi} \tag{4.74}$$

which gives either one or two relations, depending on the number of nonzero elements in L. To complete the set of equations, some variables must be specified to be constant. Now define a vector Ω of the variables to be held constant such that:

$$\frac{\partial \Omega}{\partial \tau} = 0 \rightarrow \frac{\partial \Omega}{\partial D}\frac{\partial D}{\partial \tau} = 0 \rightarrow \frac{\partial \Omega}{\partial D}\left(D^{n+1} - D^n\right) = 0 \tag{4.75}$$

Combining Equations (4.74) and (4.75), we obtain:

$$\left(\frac{LX^{-1}}{J\Delta \tau} + L\Lambda X^{-1}\frac{\partial}{\partial \xi} + \frac{\partial \Omega}{\partial D}\right)\left(D^{n+1} - D^n\right) = -L\Lambda X^{-1}\frac{\partial D^n}{\partial \xi} \tag{4.76}$$

Equation (4.76) can be used to update the variables implicitly at any of the inflow or outflow boundaries with the proper choice of L and Ω.

For inflow boundaries, two different sets of specified variables have been used successfully. One set consists of the total pressure and the cross-flow velocity. This set is useful for problems in which the inflow velocity profile is unknown. For outflow boundaries, static pressures have been specified for computations presented later in this monograph. The algorithm discussed in this section can also be used for obtaining steady-state solutions. The only difference is that for steady-state calculations, only one-time-level iteration is needed. Further discussions can be found in Rogers and Kwak (1990).

4.6 Validation of Solution Procedure

The physical interpretation of the artificial compressibility method is given in Section 4.1, which explains how the artificial wave, brought in by the introduction of artificial compressibility, interacts with vorticity transport. Therefore, it is of interest to verify the validity of the guideline given by Equations (4.11) and (4.30) on the permissible range of the artificial compressibility parameter, β. To do this, a validation calculation is performed over a range of β using a simple test problem. This study is done using the steady-state algorithm as explained in Section 4.3.2.

4.6.1 Two-Dimensional (2-D) Channel Flow

The channel flow is perhaps the simplest internal-flow test problem, where the pressure wave propagates between the in- and out-flow boundaries while the viscous effect spreads inward from two walls of the channel. The coordinate system of a 2-D straight channel with a width of 1 and length of 15 is illustrated in Fig. 4.1, which also shows velocity vectors for a converged solution.

To obtain fully developed velocity profiles within a reasonable channel length, a partially developed boundary layer profile can be imposed at the inflow boundary. In our numerical experiment, a uniform inlet flow is prescribed, the Reynolds number based on the duct width and the average velocity is 1,000, and the pseudo-time step is $\Delta\tau = 0.1$. Then, the recommended range of β is estimated to be $0.12 < \beta < 10$. To illustrate the pressure wave propagation phenomena and its effect on the convergence property, the channel is impulsively started. Here, five different values of β (0.1, 1, 5, 10 and 50) were chosen such that two cases are outside the recommended range and three values are kept within range.

Pressure contours at three different time levels are shown in Fig. 4.2a for $\beta = 5$, which is within the recommended range. The expansion wave from the exit plane propagates upstream sufficiently fast to balance the spreading of the viscous effect. The solution converges nicely, in this case. However, as shown in Fig. 4.2b, for $\beta = 0.1$, which is lower than the recommended lower bound of β, the speed of the upstream propagating pressure wave from the exit plane is very low. Therefore, the expansion wave is confined near the exit plane while the viscous effect spreads into the flow field from both upper and lower surfaces. The viscous field is not properly balanced by the physically correct pressure gradient. This causes the spurious fluctuations to amplify, as shown at $\tau = 2.5$—and eventually the computation blows

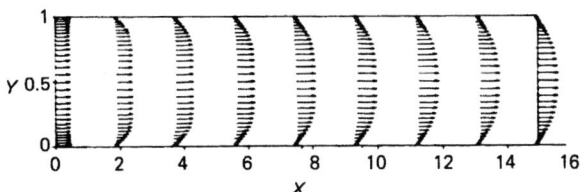

Fig. 4.1 Channel flow: velocity profile at Re $= 1000$, $\beta = 5$, $\tau = 0.1$

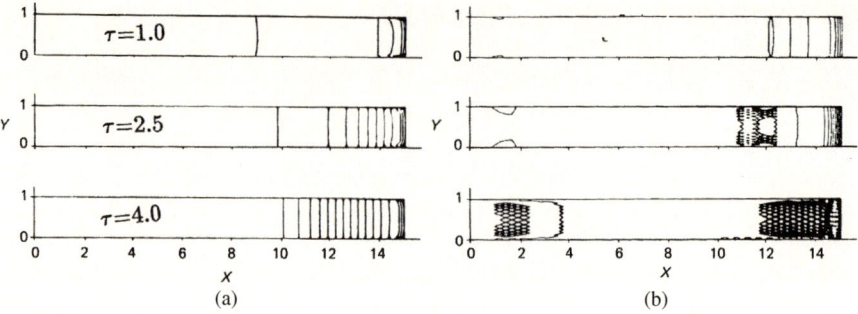

Fig. 4.2 Pressure contours for developing channel flow during the initial iteration process:
a $\beta = 5$, **b** $\beta = 0.1$

up—as shown at $\tau = 4.0$. In Fig. 4.2a, b, the pressure contours are shown during
the initial iteration process until $\beta = 0.1$ case diverges. When fully converged, the
pressure contours are straight lines for $\beta = 5$ case.

The convergence history for these cases is shown in Fig. 4.3. The log of the root-
mean-square of the change in pressure and velocities (RMSDQ) is plotted against
the computation time τ. It can be seen that calculations for $\beta = 0.1$ and 50 become
unstable within 50 steps and start to diverge, whereas other cases converge to a
stable solution. The effect of β values on the incompressibility of the fluid is shown
in Fig. 4.3b in the form of the log of the root-mean-square of the divergence of the
velocity field (RMSDIV) plotted against the pseudo-time τ.

Internal flow, especially the current 2-D channel flow, is an excellent example of
an instance where interaction between upstream propagating pressure wave and vorticity transport is visible. For external flows, such as flow over a circular cylinder for

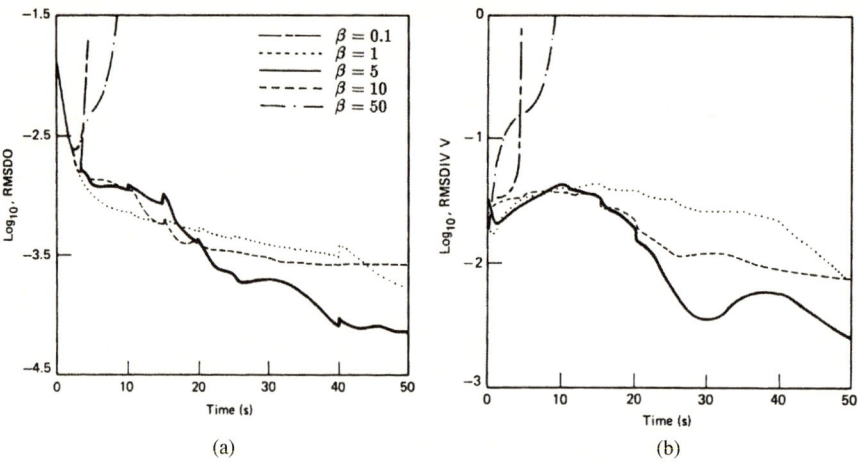

Fig. 4.3 Convergence history for channel flow at Re $= 1000$: **a** RMSDQ, **b** RMSDIV

example, the effects of these two are not very discernable. However, the guideline for the selection of artificial compressibility can still be applied, as will be shown in the next chapter.

4.6.2 Flow over a Backward-Facing Step

The flow over a 2-D backward-facing step is simple in geometry but offers rich fluid dynamics phenomena with recirculating zones and separation bubbles on the opposite wall. Maintaining two-dimensional flow is a challenge in conducting experiment. Even when the cross-flow direction is large using a constant cross section of backward-facing step like Fig. 4.4, three-dimensional flow can be developed for high Reynolds numbers. Therefore, for the current study of numerical methods, laminar ranges of Reynolds numbers are computed and compared with experiments. Figure 4.4 shows the schematic of the problem, where the step height is equal to the inlet height and the Reynolds number is based on twice the step height. The upstream boundary is located at the step and a fully developed channel flow is imposed at the inlet. This problem is very challenging computationally as it involves a primary and a secondary separation bubble. The size and location of these separation zones are very sensitive to the pressure gradient, providing a good viscous flow validation case.

Results obtained using an approximate factorization scheme have been reported previously by Rogers et al. (1985). The separation lengths were found to be sensitive to the magnitude of the numerical dissipation coefficient at higher Reynolds numbers. In essence, this is equivalent to changing the effective Reynolds number of the flow as the Reynolds number approaches 1,000. So, it is quite desirable to remove the dissipation model dependence on the solution at a high laminar range of Reynolds numbers. When the flow becomes turbulent, the Reynolds number based

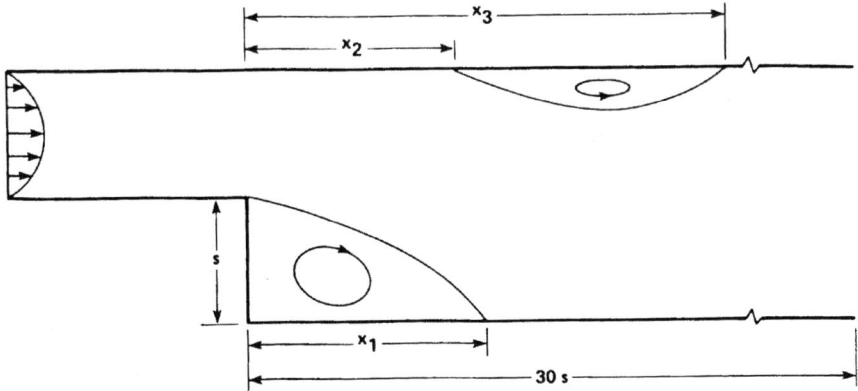

Fig. 4.4 Geometry of a backward-facing step flow problem

Fig. 4.5 Separation and
reattachment lengths for the
flow over a backward-facing
step

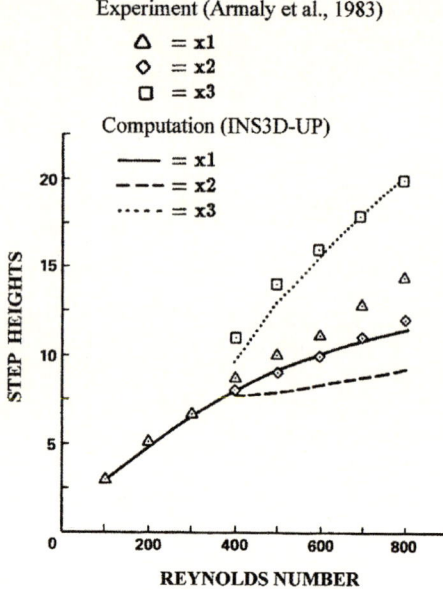

on turbulent eddy viscosity is on the order of a few hundred, and thus dissipation model presented here perform adequately.

More extensive validation has been done by using the upwind discussed in Section 4.5 and a line relaxation scheme by Rogers (1990). The computed separation and reattachment locations are then compared to experimental values given by Armaly et al. (1983) in Fig. 4.5. For the primary reattachment length, ×1, good agreement is observed between the experiments and the computation—until the secondary separation appears at a Reynolds number of about 400. At a higher Reynolds number, the primary separation length, ×1, and the secondary separation point, ×2, deviates from the experimental data. Armaly et al. reported that 3-D flow was observed near the step when the Reynolds number is greater than 400.

For more complete validation, the numerical simulation needs to contain the entire experimental configuration, including the possible 3-D effect. In Fig. 4.6, the convergence history is shown. Using the grid of 100 points in the stream-wise direction and 53 points in the cross-flow direction, a good converged solution is obtained within 100 iterations and with less than 11.5 s of computing time on the Cray 2, a number cruncher during the 1980s.

The two examples shown in this section are used to demonstrate the artificial compressibility procedure, and are not intended to validate the entire spectrum of flow problems we are likely to encounter in basic and applied work. In Chapter 5, more computed cases are presented for the purpose of evaluating capabilities and performances of different approaches.

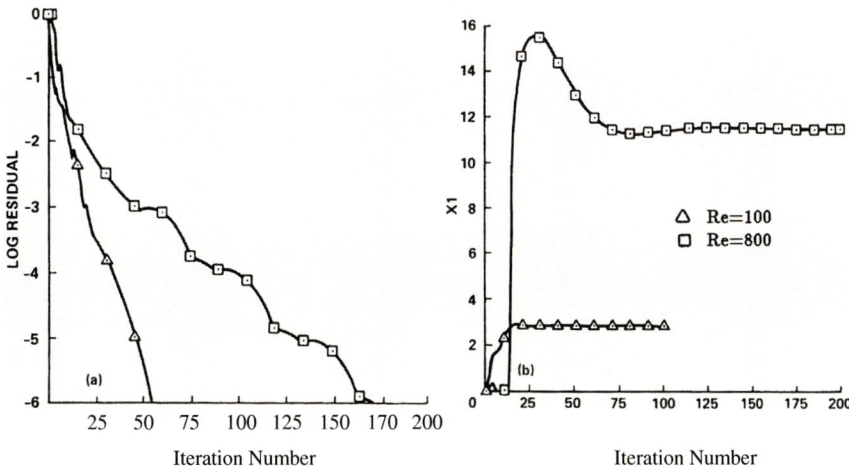

Fig. 4.6 Convergence history for the flow over a backward-facing step: (**a**) residual, (**b**) primary attachment length

4.7 Unified Formulation

Previously, we presented implicit methods for incompressible flow computations. When the flow field contains a wide range of speed regimes, Mach numbers can vary from almost zero to the transonic range. In such cases, it will be of practical interest to have a flow solution method that can cover both the incompressible and compressible flow regimes. In this section, we discuss a single unified solution approach that can be applied to flows of both regimes.

The need for a unified formulation for "all-speed" flow becomes apparent when compressible flow codes break down at low speed regime (see Hafez, 2001). For example, time-marching methods developed for solving compressible flow problems become inefficient and lose accuracy when applied to low speed flows.

One idea for designing a unified scheme came from the artificial compressibility method. Since the artificial compressibility formulation resembles the compressible Navier-Stokes solver, these two can be combined in such a way that the compressible Navier-Stokes algorithm behaves similarly to the artificial compressibility approach at a low speed. A series of development work appeared starting in the late 1980s (for example see reviews by, Merkle, 1995; Venkataswaran and Merkle, 1999, 2002).

We show a method using preconditioning that can be applied to compressible flow solvers to overcome difficulties in the low-Mach regime. One such approach is to extend the time-derivative preconditioning method, used in the artificial compressibility formulation to the compressible flow equations. The time-derivative preconditioning method developed by Housman et al. (2009), investigates both conservative and non-conservative discretizations, where the non-conservative approach follows the split coefficient matrix (SCM) method of Chakravarthy et al. (1980).

4.7.1 Time-Derivative Preconditioning Method

The unified formulation can be applied to both gases and liquids, so no explicit equation of state is assumed in the derivation. In the present formulation we assume that the state equations can be expressed as:

$$\rho \equiv \rho(p, T) \text{ and } h \equiv h(p, T) \tag{4.77}$$

where ρ is the fluid density, h is the specific enthalpy, p is pressure, and T is temperature. The only other restriction on the equation of state is that the inviscid system remains hyperbolic in time.

The time-derivative preconditioned system of equations written in strong conservation law form for a non-orthogonal curvilinear coordinate system are written as:

$$\Gamma_p \frac{\partial \hat{Q}}{\partial \tau} + \frac{\partial \hat{E}}{\partial \xi} + \frac{\partial \hat{F}}{\partial \eta} + \frac{\partial \hat{G}}{\partial \zeta} = 0 \tag{4.78}$$

where

$$\hat{Q} = J^{-1} \begin{bmatrix} p \\ u \\ v \\ w \\ T \end{bmatrix}, \hat{E} = \begin{bmatrix} \rho \hat{U} \\ \rho \hat{U}u + \hat{\xi}_x p \\ \rho \hat{U}v + \hat{\xi}_y p \\ \rho \hat{U}w + \hat{\xi}_z p \\ \rho \hat{U}H - \hat{\xi}_t p \end{bmatrix} \hat{F} = \begin{bmatrix} \rho \hat{V} \\ \rho \hat{V}u + \hat{\eta}_x p \\ \rho \hat{V}v + \hat{\eta}_y p \\ \rho \hat{V}w + \hat{\eta}_z p \\ \rho \hat{V}H - \hat{\eta}_t p \end{bmatrix} \hat{G} = \begin{bmatrix} \rho \hat{W} \\ \rho \hat{W}u + \hat{\zeta}_x p \\ \rho \hat{W}v + \hat{\zeta}_y p \\ \rho \hat{W}w + \hat{\zeta}_z p \\ \rho \hat{W}H - \hat{\zeta}_t p \end{bmatrix}$$

$$\tag{4.79}$$

In the system of equations above, (u, v, w) are the Cartesian velocity components, $(\hat{\xi}, \hat{\eta}, \hat{\zeta})$ are the inverse Jacobian scaled metric terms, $(\hat{U}, \hat{V}, \hat{W})$ are the scaled contravariant velocities, and $H = h + (u^2 + v^2)$ is the total enthalpy. The time-derivative preconditioning matrix is:

$$\Gamma_p = \begin{bmatrix} \rho'_p & 0 & 0 & 0 & \rho_T \\ u\rho'_p & \rho & 0 & 0 & u\rho_T \\ v\rho'_p & 0 & \rho & 0 & v\rho_T \\ w\rho'_p & 0 & 0 & \rho & w\rho_T \\ H\rho'_p + \rho h_p - 1 & \rho u & \rho v & \rho w & H\rho_T + \rho h_T \end{bmatrix}$$

where the physical material derivatives, (ρ_T, h_p, h_T) are defined by the equation of state and the preconditioning parameter ρ'_p is defined locally by:

$$\rho'_p = \frac{1}{V_p^2} - \frac{\rho_T(1 - \rho h_P)}{\rho h_T}. \tag{4.80}$$

The characteristic velocity scale is:

$$V_p = \min\left(c, \ \max\left(\sqrt{u^2 + v^2 + w^2}, \beta\right)\right) \tag{4.81}$$

where $\beta > 0$ is a problem-dependent constant, which avoids division by zero in the evaluation of ρ'_p. This is equivalent to the definition of β in the artificial compressibility method. The physical isentropic speed of sound is defined by:

$$c^2 = \frac{\rho h_T}{\rho h_T \rho_p + \rho_T(1 - \rho h_p)} \tag{4.82}$$

Note that when the characteristic velocity scale approaches the isentropic speed of sound, the preconditioned equations converge to the non-preconditioned system. This is the preferred behavior since the standard time-marching system is well conditioned in the transonic limit. A complete derivation of the preconditioned system is given in Housman et al. (2009).

4.7.2 Numerical Results

The unified formulation is tested using steady-state flow problems over a large range of Mach numbers by Housman et al. (2009). Here, low-speed liquid flow through a channel containing a hydrofoil is presented to illustrate the capability of the preconditioned approach. In this case, the conservative and non-conservative approaches both agree reasonably well with the experiment and are indistinguishable from one another.

4.7.2.1 Liquid Flow over a NACA 0015 Hydrofoil

The test case considers the flow of water through a channel containing a NACA 0015 hydrofoil. This case was proposed as a benchmark problem for low-Mach compressible flow solvers by Salvetti and Beux (2004) as part of a numerical workshop conducted for low Mach number flows.

The purpose of this case is to test the capability of the unified formulation to compute nearly incompressible flow. An experimental study on the same problem has been performed by Rapposelli et al. (2003). An outline of the geometry with a structured overset grid arrangement is shown in Fig. 4.7.

Liquid water is assumed to obey a stiffened gas equation of state of the form:

$$\rho = \frac{p + p_\infty}{RT} \text{ and } h = C_p T$$

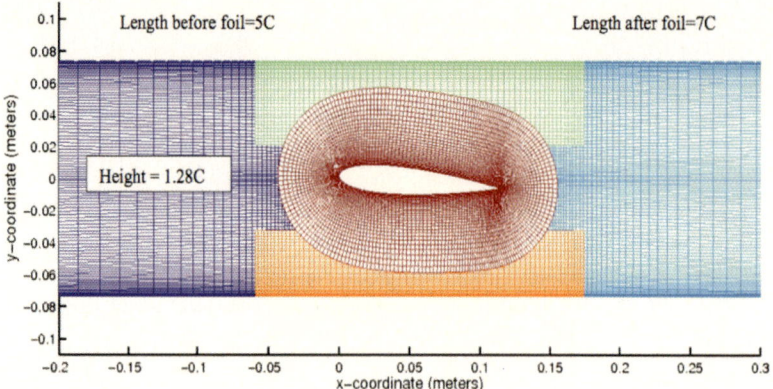

Fig. 4.7 Structured overset grid for NACA 0015 hydrofoil at the center of a channel

where the gas constant is defined by $R = C_p(\gamma - 1)/\gamma$. The material properties are:

$$\gamma = 1.9276, \ C_p = 8,076.73 \ J/kg/K,$$

$$p_\infty = 1.137279 \times 10^9 \ Pa$$

The inlet conditions $P = 59,000$ Pa, $U = 3.11$ m/s, and $T = 298$ K, correspond to an inlet Mach number of 0.0021. Figure 4.8a shows the pressure coefficient on the upper and lower surfaces of the NACA 0015 hydrofoil, comparing the pre-conditioned Roe (PROE) method and the non-conservative preconditioned split coefficient matrix (PSCM) method, a potential flow solution, and experimental data.

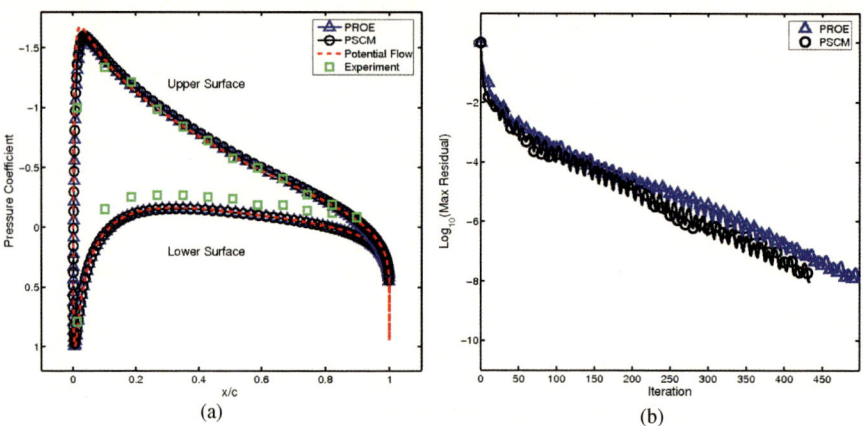

Fig. 4.8 Comparison of computed (PROE, PSCM and potential flow solution) and experimental results for liquid flow over a NACA 0015 hydrofoil: (**a**) C_p on upper and lower surfaces of NACA 0015 hydrofoil; (**b**) maximum residual convergence history

The PROE and PSCM methods converge to identical solutions that match the potential flow solution up to the trailing edge. At the trailing edge, the O-grid boundary condition causes additional numerical dissipation, resulting in C_p not dropping completely. The computed results match the experiment on the upper surface, while the lower surface is not captured. This is most likely due to the inviscid assumption used for computing this case.

In Fig. 4.8b convergence of the maximum residual is plotted versus iteration number for both the conservative PROE and the non-conservative PSCM methods. Nearly identical convergence rates are obtained and the overall number of iterations is less than 500. The results of these two test cases suggest that the unified formulation with the alternating line Jacobi relaxation algorithm is Mach number-independent regardless of the fluid medium. More comprehensive validation is necessary to make a definitive conclusion of this approach. However, in general, a unified approach can be very useful for many engineering applications where flow speed is in a wide range or where both incompressible and compressible flow coexist, such as in a multi-material or multi-phase flow.

Further validations of the artificial compressibility method are presented in Chapter 5.

Chapter 5
Flow Solvers and Validation

Up to this point, we have reviewed numerical algorithms for computing viscous incompressible flows, primarily using primitive variables along with finite difference and finite volume frameworks. The solution methods for incompressible flows are based on the assumption that the flow can be approximated by incompressible Navier–Stokes equations. Once a solution algorithm is developed, flow solvers and software procedures need to be developed to compute fluid dynamic problems. This process includes setting up the problem, solving the flow with the proper initial and boundary conditions, and then post-processing the computed results. These solutions include several levels of approximations including algorithmic, geometry-related and physical-modeling related approximations.

The methods we have chosen for developing flow solvers and solution procedures, namely, pressure projection and artificial compressibility, are discussed in Chapters 3 and 4, respectively. Our discussions on algorithms in these chapters are given from a practical utility viewpoint and, of course, other methods and algorithms are available in the literature.

Once the solution procedure has been developed, numerical simulation can be used to study fundamental fluid dynamics problems and/or to utilize the software as a tool for fluid engineering. Geometry definition and grid generation can be relatively simple for fundamental problems. However, for engineering applications, this step could be very involved and require much human time, as the surface definition of geometry is often not very well defined and variations in format are diverse. Depending on the geometry complexity, the grid resolution requirement will vary, affecting the computational strategy regarding how to divide the computational domain, what grid resolution is required to capture the correct flow phenomena, how to utilize particular computer architecture for implementing parallel processing and data management, and so on.

As computing power has increased, has now become feasible to model more complete geometries at a very high resolution. So, the programming and computer science aspects of the flow solvers are important for accomplishing computational efficiency and decreased solution time. When the problem size is large, data management such as communication among processors and data transmittal time in and out of the memory become significant.

D. Kwak, C.C. Kiris, *Computation of Viscous Incompressible Flows*, Scientific Computation, DOI 10.1007/978-94-007-0193-9_5, © US Government 2011

In this chapter, the flow solver validation process will be discussed using several test problems of a fundamental nature. For these problems, the geometry is simple and we hone in on the accuracy of various algorithms presented earlier. These test problems cover internal and external flows and steady and time-dependent flows, selected as building-block validation cases representing flow characteristics encountered in many real-world applications. Some of these flow features will be illustrated later when we describe computational examples for supporting specific missions. The first class of validation cases were chosen in the laminar range of Reynolds numbers, followed by turbulent flow cases to discuss issues related to physical modeling.

5.1 Scope of Validation

Computational performance depends not only on the methods implemented but also on how solvers are coded. The flow solution codes selected here have been developed and used by the authors and their colleagues at NASA Ames Research Center over the years, so we trust that solvers used in this chapter are reasonably well coded, and as such can be used to represent the algorithms explained earlier in Chapters 3 and 4. For convenience, these codes are identified as below.

5.1.1 Artificial Compressibility Codes

5.1.1.1 INS3D

Historically, a flow solver code named INS3D was developed first based on the steady-state algorithm described in Section 4.3.2. In INS3D, the artificial compressibility approach is implemented using an approximate factorization scheme. This takes advantage of the advances made in conjunction with compressible flow computations. The spatial discretization utilizes second-order central differencing with additional numerical dissipation terms. The code name is derived from the Incompressible Navier–Stokes code in 3-D generalized coordinates. This code was developed primarily to obtain steady-state solutions in conjunction with an early task in the 1980s of redesigning the Space Shuttle main engine. The experience gained in that project was then extended to the following version using the same line of approach.

5.1.1.2 INS3D-UP

To obtain time-accurate solutions using the artificial compressibility formulation, the continuity equation must be satisfied at each time step by sub-iteration in pseudo-time. In order to use a large time step in the pseudo-time iteration, an upwind differencing scheme based on flux-difference splitting is used in combination with an implicit line relaxation scheme. This removes the factorization error and the need for specifying a numerical dissipation amount. To characterize upwind differencing,

"-UP" was added to the code. After successful validation of this version, the code has been used in most applications at NASA and other organizations.

Other variants of INS3D have been tried using the artificial compressibility approach, including one called INS3D-LU. This code is mainly used to investigate the LU-SGS scheme for computing incompressible flow, as discussed in Section 4.3.3. For spatial discretization, a finite volume scheme in conjunction with either central or upwind differencing is implemented. An LU-SGS implicit algorithm is employed for temporal discretization. Results for steady-state solutions are compared with other codes in this chapter, but only for limited cases. For time-accurate solutions, this code has similar characteristics to the original three-factored scheme. INS3D-LU was created to investigate various algorithmic options, but has not been much utilized in actual engineering applications.

5.1.2 Pressure Projection Code

5.1.2.1 INS3D-FS

A generalized flow solver based on a pressure projection method using a fractional-step approach has been developed for time-dependent computations of the incompressible Navier–Stokes equations. The governing equations are discretized conservatively using a finite-volume approach on a staggered grid. Here, the discretized equations are advanced in time by decoupling the solution of the momentum equation from that of the continuity equation. This procedure, combined with accurate and consistent approximations of the geometric quantities, satisfies the discretized mass conservation equation exactly in a discrete sense. The addition of "-FS" in the code name represents the fractional step method implemented in the pressure projection approach. As in the case of "-LU", this version was primarily used to compare different approaches.

5.2 Selection of Codes for Engineering Applications

Historically, many research versions of incompressible flow solvers have been developed at Ames. Practically speaking, not all the results from those codes can be presented here. Those benchmark problems presented are fundamental fluid dynamics problems in nature. These problems are geometrically simple, and approximations related to geometry and grid do not cause significant issues—so, these problems offer good validation cases for characterizing algorithms and for comparing the codes' capabilities in predicting various flow features.

Based on our benchmark test runs and experience gained from other applications, we have observed the following:

For obtaining steady-state solutions, the artificial compressibility approach (INS3D-UP) is very effective. Combined with overset-grid topology, this approach offers the greatest flexibility in modeling complex geometry problems.

1. For obtaining time-accurate solutions, the artificial compressibility approach with sub-iterations at the pseudo-time level works very well and offers great flexibility in modeling complex geometry problems. Even though the implicit sub-iteration is computationally expensive, since the method does not require strict enforcement of a divergence-free velocity field in each time step, one can advance the time without tight convergence in the continuity equation.

2. The pressure projection method (INS3D-FS) offers an alternative to the artificial compressibility code when the flow physics require a small time step. Even though the pressure Poisson equation is expensive, when the time step is small, iterations can be less expensive. Combined with multi-grid acceleration, this can be a competitive approach.

3. In addition to benchmark problems, the INS3D-UP artificial compressibility code has been used in engineering applications presented in Chapters 6, 7 and 8. The above observations are presented here first so that readers may understand why we chose a particular solver in conjunction with applications for engineering. Several basic validation cases will be presented next with some detailed discussion.

5.3 Steady Internal Flow: Curved Duct with Square Cross Section

The flow through a square duct with a 90° bend offers a good test case for a three-dimensional Navier–Stokes solver. This flow is rich in secondary flow phenomena, both in the corner regions and through the curvature in a streamwise direction. Flow through this geometry was studied experimentally by Humphrey et al. (1977) and Taylor et al. (1981, 1982), and extensive laminar flow data are available. This particular geometry was used as a steady-state test case for both the artificial compressibility and pressure projection methods discussed in Chapters 3 and 4. The geometry is shown in Fig. 5.1; the Reynolds number of the flow is set to 790 based on the unit length and average velocity, which is identical to the experimental cases mentioned. The problem was non-dimensionalized using the side length, H, of the square cross section.

Since the flow considered here is laminar, accuracy of the computed results in this case depends primarily on the algorithm selected, including numerical dissipation associated with the procedure and grid resolution. For turbulent cases, issues related to turbulence modeling have to be added to the uncertainties of the results. Even in this seemingly simple example, the variations in the numerical results coming from formulations and computational modeling are clear. Two aspects affecting the computed results will be examined next; namely, grid resolution and algorithm.

Next, we present the study on the effect of grid resolution from the work by Rogers and Kwak (1989), where the INS3D-UP artificial compressibility code is used. However, similar variations in numerical results can be observed with other codes. For the inflow boundary condition, the velocity was specified to be that of the fully developed, laminar, straight square duct (see White, 1974).

Fig. 5.1 Geometry of a
square duct with a 90° bend,
with a grid of 31 × 11 × 11

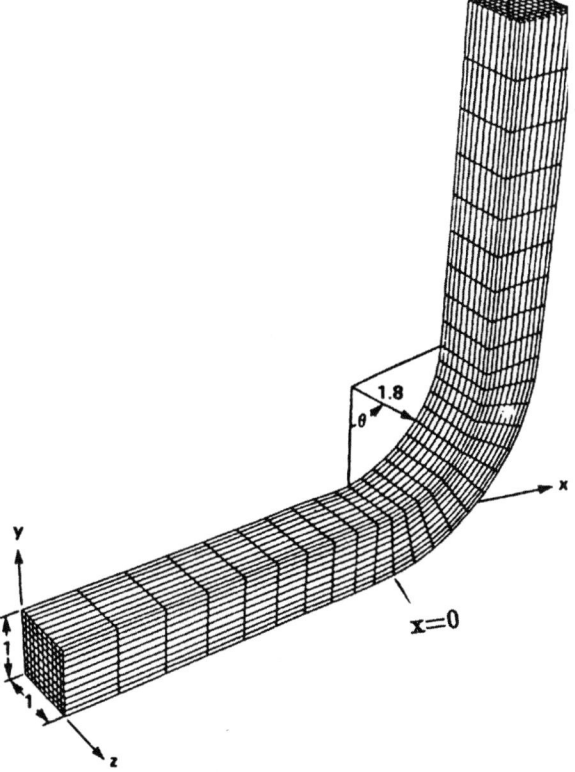

The velocity is normalized by the average inflow velocity. The computed results
are compared to the experimental results of Humphrey et al. (1977). Three different
grids are used, each with dimensions of 31 × 11 × 11, 61 × 21 × 21, and 121 ×
41 × 41. The coarsest grid (31 × 11 × 11) is shown in Fig. 5.1. Both the straight
inflow section before the bend and the outflow section after the bend were set to
a length of five. Computational experiments (Rogers et al., 1991a) show that the
solutions are insensitive to the downstream boundary locations and conditions. As
shown in the figure, the radius of curvature for the inner wall of the bend is set to
be 1.8 units.

To study how the artificial compressibility affects the convergence, the 31 × 11 ×
11 grid problem is computed using β values ranging from 0.1 to 10,000. The convergence of the maximum residual values is plotted against iteration numbers,
shown in Fig. 5.2a. The values of β, ranging from 1 to 100, leads to excellent
convergence. Therefore, for the remaining computations for this problem, the β
value of 10 is used.

The convergence history of three grid cases is compared in Fig. 5.2b. The maximum residual over all grid points vs. iteration number is plotted. The convergence
is shown to be very fast, although it is somewhat slower for the finest grid. It is
expected that finer grid takes longer to converge because the information has to
propagate through a greater distance in computational space.

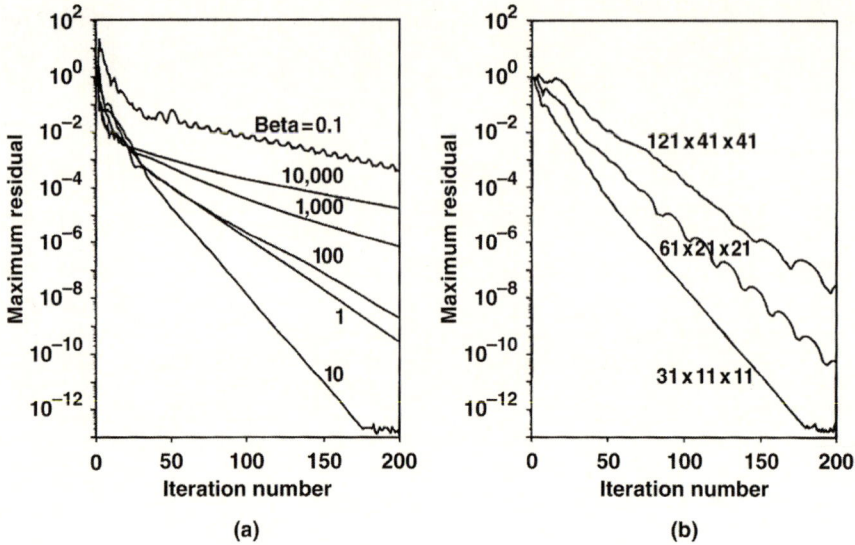

Fig. 5.2 Convergence rate for flow through a square duct with a 90° bend at Re = 790: (**a**) effect of β on convergence on 31 × 11 × 11 grid; (**b**) using different grids with $\beta = 10$

Comparing computed results to experimental data brings up a question on the location for plotting. Since secondary flow is generated through the bend, comparing the results on one cross-section could exhibit misleading discrepancies, while over-all flow quality is fairly well captured by the computations. This will be explained in more detail later in this section. For the moment, computed results for two grids are shown in Fig. 5.3 using velocity magnitude contours at the 90-degree cross section at the end of the bend. As the figure shows, there is very good comparison between the medium- and fine-grid solutions throughout most of this cross section. In par-ticular, the location and value of the maximum velocity magnitude agree very well. Some minor difference occurs in the swirling flow in the region close to the inner wall where the flow is more dissipated in the medium grid compared to fine grid solution.

The computed streamwise velocity profiles at various streamwise stations are then plotted in Fig. 5.4. The plots on the left are from $z = 0.25$, which is halfway between the x-y plane wall and the x-y symmetry plane. The right-hand side plots are from the x-y plane at $z = 0.5$, which is the x-y symmetry plane. In both parts of the figure, the profiles are shown at four positions in the curved section correspond-ing to θ equal to 0, 30, 60, and 90°. The symbols represent the experimental results, and the lines represent computed results.

The results generally follow closely to one another. One exception to that general trend is the formation of the second maximum in velocity on the inner wall side. This second maximum is developed farther upstream in the computation than in the experiment, causing discrepancy at the $\theta = 60°$ location in both plots. At the $\theta = 90°$

Fig. 5.3 Velocity magnitude contours for fine-grid (121 × 41 × 41) and medium-grid (61 × 21 × 21) computations at $\theta = 90°$

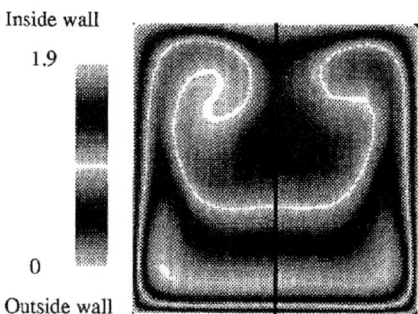

location for $z = 0.25$ plot, three different maxima occur in the computations. Details of the dynamics can be seen more clearly by observing the entire cross section rather than from line plots alone. It would be even more illuminating if there were a convenient way to observe the entire three-dimensional flow. A series of 2-D cross-sectional views from the fine-grid computation, shown in Fig. 5.5, provides a better picture for understanding these multiple peaks.

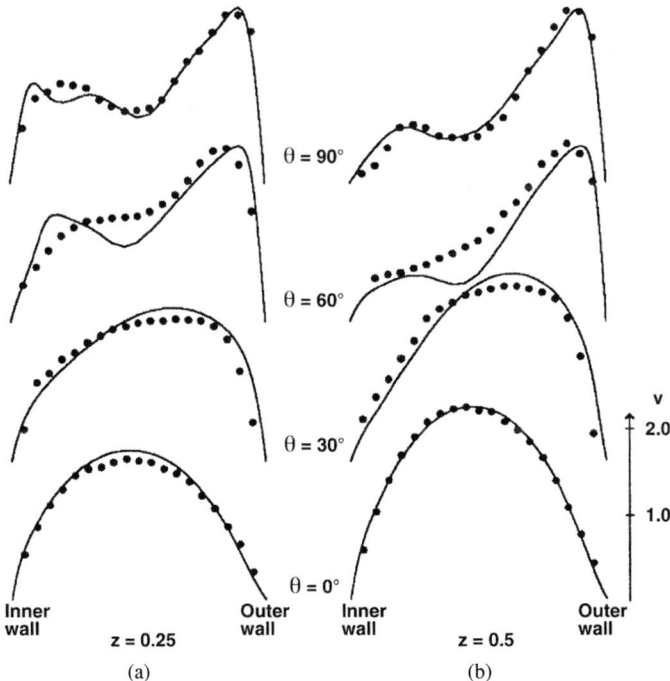

Fig. 5.4 Streamwise velocity profile for flow through a square duct with a 90° bend at Re = 790: (**a**) x-y plane at $z = 0.25$; (**b**) x-y plane at $z = 0.5$; $r = 0$ represents inner wall and $r = 1$ represents outer wall

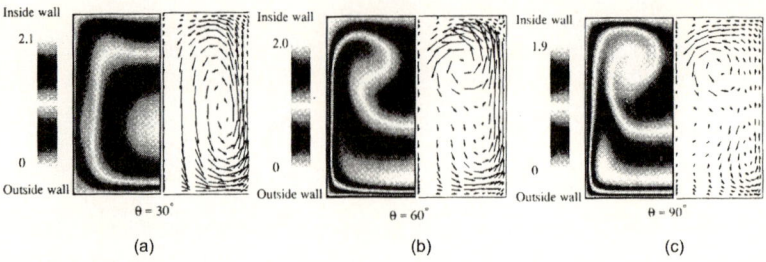

Fig. 5.5 Velocity magnitude contour and cross-sectional velocity vectors at three sections for flow through a square duct with a 90° bend at Re = 790

First, the high-velocity fluid moves toward the outside wall (Fig. 5.5a) as the flow turns the bend, bringing some of this high-velocity flow toward the inner wall (Fig. 5.5b), which later forms multiple peaks in velocity near the inner wall. Then, as seen at the 90-degree location (Fig. 5.5c), the swirl has wrapped the region of higher velocity around toward the middle. Therefore, any small change in flow conditions, such as Reynolds number, grid resolution, and dissipation terms, can change the magnitude and the exact location of this swirl. Even though the overall comparison is quite satisfactory, some of the details can best be compared by viewing at least the entire cross-sectional results, which also requires detailed measured data.

To further illustrate this point, the computed results on the square duct with a 90° bend by McConnaughey et al. (1989) is presented next. In their work, extensive validation of the artificial compressibility method was performed using the original version of INS3D. Utilizing the detailed inflow measurement by Taylor et al. (1981, 1982), they investigated various aspect of the flow solver, including a grid refinement study and a sensitivity study of the numerical dissipation terms. Computed results on a 90° bend and an S-bend are extensively compared in the same report.

From the grid resolution study using four successively refined grids, the finest grid results are shown in Fig. 5.6, using grid dimension of 28 × 52 × 121 for one-half height by width by length. The figure shows predicted axial flow contours compared to experimental data, while Fig. 5.7 shows predicted radial flow contours compared with experimental data. These contours on a cross-sectional plane illustrate how the numerical solutions can be compared with experiments. The flow field details from the cross-sectional view shed more light than studying line plots alone. At the final plane of data at +2.5H past the bend, the experimental data exhibit more dissipation than that predicted by computations.

The source of this difference needs further investigation. However, this suggests that computations can be performed for planning experiments. Overall, both computations and experiments compare very well. As noted by those authors, the square duct problem offers a good test case for validating secondary flow prediction capability, even though the geometry is simple. Correct prediction of this phenomenon can play an important role in engineering problems involving complex internal flow geometry such as that encountered in advanced rocket propulsion systems.

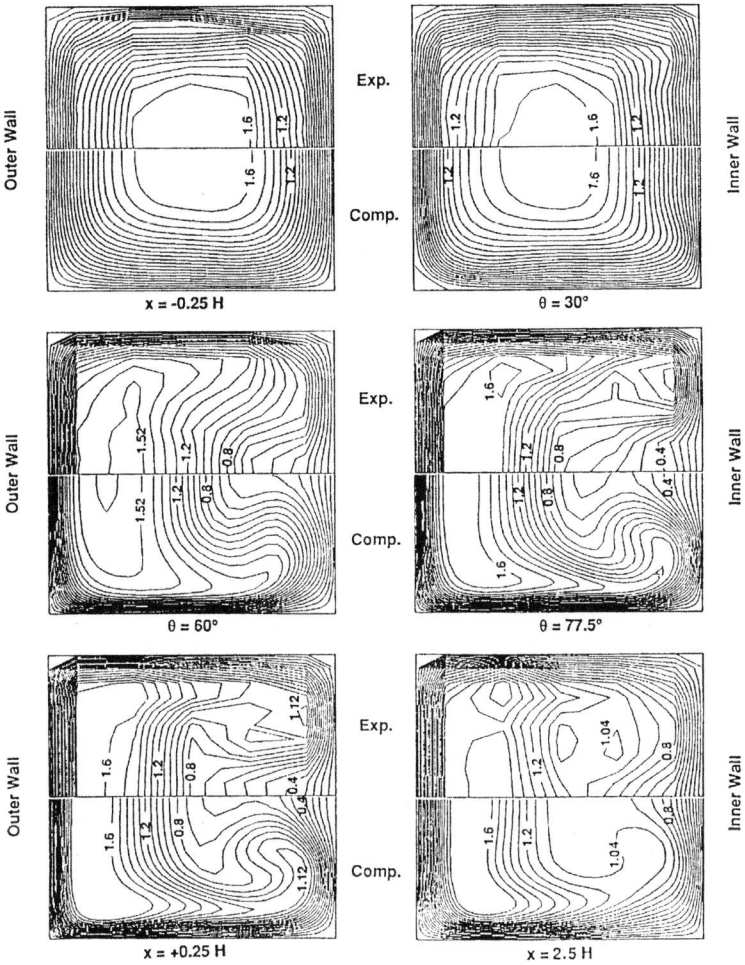

Fig. 5.6 Predicted axial flow in a 90° bend (McConnaughey et al., 1989, using INS3D) compared with the experimental data of Taylor et al. (1981)

We have presented how grid resolution and artificial compressibility affect the quality of solutions, as well as the issue related to post-processing the results. As explained in Section 5.1, three different flow solver codes have been developed based on different algorithms and discretization. The performance of these three approaches is compared in Fig. 5.8 using a medium grid resolution. As the figure shows, results vary somewhat depending on differencing schemes and the smoothing applied. Again, as explained above, the magnitude of differences in line plots can be somewhat bigger than magnitude observed in contour plots. However, this illustrates differences among codes that may be expected in practice. Quantification of uncertainties or errors as illustrated here is a challenging issue, and generally accepted criteria have yet to be developed.

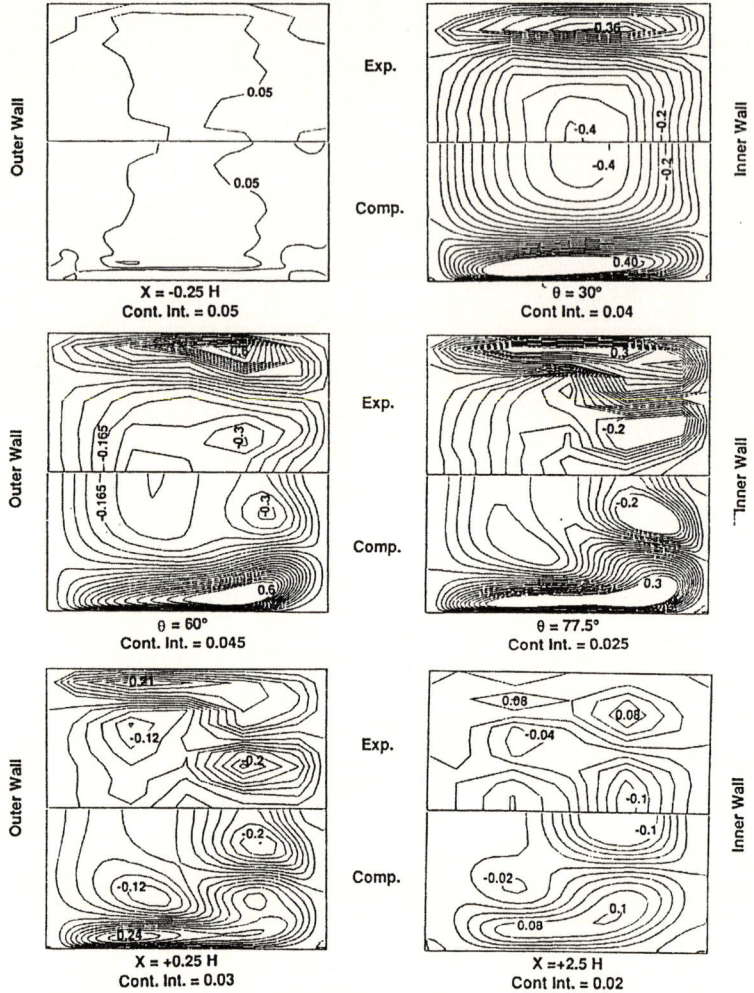

Fig. 5.7 Predicted radial flow in a 90° bend (McConnaughey et al., 1989, using INS3D) compared with the experimental data of Taylor et al. (1982)

5.4 Time-Dependent Flow

For quantification of fluid dynamic characteristics such as forces and moments on a vehicle, steady-state solutions are used. For practical applications, using steady-state solutions as ensemble-averaged quantities is in many cases the most reasonable approach in design and analysis. However, in many realistic situations, flow becomes time dependent, either in transitional mode such as in impulsively started flow or in unsteady fluctuating mode. For example, for the analysis of startup conditions as encountered in an impulsively started vehicle and for determining

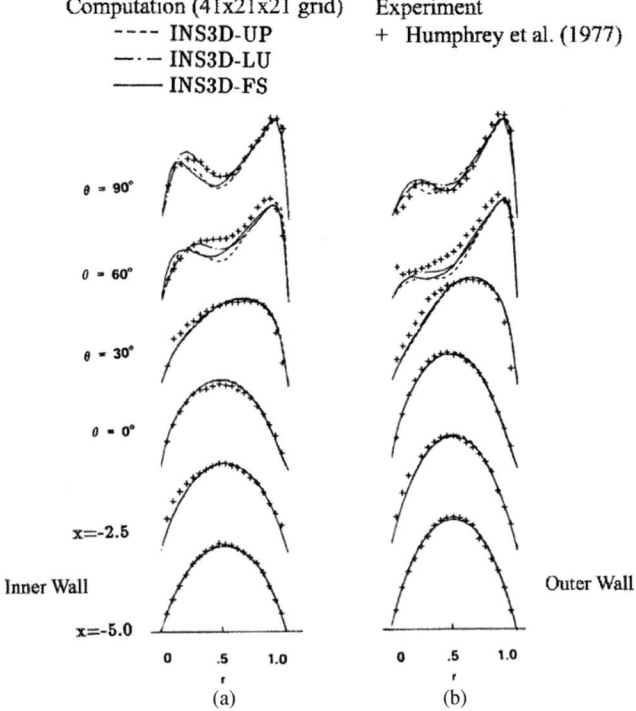

Computation (41x21x21 grid)   Experiment
   ---- INS3D-UP   + Humphrey et al. (1977)
   —·— INS3D-LU
   —— INS3D-FS

Fig. 5.8 Comparison of streamwise velocity distribution along (**a**) z = 0.25, and (**b**) z = 0.5

vibration load due to fluctuating flow, time-accurate computations are necessary, which are much more expensive than steady-state computations—usually at least one order of magnitude higher. For the purpose of validating the time accuracy of the algorithms and processes discussed earlier, a few building block problems are studied next.

5.4.1 Flow Over a Circular Cylinder

Flow over a circular cylinder has been of interest for many decades, as it offers a full range of phenomena from laminar, periodic shedding of vortices, transition to turbulent, and fully turbulent flow regimes (see Morkovin, 1964)—making it a challenging problem for computational simulation. Computational studies of flow over a circular cylinder began as early as the 1930s (e.g., Thom, 1933), and this continues to be a popular subject.

In this section, an impulsively started circular cylinder at Re = 40 is first presented, followed by vortex shedding cases with higher Reynolds numbers up to 1,000. This represents a simple case for external flows, for the purpose of comparing artificial compressibility and pressure projection formulations without involving transition and turbulence modeling.

For external flows in which the computational domain extends a large distance from the body, the pressure waves originating from the body surface propagate into the far field. Therefore, to obtain the near-field solution using only an artificial compressibility code, the distance traveled by the waves and the spreading of vorticity can be considered approximately the same in magnitude. The range of β in INS3D can then be estimated based on this reasoning. For example, at Re $= 40$, if the viscous region is taken to be approximately two diameters away from the body, one can estimate the following range for β using $\Delta\tau = 0.1$:

$$0.1 << \beta < 10$$

Physically, for external flows, the pressure wave can quickly travel a short distance to balance the viscous region close to the body. Therefore, the magnitude of β is less restrictive than for internal flow cases such as in the channel flow described in Chapter 4. Results from computations using this range of artificial compressibility produce similar data found in the literature for steady-state computation at Re $= 40$ (Kwak et al., 1986, INS3D). Far more extensive validation computations were performed at Re $= 5$, 10, 20, and 40 by Rogers and Kwak (1988) using INS3D-UP, where computed data showed good agreement with experiments and other computations (results not illustrated here).

To validate the time accuracy of the pressure projection method, the near-field detail of transient flow for impulsively started circular cylinders at Re $= 40$ and 200 is computed using the INS3D-FS code (Rosenfeld et al., 1988; Rosenfeld and Kwak, 1989). In Fig. 5.9, the computed time evolution of the separation length is compared with experiments by Coutanceau and Bouard (1977). Also plotted are the computed result by Collins and Dennis (1973). During the initial, short start-up time period, the flow development is viscous dominated and both methods produced equivalent results in capturing time accuracy (see Rogers et al., 1985, for INS3D-UP results).

As the Reynolds number increases above 40, a non-symmetric wake develops and periodic vortex shedding sets in. Both the artificial compressibility approach (INS3D-UP) and pressure projection method (INS3D-FS) are validated using this simple yet challenging problem. In Fig. 5.10, these computations are compared with other numerical and experimental results.

The calculations using both codes were performed using an O-type grid clustered near the body. To obtain time-accurate solutions from INS3D-UP, sub-iterations were carried out at each physical time step. Starting impulsively from rest, over 20 sub-iterations were required during the transient phase. A non-symmetric wake develops spontaneously, followed by shedding of vortices, without introducing any artificial disturbance—probably due to biasing in the upwind scheme, which in turn may have introduced enough disturbance into the flow field. When a central differencing scheme is used (as in INS3D-FS), it is necessary to introduce asymmetric disturbance to initiate the shedding within a reasonable time. This technique is

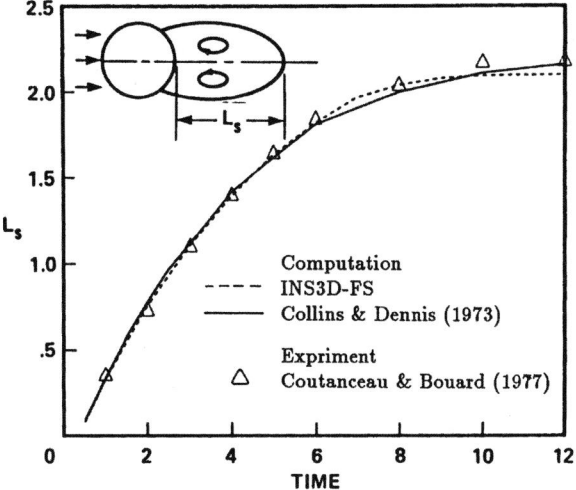

Fig. 5.9   Time evolution of separation length for flow over a circular cylinder at Re = 40

consistent with the natural process where some sort of disturbances trigger vortex shedding. Figure 5.10 plots a Strouhal number versus a Reynolds number from computed results compared to experiments. Both methods produced comparable solutions. The Strouhal number is perhaps relatively easier to predict. However, the time accuracy of the code itself has been validated using other problems with exact solutions (see Rogers and Kwak, 1990).

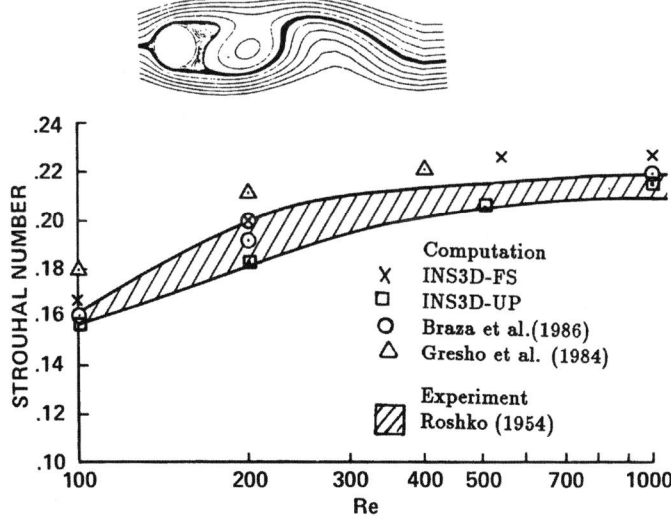

Fig. 5.10   Vortex shedding from a circular cylinder

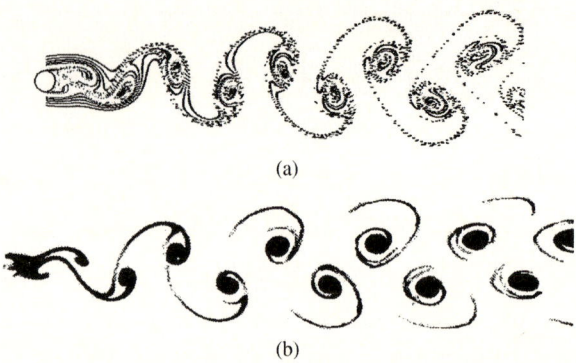

Fig. 5.11 Karman vortex street behind a circular cylinder at Re = 105: (**a**) computed result using the artificial compressibility method (INS3D-UP code); (**b**) experimental visualization by Taneda (see Van Dyke, 1982)

The staggered pattern of the vortex shedding known as Karman's vortex street has been a subject of many flow visualization studies. For the purpose of comparison, particle traces are generated from the time-dependent solution of flow around a circular cylinder at a Reynolds number of 105 using INS3D-UP. Figure 5.11a shows this computed vortex street, while Fig. 5.10b shows an experimental photograph of the same conditions taken by Taneda in 1972 and reproduced from Van Dyke (1982). The streaklines in the experiment are shown by electrolytic precipitation in water. As can be seen, the vortex structure is very similar between the two. The experimental picture is digitized and displayed on a workstation along with the computationally generated flow visualization image. This example illustrates the potential of using post-processing of the CFD results for studying fundamental fluid dynamics phenomena.

5.4.2 Impulsively Started Flat Plate at 90°

To further investigate various algorithm features of the two primary methods designed for applications, namely the artificial compressibility method (INS3D-UP) and the pressure projection method (INS3D-FS), an impulsively started flat plate at 90° to the flow direction is solved next. Even though the unsteady flow encountered in two-dimensional problems does not encompass all the features observed in three-dimensional problems, such as vortex stretching, current numerical experiments can provide some basis for selecting methods for real-world applications.

Computed results from both of these methods are compared with the experimental data by Taneda and Honji (1971). The experiment has been carried out in a water tank 40 cm wide, where a thin, 3-cm high flat plate was immersed. The flow started from rest impulsively at the velocity u = 0.495 cm/s. The Reynolds number for this case is 126 based on the plate height; the computational grid size is 181 × 81 in

Fig. 5.12 Computational grid for the impulsively started flow past a 90-degree flat plate: plate thickness = 0.03 H, Reynolds number based on plate height = 126

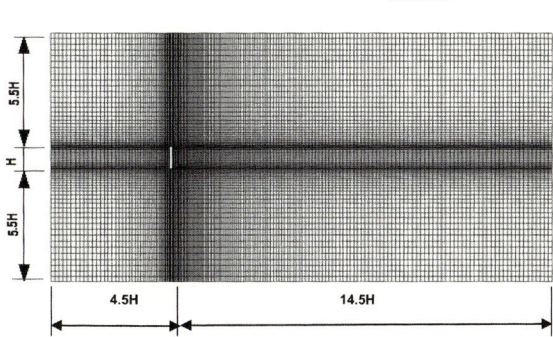

the flow and vertical directions, respectively (Fig. 5.12). Recalling that INS3D-FS is written in a finite volume staggered-grid formulation, one additional ghost cell is required in each direction.

To visualize the time evolution of the flow, velocity vectors at various non-dimensional times are plotted in Fig. 5.13. The flow separates at the edge of the plate and forms a vortex pair. The twin vortices are elongated in flow direction as time progresses. To quantify the time history, the separation bubble lengths from computations and experiment (Taneta and Honji, 1971) are compared in Fig. 5.14. The separation bubble length is defined in the figure. Computations using the artificial compressibility, pressure projection, and a finite element method (Yoshida and Nomura, 1985) all produced very similar results.

Although the plate started impulsively in the experiment, it was started slowly in the computation. Two different ways of starting the flow are used in the computations, as illustrated in Fig. 5.15. The computed results plotted in Fig. 5.14 used a slow start procedure with the time interval of 0.5 s, which corresponds to a non-dimensional time step size of 0.0825. For the finite element computations plotted in the same figure, Yoshida and Nomura (1985) used the same slow start procedure.

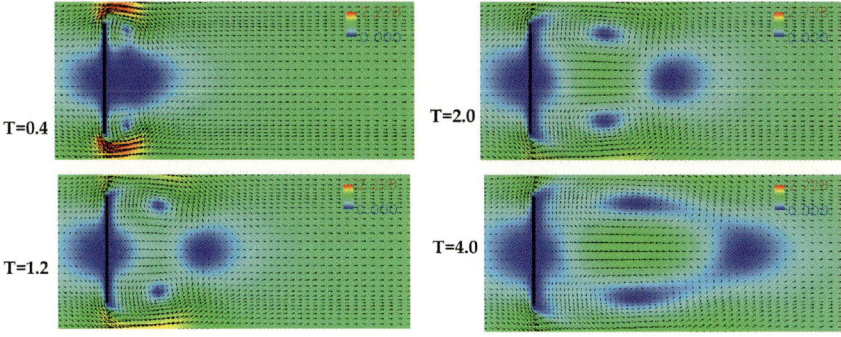

Fig. 5.13 Computed velocity vectors at various non-dimensional times (INS3D-FS)

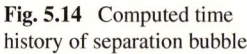

Fig. 5.14 Computed time history of separation bubble

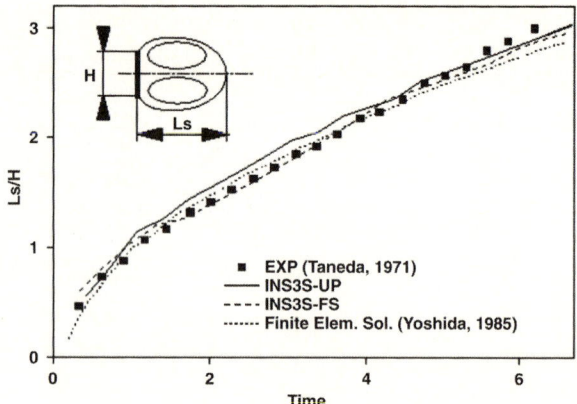

The velocity profile for the slow start case is prescribed as shown in Fig. 5.15b and the starting time of calculation is shifted to match that of the experiment.

Next, computed results using the INS3D-FS pressure projection code are presented in some detail. The effect of the starting procedure on the flow development is shown in Fig. 5.16a. Measurable differences between the two procedures can be seen in the resulting flows. In Fig. 5.16b, results using different time step sizes and grid resolution are plotted. Increasing the spatial resolution does not improve the results significantly, while decreasing the time step size improves agreement with experiment.

For the artificial compressibility method using INS3D-UP, two important parameters affecting time accuracy are the artificial compressibility parameter, β (BETA), and the number of sub-iterations at each time level to recover the incompressibility condition. Two different β and sub-iteration numbers have been tested, as shown in Fig. 5.17. Since a slow start produces more favorable results, as studied above, that procedure is employed here.

This experiment shows the importance of satisfying incompressibility for time accurate computations using the artificial compressibility approach. However, a large number of sub-iterations can impose a heavy burden on computational resources, in contrast to the small number of time steps required for the pressure projection approach. In reality, one can limit the sub-iterations to a fixed number. In that case, users have to determine the level of time accuracy needed for the analysis at hand, in light of the available computing resources.

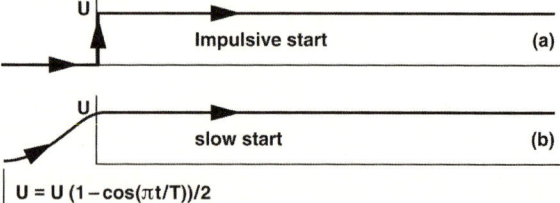

Fig. 5.15 Prescribed velocity for an impulsive started (a) and a slow start (b) procedure

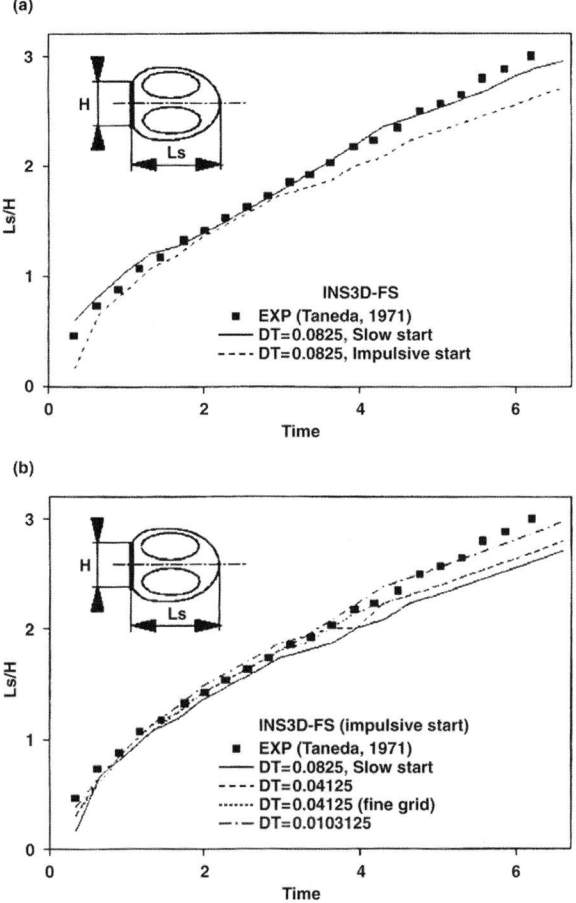

Fig. 5.16 Effects of starting procedure and time step size for pressure projection method (INS3D-FS code)

5.4.3 Pulsatile Flow Through A Constricted 2-D Channel

The next validation cases are designed to test time accuracy for internal flow using INS3D-FS and INS3D-UP, representing the pressure projection and artificial compressibility methods, respectively.

5.4.3.1 Oscillating Wall

The first case is intended to model the large-amplitude self-excited oscillations generated when fluid flows through a collapsible tube, such as blood flow through a vein. Stephanoff et al. (1983) and Pedley and Stephanoff (1985) performed a series of flow visualization experiments in which the channel walls were rigid except for the indented region. The length of an indentation is 10, measured by an unindented

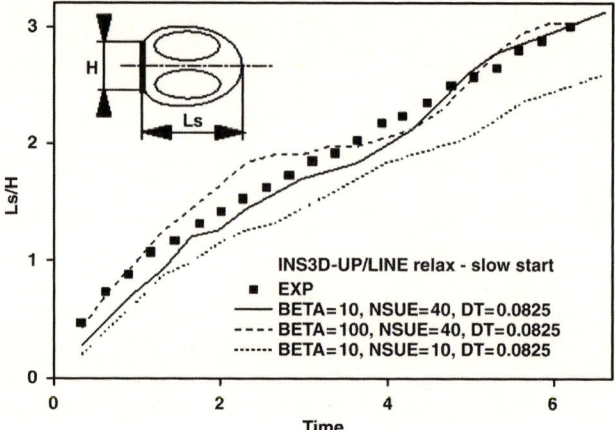

Fig. 5.17 Effects of starting procedure and effects of time step size

channel height as the reference unit length. The channel starts at a distance of 120 units upstream of the oscillating constriction and is 250 units long. The indentation is made of a thick rubber membrane and is driven by a piston with a sinusoidal motion in time and a maximum indentation of 0.38 units. At the beginning of each cycle the indentation is flush with the wall of the channel.

A computational model for this experiment was constructed with the upstream boundary placed at 5 units from the oscillating constriction and the downstream boundary placed at 30 units from the upstream boundary. A grid for this model is illustrated in Fig. 5.18, with the grid dimension of 31×251 in cross-stream and flow directions, respectively. The grid in the indentation region stretches and compresses linearly with the moving indentation. Every other grid point is plotted in the figure. The shape of the indentation is approximated by a hyperbolic tangent, as suggested by Pedley and Stephanoff (1985). The grid for the upstream side of the indentation is much coarser compared to the downstream side, making the computational geometry asymmetric. However, numerical experiments show that flow is insensitive to the upstream geometry of the indentation, so the grid points are more clustered near the downstream side of the indentation.

A fully developed flow profile is given at the initial and upstream boundaries. At the downstream boundary, a non-reflecting condition is imposed. In Fig. 5.19,

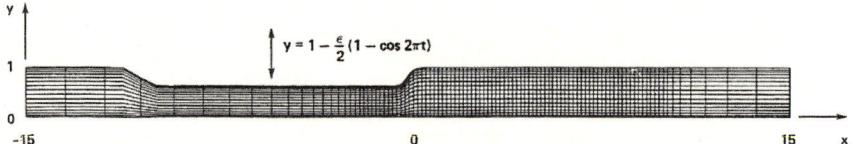

Fig. 5.18 Grid for channel with an oscillating indentation

Fig. 5.19 Comparison of
instantaneous streamlines at
$t = 0.55$, $Re = 610$: (**a**)
experimental results; (**b**)
computed results (INS3D-FS)

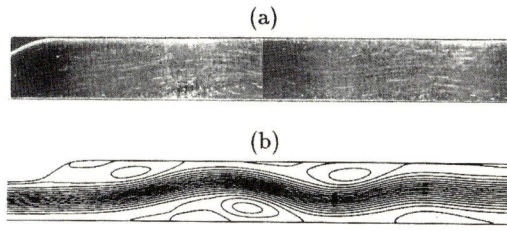

(a)

(b)

instantaneous streamlines from computations are compared to the experimental visualization by Stephanoff et al. (1983) at the non-dimensional time of $t = 0.55$, based on one period. The Strouhal number $St = 0.038$ and the Reynolds number $Re = 610$ are based on the channel height and average velocity.

The first separation occurs downstream of the sloping wall, followed by a second large eddy formed on the opposite wall with a secondary separation bubble buried inside the primary bubble. A similar pattern of weaker vortex pairs repeats as flow goes downstream. The computed separation length of the first eddy at the upper wall is under-predicted compared to experiments. However, the distance between vortices, which are related to the wavelength of the core flow, compare favorably. Overall, the results are shifted about 0.4 units of the channel height.

To identify the cause of this discrepancy, we conducted a grid refinement study, and applied the artificial compressibility code INS3D-UP to compute the same case. Both produced almost identical grid-independent results. This indicates that it is likely that the modeling of the experimental set up does not exactly match the experimental conditions.

Evolution of the flow over one complete cycle is shown in Fig. 5.20 for $Re = 600$ and a Strouhal number, $St = 0.057$. The instantaneous streamlines are plotted at several instances over one complete cycle of the indentation. The flow development is very similar to that observed experimentally by Pedley and Stephanoff (1985). At the beginning of the cycle, during which the indentation moves downward, the downstream flow is accelerating and a single separation bubble forms on the sloping wall behind the indentation, as shown in Fig. 5.20a. As time progresses, the separation length increases and a second counter-rotating eddy appears on the opposite wall downstream of the primary eddy, as shown in Fig. 5.20b. The flow field significantly changes during the second half of the period, when the indentation moves upwards and causes deceleration of the flow downstream. A third eddy is formed at the upper wall further downstream at $t = 0.55$, as shown in Fig. 5.20c. In the third quarter of the period, the core flow becomes wavy and a series of eddies appear along the walls. The amplitude of the core flow increases with time up to $t = 0.75$, as shown in Fig. 5.20g, which corresponds to the maximum deceleration. During the last quarter of the period, eddies shrink in size and strength and wash downstream. By the end of the cycle, the residual eddies are quite small. and were found not to affect the next cycle. Following the experiments by Pedley and Stephnoff (1985), eddies are labeled alphabetically, as shown in Fig. 5.20d.

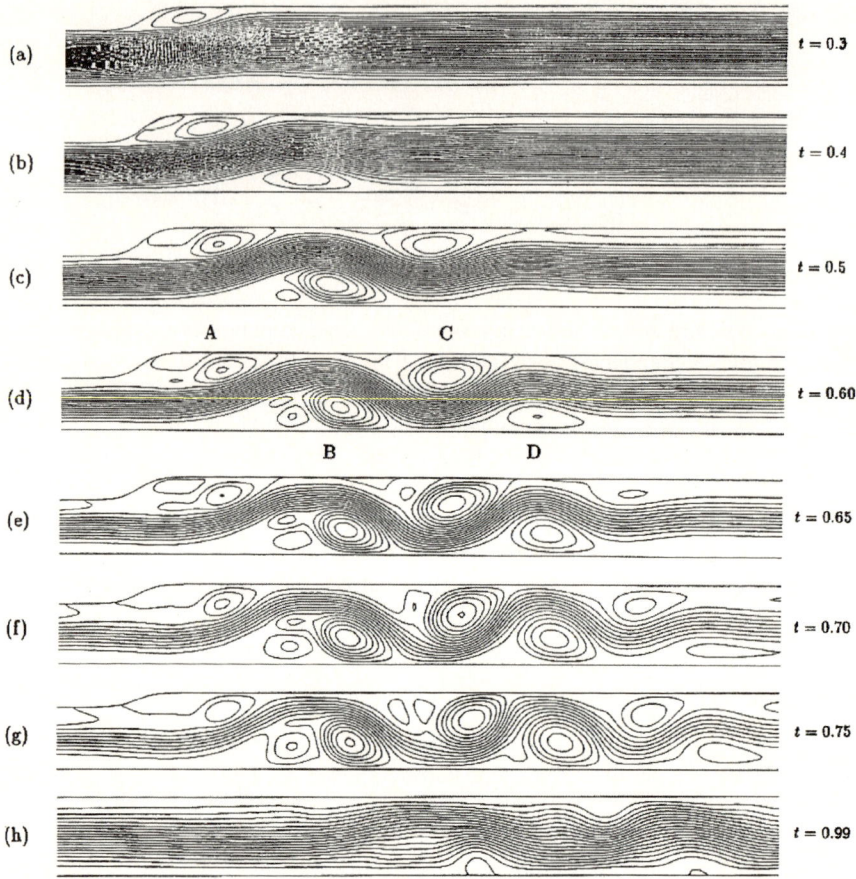

Fig. 5.20 Evolution of instantaneous streamlines computed using INS3D-FS: St = 0.55, Re = 600

It is difficult to understand the entire flow field dynamics of this case, especially from a validation point of view. As a measure of validating time accuracy, the time evolution of the center of vortices A, B, C, and D is plotted in Fig. 5.21. Computed results from the two methods and experimental measurements are compared. Even though this plot is somewhat qualitative, dynamics of eddies generally follow the experimental trend. For full validation, a grid resolution study along with more quantitative measurements will be necessary.

5.4.3.2 Oscillating Inflow

In the second case, we model pulsating flow through a constricted channel, such as in the case of blood flow through stenosed arteries. The computational geometry is consistent with the experimental setup of Park (1989), and is shown in Fig. 5.22. The

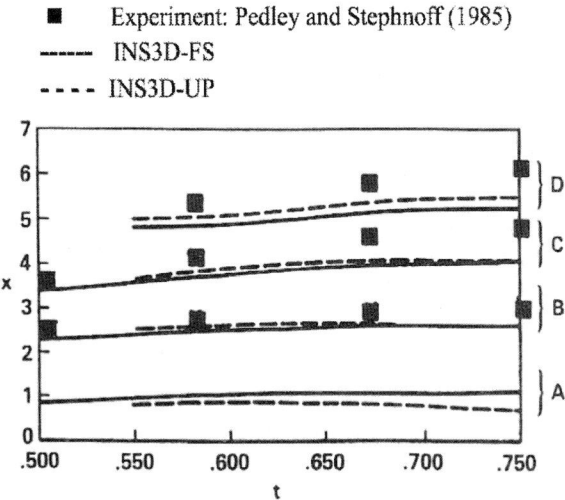

Fig. 5.21   Time evolution of center of vortices

channel height is h and has been normalized to unity. The height of the constriction, a, is 0.57, which is the distance from the top of the wall to the lowest point in the constriction. The length of the channel upstream of the constriction is given by $L_u = 7$. The length of the channel is $L_c = 4.66$ and the length of the straight channel downstream of the constriction is $L_d = 15.34$ (Fig. 5.22).

The inflow boundary for the experimental setup was at 100 channel heights upstream of the constriction. For computational efficiency, the boundary for the inflow condition was placed at 7 channel heights upstream of the constriction. The inflow boundary condition is designed to match with experiment (Rosenfeld et al., 1991b). A parabolic profile was imposed and scaled to match the mass flow from the experimental setup.

The inflow velocity profile is periodic and is given by the shape function given in Fig. 5.23. This shape is defined analytically by:

$$U(t) = U_s \qquad 0 < t/T < 1/2$$
$$U(t) = U_s - U_p \sin(2\pi t/T) \quad 1/2 < t/T < 1$$

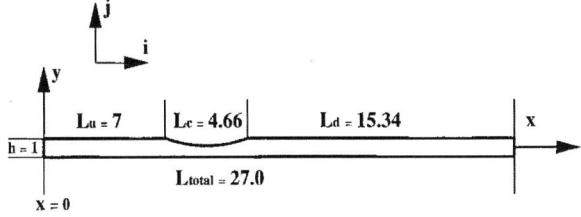

Fig. 5.22   Computational model of 2-D constricted channel

Fig. 5.23 Inflow velocity
versus time for one period

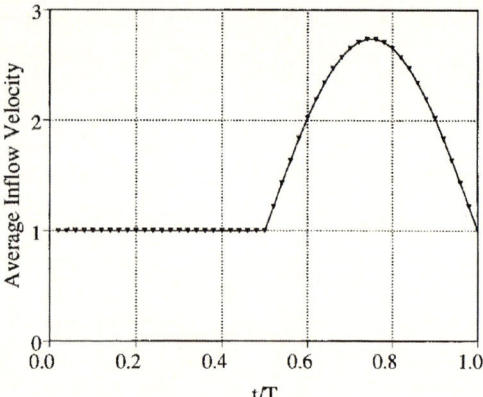

where U_s is the non-dimensional steady component of the average velocity, U_p is the pulsatile component and T is the period. This wave form was created to represent diastole and systole in a mammalian blood circulatory system and could generically represent conditions encountered for modeling arteriosclerosis.

Since the wall geometry is fixed, this case offers an opportunity to compare the two methods in achieving time accuracy. Both methods use pressure as a mapping parameter to obtain incompressibility after advancing each time step, and the iterative processes are very similar. However, it is interesting to observe a subtle difference between the artificial compressibility method (IS3D-UP) and the pressure projection method (INS3D-FS). The former shows upstream propagating waves from the downstream boundary, while the latter changes the pressure in time. Figure 5.24 shows the evolution of pressure contours during the iteration process

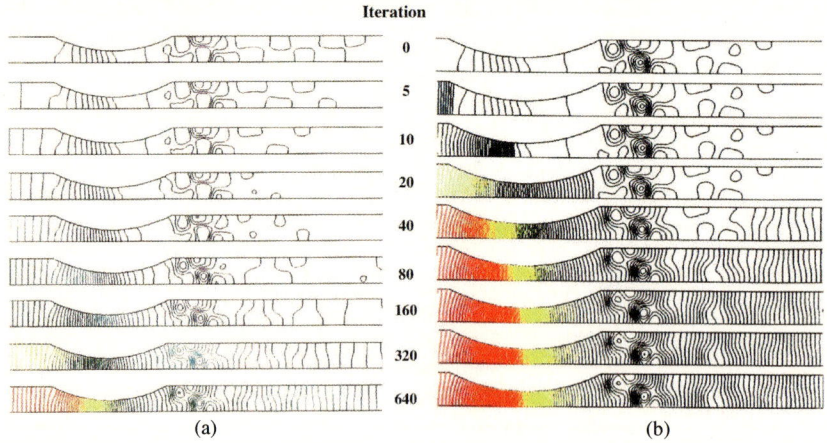

Fig. 5.24 Evolution of pressure contour during iteration within one time step at $t/T = 0.5$ using $\Delta t/T = 0.01$: (**a**) sub-iteration using INS3D-UP with $\beta = 100$; (**b**) Poisson iteration using INS3D-FS

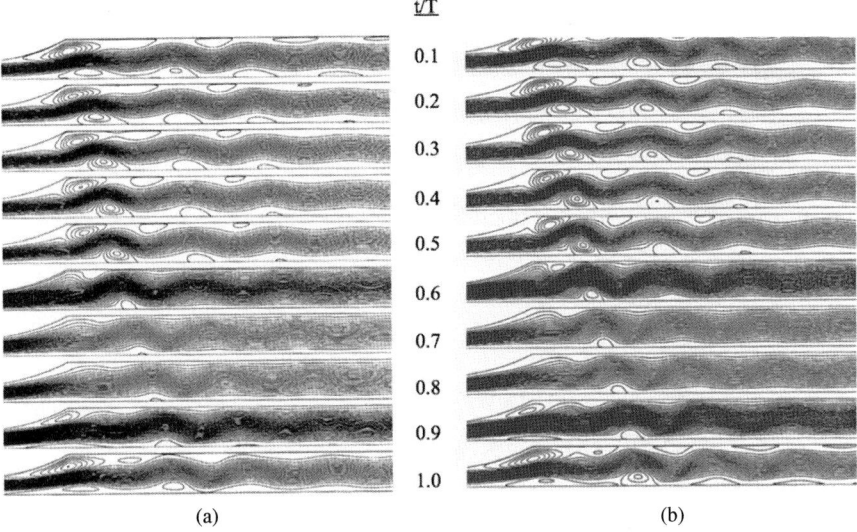

Fig. 5.25 Evolution of pressure contour for one period: computed results using (**a**) INS3D-UP, and (**b**) INS3D-FS

for one time-step advancement. The plots are obtained from two separate computations and contours are not in the same scale. However, the two show the pressure propagation phenomena qualitatively during the iteration within one time step.

To compare the time accuracy of results from the two methods, a train of vortices propagating downstream from the constriction is plotted for one period. This comparison, as shown in Fig. 5.25, is qualitative, and also shows how the flow develops as the vortices convect downstream.

One easily quantifiable physical result is the location of the center of the B-vortex, defined to be the vortex along the bottom wall of the channel, as shown in Fig. 5.26; it grows immediately behind the end of the constriction and is shed downstream. The location of this vortex was examined as function of time. Results from both the INS3D-UP and INS3D-FS methods are compared to the experimental results. Both methods produce equivalent results and agreement with experiment is good for $t/T < 2.0$, but deviates as it goes further downstream after $t/T > 2.0$. At that point the vortex strength is considerably weaker than at the upstream location, and thus the uncertainty of measuring the center of the B-vortex increases. This suggests that more precise measurements can be useful for validating the time accuracy of these numerical methods and simulation procedures.

5.4.4 Flapping Foil in a Duct

In this section, the time-accurate procedure of the artificial compressibility method described in Chapter 4 will be validated using a flapping airfoil problem. This next

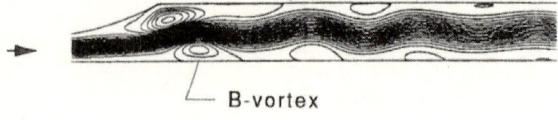

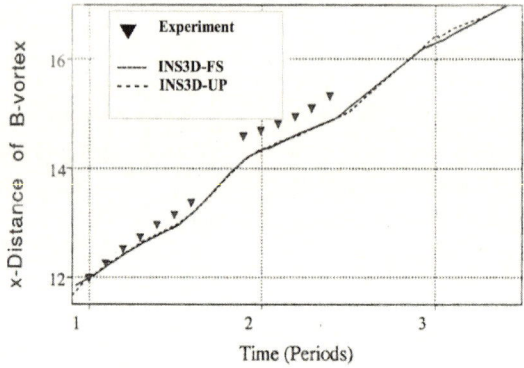

Fig. 5.26 Location of B-vortex as it goes downstream in time

validation case has been selected to represent time-dependent flow generated by
bodies in motion relative to stationary components—a class of unsteady flow often
encountered in engineering. For example, the next generation of fluid engineers
may be required to analyze advanced components such as high-lift devices, marine
propulsion systems, turbopumps in liquid-propellant rocket engines, and mechanical
heart valves and assist devices. These advanced devices are likely to require more
efficient and simpler designs with lower manufacturing costs. Accurate and detailed
knowledge of the flow field obtained by unsteady flow calculations can greatly help
designers to reduce cost and improve the reliability of such advanced systems. In
addition to geometric complexities the challenges in these numerical simulations
include turbulent boundary layer separation, wakes, transition, tip vortex resolution,
Reynolds number effects, and moving boundaries.

An ideal validation case for the time-accurate artificial compressibility method
was presented by the Office of Naval Research (ONR) and Massachusetts Institute
of Technology (MIT) at the Unsteady Flow Workshop held March 29–30, 1993.
The ONR/MIT designed a flapping foil experiment (FFX) as a two-dimensional
representation of the interaction between the propeller blade and wake flows. One
purpose of the experiment was to provide detailed experimental data to be used
to evaluate computational methods for marine propulsors. At the ONR/MIT workshop, the computed results obtained by various research groups were compared with
experimental measurements. The flappers in the FFX generate high frequency periodic wakes, which impose an unsteady loading on the stationary foil. In addition to
the complexity of the flow physics, the numerical simulation of the FFX requires

proper domain decomposition and moving boundary procedures. This makes the FFX a good validation case for a time-accurate numerical procedure. Kiris et al. (1994a) reported time-accurate computations using the artificial compressibility approach. Key features of that report are summarized here, from a validation perspective.

5.4.4.1 Experimental and Computational Models

A schematic of the experimental setup is shown in Fig. 5.27. The stationary foil, which has an 18-inch chord and a 1.18-degree angle-of-attack, represents a propeller blade embedded in the wake generated by upstream pitching foils. The upstream flapping foils are NACA 0025 foils of 3-inch chord. The flappers perform synchronized sinusoidal motions of 6-degree amplitude at a reduced frequency of 3.62.

The flappers in the FFX generate periodic wake, which imposes an unsteady condition on the stationary foil. Velocity and pressure measurements at a Reynolds number (based on the stationary foil chord and the in-flow mean velocity) of 3.7×10^6 were taken on and around the stationary foil inside the measurement box shown by the dashed line in Fig. 5.27. The measurements in this box are given to provide the upstream, downstream, and outer boundary conditions for calculations of the stationary foil alone. Since the purpose of the current numerical study is to investigate the moving boundary capability, the computational model includes the entire domain shown in Fig. 5.27, where experimental inflow and exit conditions are provided for computation.

Various grid topologies are created for computations, and the computed results obtained from a time-accurate artificial compressibility formulation are compared with the experimental data. To simulate the entire configuration, two commonly used grid topologies in structured-grid approaches are tested, i.e., multi-block patched and Chimera overlapped grids.

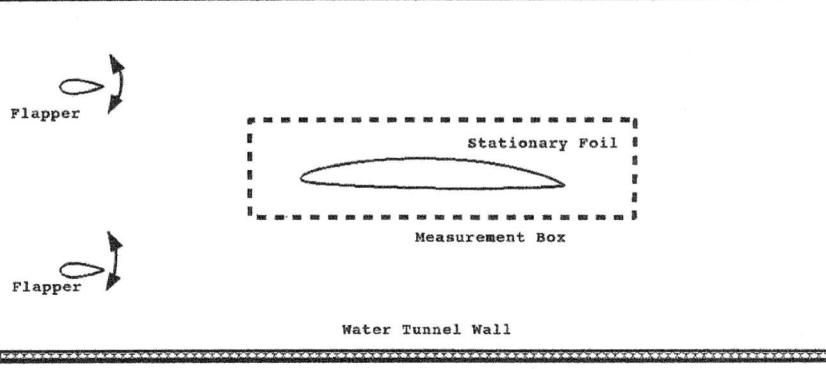

Fig. 5.27 Schematic of MIT flapping foil experiment (FFX)

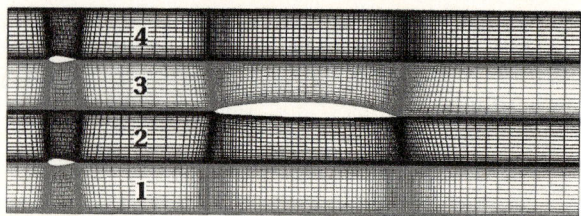

Fig. 5.28 Multi-block patched grid topology for the flapping foil computation

A multi-block patched grid topology applied to the FFX geometry, shown in Fig. 5.28, consists of four H-grids. The patched grids are point-wise continuous at the zonal boundaries, and the interfaces have two points of overlap. Each grid has the dimension of 319×63, resulting in a total number of grid points of 80,388. Alternate grid lines were plotted in all grid-related figures in this section. Grid 1 covers the region between the lower tunnel wall and the lower flapper surface; grid 2 extends between the upper surface of the lower flapper and the pressure side of the main foil; grid 3 is located between the suction side of the main foil and the bottom surface of the upper flapper; and grid 4 extends between the top surface of the upper flapper and the upper tunnel wall.

The advantage of a multi-block patched grid scheme is that the grids remain point-wise continuous as the bodies move, avoiding any interpolation error at the interface boundaries. However, the grid does have to be regenerated at each physical time step to account for the flapper motion. For the FFX, the interface boundaries between zones move up or down with the flappers and each zone contracts and expands during the cyclic motion.

An alternative to the multi-block patched grid scheme is the Chimera overlapped grid scheme. The overlapped grid topology for the FFX is shown in Fig. 5.29a. An H-grid with a dimension of 253×191 (Grid 1) occupies the water tunnel without considering the foils. Three C-grids are generated for the foils and are overlapped with the tunnel grid. Grid 2 is generated for the stationary foil with a grid dimension of 337×61. Grids 3 and 4 wrap around the flappers with grid dimensions of 215×40 each; these rotate with the flappers. The total number of grid points for this grid system is 86,080.

The advantage of the overlapped-grid scheme as a moving boundary procedure is a simplified grid-generation procedure. For the FFX, the grids are generated, then the flapper grids are rotated relative to the tunnel grid. However, additional numerical boundary conditions and the data management for time-dependent interpolation stencils are introduced. The overlapped grid regions in the near field of the flapper and in the near field of the stationary foil leading edge are plotted in Fig. 5.29b.

The individual grids receive information from each other by interpolating the dependent variable. The grids around the foils have outer boundaries overlapping the interior region of the tunnel grid, which has an interior boundary surrounding a hole. A hole-point is a mesh point that is removed from the solution procedure. The

(a)

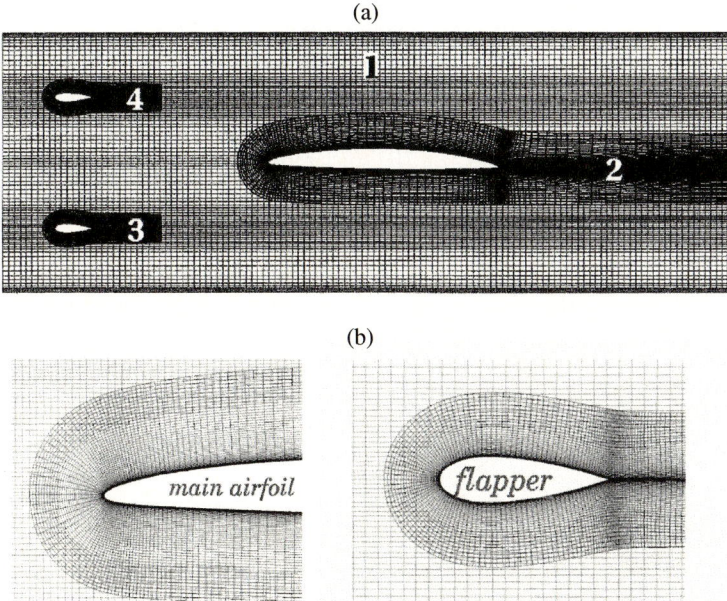

(b)

Fig. 5.29 (**a**) A Chimera overlapped grid topology for the flapping foil computation; (**b**) Overlapped grid in the near field of the stationary foil and the flapper

immediate neighboring points of the hole-points, called "fringe points," are updated from the interpolation procedure. For all computations presented in this section, two layers of fringe points are used for interior and outer boundaries. The interpolation of data between the flapper grid and the tunnel grid is time dependent, while the interpolation of data between the stationary foil grid and the tunnel grid remains steady in time.

The major differences between the patched and overlapped grid approaches are the amount of effort required to generate the time-varying grids and the amount of computation required for interpolation between zonal boundaries. Generating the time-varying patched grid system for the FFX using the elliptic grid genera-tor requires an order of magnitude more work than that required to generate the overlapped grid system and the interpolation database. The overlapped grid sys-tem provides the flexibility of choosing grid topologies since the grids do not have boundary constraints. Therefore, the C-type hyperbolic grids can be easily used around the foils, in which very fine grid resolution is required near the boundary layer. The patched grid system designed for the FFX requires use of the H-type grid with constraints at the boundaries. The elliptic grid generator was used for these H-grids. Obtaining the preferred grid density near the boundary layer is the most time-consuming part in this procedure, and this process has to be repeated at every boundary movement.

Fig. 5.30 Composite grid topology for the flapping foil computation

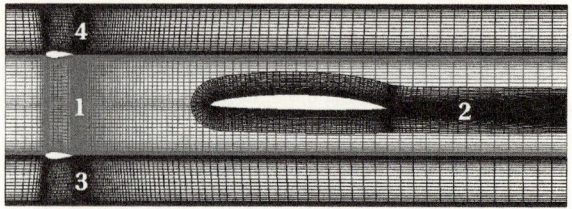

The third grid topology studied is illustrated in Fig. 5.30. This composite grid combines both patched and overlapped grid schemes. Three H-grids (grids 1, 3 and 4) are patched around the flappers. A C-grid is generated around the stationary foil (grid 2), and a hole is cut in grid 1 to accommodate this stationary foil. Grids 1 and 2 communicate with each other through the Chimera interpolation procedure. The total number of grid points for this composite grid system is 77,932; the grids dimensions are $255 \times 99, 337 \times 61, 255 \times 63$ and 255×63, for grids 1, 2, 3, and 4, respectively.

5.4.4.2 Computed Results

We have compared the computational procedures for obtaining both steady-state and time-accurate solutions using these three grid topologies.

Steady-State Solutions

Because experimentally measured data was available for steady flow with stationary flappers with $0°$ angle of attack, we first carried out steady-state calculations to validate the computational procedures not involving moving grid. The artificial compressibility coefficient, β, is set to be 10 for all computations in this section.

In Fig. 5.31, the measured and calculated static pressure coefficient, Cp, on the stationary foil surface, are compared. Symbols represent the experimental measurement. Computed results using three different grid topologies are compared: the dashed line represents the patched grid results; the solid line represents the overlapped grid results; and the chain-dotted line represents the composite grid results. All results compare well with the measured data. The composite grid system Cp results are nearly identical to the overlapped grid results.

Total velocity magnitude contours from the overlapped grid calculations are shown in Fig. 5.32. The wakes from the stationary foil and the flappers are clearly seen. Contours obtained from two different grids in the overlapped regions match quite well, indicating that the solution has been converged.

The convergence history for this case is shown in Fig. 5.33. Converged solutions were obtained after 250 iterations. Very similar convergence behavior was observed for all grid topologies. Even though the results did not change significantly after 250 iterations, the computation was continued to 600 iterations in order to verify the convergence characteristics. The solid line in Fig. 5.33 shows the history of the

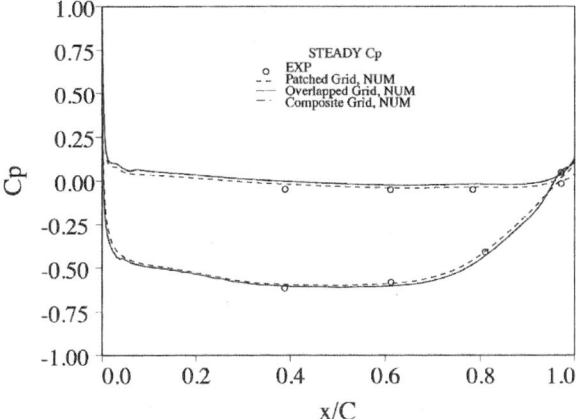

Fig. 5.31 Steady-state pressure coefficient (Cp) distribution on the stationary foil

maximum residual of the flow equations. The dashed line shows the history of the maximum of the divergence of velocity, and the chain-dotted line shows the history of the RMS of the Baldwin–Barth one-equation turbulence model (1991). This result indicates that all three measures of convergence behave in a similar fashion.

In Fig. 5.34, the velocity profiles on the suction side boundary layer region of the stationary foil at the streamwise station of $x/c = 0.612$ are plotted. These streamwise velocity plots compare the effects of grid resolution for three different grid topologies. The dashed and chain-dashed lines represent the results from the patched grid system, in which each zone has an H-type grid. The elliptic grid generator used for this grid does not provide the capability to control the grid spacing exactly at the solid wall boundary. Even though spacing on the order of 10^{-5} was specified as input to the elliptic grid generator, the resulting wall spacing was typically on the order of 10^{-3}. As a result, the grid resolution near the stationary foil wall is poor in this calculation. The velocity profile shown with the dashed line does not compare well with the experimental data.

In order to improve the grid resolution near the wall region, grid points near the wall are prevented from moving away from the wall in the elliptic grid generator.

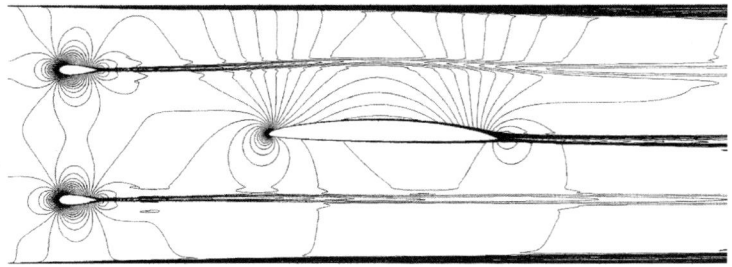

Fig. 5.32 Total velocity magnitude contour for steady-state solution

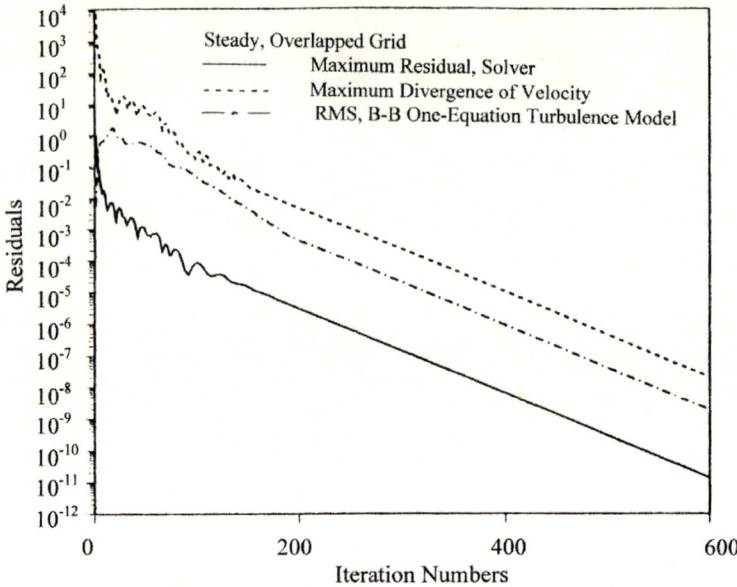

Fig. 5.33 Convergence history for the overlapped grid topology

The result from this modified grid is shown by the chain-dashed line. Although the velocity profile shown with the chain-dashed line is improved compared to that of the dashed line, it still does not compare well with the experimental data.

Next, we performed the overlapped grid computations for the three levels of grid density. The total number of grid points for the coarse grid system was 38,607 with grid dimension of 127 × 96, 337 × 35, 215 × 34, and 214 × 34 for the

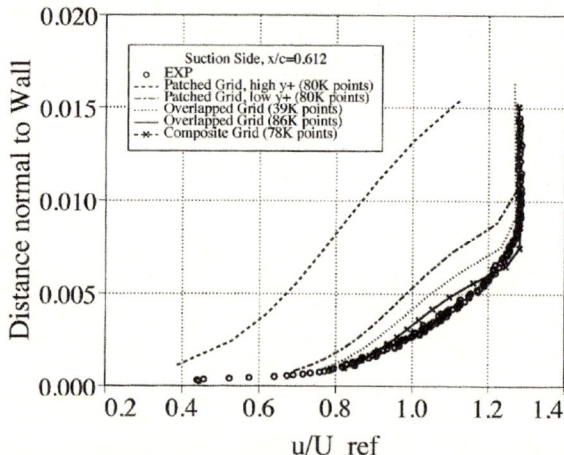

Fig. 5.34 Velocity profile on the upper surface of the stationary foil at x/c = 0.612

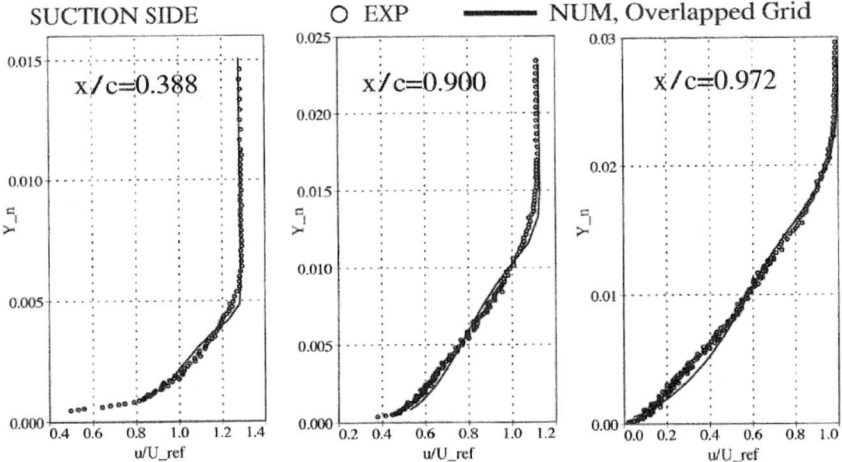

Fig. 5.35 Velocity profile on the upper surface of the stationary foil

four regions. The total number of grid points for the finest overlapped grid system was 97,480 with the grid dimensions of 253×191, 337×91, 215×43, and 215×43 for the four regions. The dotted line in Fig. 5.34 represent overlapped grid results from the coarse grid. The grid spacing near the wall for the coarse grid is 2.5×10^{-4}. The overlapped-grid result using the coarse grid (dotted line) shows better agreement with the experimental data than the patched grid results. The solid line represent the result obtained by using the finer grid system, and the dashed line with the x-symbols represent the result obtained by using the composite grid topology. The grid spacing near the stationary foil wall for these fine grids is 5.0×10^{-5}.

Both velocity profiles are virtually identical and compare fairly well with the measured data. These calculations using a C-type hyperbolic grid for the stationary foil show the flexibility of the overlapped grid approach, compared to the patched grid approach with an H-type grid over the stationary foil. The amount of work in generating the H-type elliptic grid increased when the grid resolution in the boundary layer region was increased. Therefore, computation results presented in the remainder of this section are primarily based on the overlapped grid computations and the composite grid topology.

Figures 5.35, 5.36 and 5.37 show the velocity profiles at several streamwise locations on the surfaces and at the stationary foil wake. Symbols represent the experimental measurements and the solid lines represent numerical results obtained by using the overlapped grid topology. The velocity profiles from the composite grid topology are not included here, as they are virtually identical to overlapped grid results (see Fig. 5.34).

Overall, the computed results compare very well with the measured data at the boundary layer and at the wake of the stationary foil. The largest discrepancy between the computed results and the measured data is seen in the wake of the foil

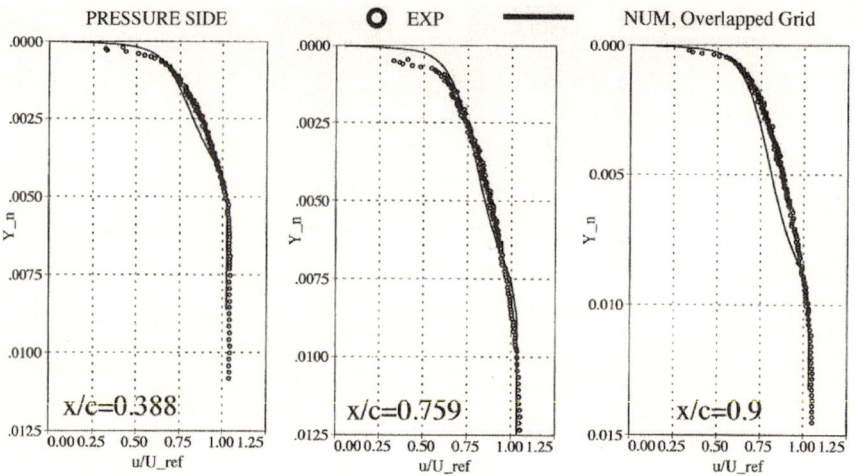

Fig. 5.36 Velocity profile on the lower surface of the stationary foil

(x/c = 1.2). The velocity at the edge of the wake is over-predicted with less than a couple of percentage points in error range.

To ensure that the steady flow results are grid independent, additional computations were performed. The order of accuracy for convective terms in coarse grid calculations was increased from third-order to fifth-order flux difference splitting. The third-order coarse grid result is indicated by the dashed line and the fifth-order coarse grid result is indicated by the chain-dashed line in Fig. 5.38. The solid line and the dotted line with x-marks represent the fine grid third-order and fifth-order results, respectively. The wake from the fine grid computations clearly shows better

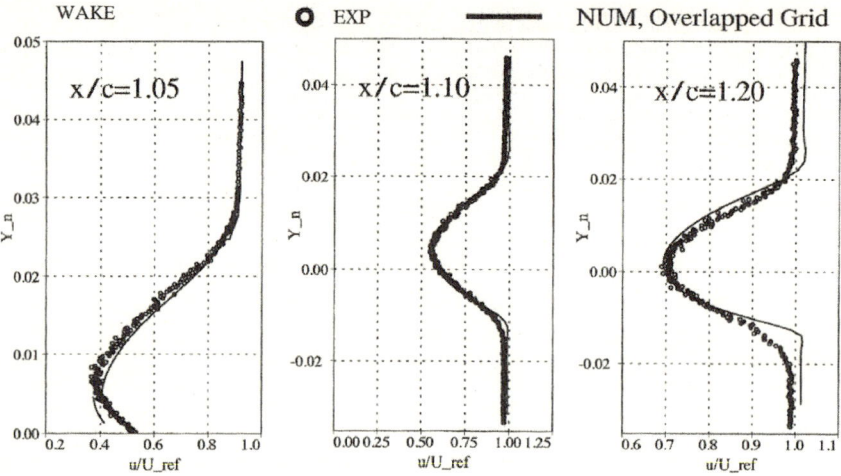

Fig. 5.37 Velocity profile on the wake of the stationary foil

Fig. 5.38 Velocity profile in
the wake of the stationary
foil: at x/c = 1.20

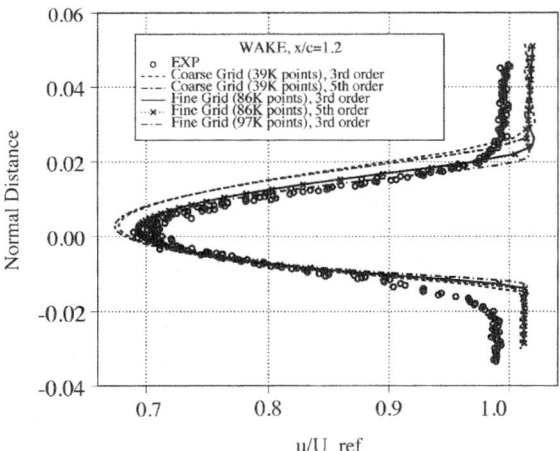

agreement with the measured data than the coarse-grid results. Note that the over-
shoot occurring at the edge of the wake in the third-order results does not occur in
the fifth-order results.

In the finest grid computations, the base grid for the stationary foil was refined
by increasing the number of grid points from 61 to 91 in the normal direction. The
velocity profile from the resulting finer grid (91K points total) with third-order differencing is plotted with the chain-dotted line. This fine-grid result is very similar to the
base grid result (86K grid points), indicating that this is close to a grid-independent
solution and that 86K-point grid will provide adequate resolution for the unsteady
calculations.

In fact, the difference between the 86K and the 97K grid results is less than the
oscillations in the measured data. Note that there is a rather large difference between
the measured and computed wake edge velocities. Since this edge velocity is shown
to be grid independent, it is thought that the experimental data used an erroneous
value for the reference velocity. In addition, all computed results have the same
velocity magnitude at the edge of the wake, and fine grid results are self-consistent.
For these reasons, the result shown with the solid line in Fig. 5.38 is considered
to be a grid-independent solution. Validation of the time-dependent procedure is
presented next.

Time-Dependent Solutions

The converged steady solutions were used as the initial conditions for the unsteady
calculations. The motion of flappers was specified as pitching about the mid-chord
point described by $\alpha = \alpha_m \sin(\omega t)$, where t is time α_m is $6°$ and $\omega = 2kU_\infty/c$ is
the angular velocity. Here, U_∞ is the reference velocity specified as 20.62 ft/s and
$c = 18$ inches is the stationary foil chord. The reduced frequency, k, is 3.62 based
on c/2. In all unsteady calculations, one cycle of the flapping foil period consisted

of 192 physical time steps, which corresponds to a discrete non-dimensional time step of 4.53×10^{-3}. We chose a time-step size that is small enough so moving mesh points in the overlapped grid system do not move more than one cell in a neighboring grid during one time step. At each physical time step, the maximum of non-dimensional divergence of velocity was dropped below 10^{-2} for all zones. This required $15-40$ sub-iterations during each time step. Numerical tests indicate that reducing the divergence of velocity further does not have measurable effects on the solution accuracy. The pseudo-time step was taken to be the same value as physical time step. The artificial compressibility coefficient, β, was set to 10. The periodic solution was obtained for both grid topologies after six flapping cycles.

Figure 5.39 shows the total velocity magnitude contours at a non-dimensional time of $t/T = 0.25$ a. Here, T denotes the period of the flapping motion. This qualitative comparison shows that the results obtained from two different grid topologies are very similar. The unsteady wake using the composite grid shows slightly more detailed features compared to the results using the overlapped grid. Since the grid boundaries are located in the middle of the oscillating wakes, we found it easier to increase the grid resolution using the patched grid. However, the difference between the two results is not easily recognizable from the contours. The quantitative comparison between the two, along with the experimental data, is presented next.

The mean Cp distributions on the stationary foil surface from the unsteady calculation are plotted in Fig. 5.40. The symbols represent the experimental mean values, the solid lines represent the overlapped grid results, and the dashed lines represent the composite grid results. The computed mean Cp values compare very well with the experimental measurements, except that there is a slight over-prediction at the 60% chord location on the pressure side of the foil. The mean Cp values from two different grid topologies are practically identical.

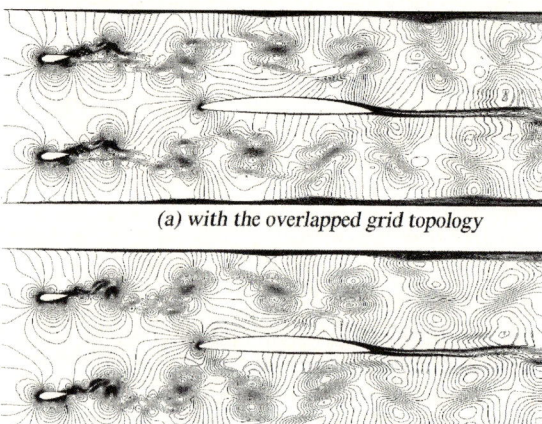

(a) with the overlapped grid topology

(b) with the composite grid topology

Fig. 5.39 Instantaneous velocity magnitude contours at $t/T = 0.25$

Fig. 5.40 Mean Cp distribution on stationary foil from unsteady calculations

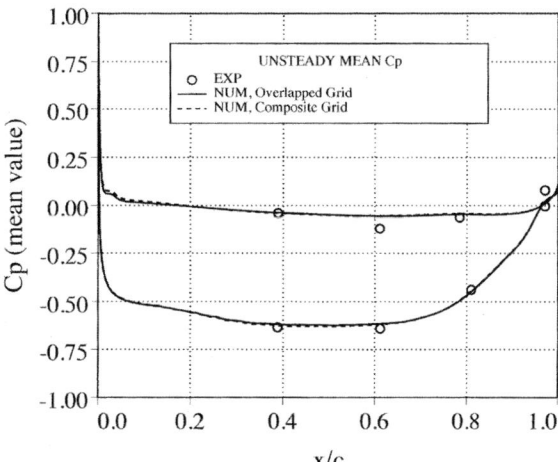

In Fig. 5.41, the time history of Cp values on the stationary foil is compared with experimental data at several streamwise locations. The Cp values obtained from both grid topologies show a very similar time history and compare fairly well with the measurement. The biggest discrepancies are seen on the pressure side of the foil at streamwise location $x/c = 0.611$, where Cp is over-predicted, and on the suction side of the foil at streamwise location $x/c = 0.972$, where Cp is under-predicted. This is consistent with the mean Cp distribution in Fig. 5.40. The major difference between the two computed results is that overlapped grid topology produces higher frequency oscillations compared to the composite grid approach. It should be noted that the grid movement and the resulting interpolations required in grid boundaries are quite different between the two grid topologies.

In the overlapped grid, as the flapper grids rotate they move through a relatively coarse region in the tunnel grid. This mismatch of the grid resolution between the two overlapped regions can lead to interpolation errors. Considering that different hole points are being cut at each time step while unsteady wakes from flappers continuously move through the region, it is difficult to obtain the same degree of accuracy between the two grids in the overlapped grid topology. In the composite grid system on the other hand, the mesh points in the tunnel grid move with the flappers, which maintains relatively fine grid resolution in the near wakes of the flappers. The FFX problem would offer a good test case for validating any improved correction schemes in the overset grid arrangement.

The sensitivity and importance of the boundary interpolation scheme for the composite grid approach are illustrated next. In Fig. 5.42, the effect of interpolation is shown when computing previous time level data for newly created fringe points. The time history of Cp on the suction side of the foil at streamwise location $x/c = 0.810$ is reported for the composite grid computation. The symbols represent the experimental measurements.

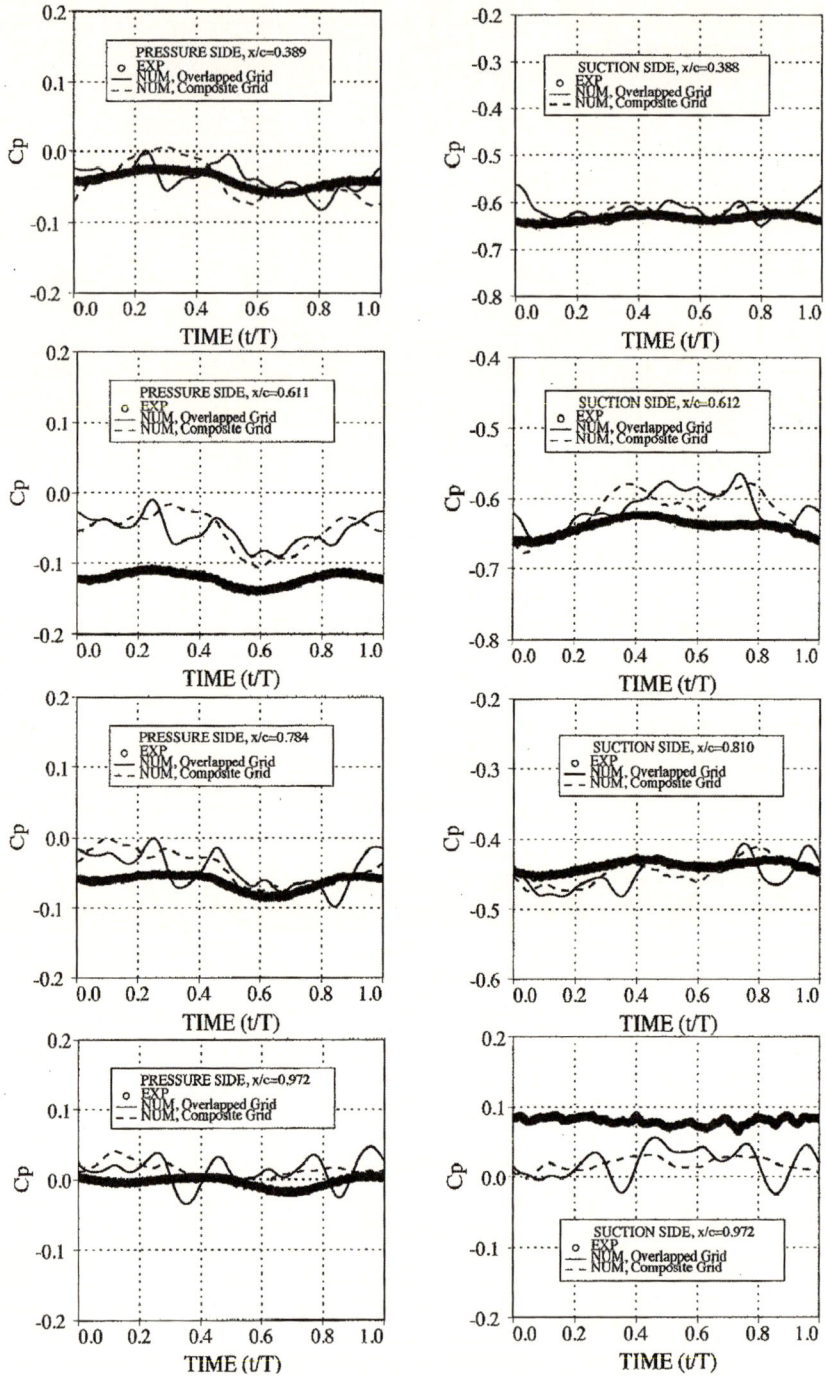

Fig. 5.41 Time history of Cp at various streamwise locations of stationary foil

Fig. 5.42 The effect of updating boundary points in the overlaid moving regions of the composite grid system

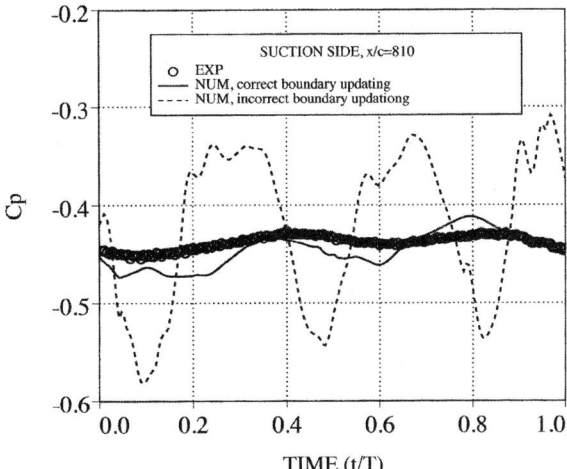

The different procedures for updating previous time level data at the fringe points are compared in Fig. 5.42. Since, in the present computations a second-order three-point backward differencing scheme is used for time discretization, information at time levels n and n–1 is needed to advance to the n+1 time-level. When a hole-point becomes a fringe point due to moving boundaries, this fringe point does not have any information from the previous time levels. One way to obtain the previous time level data is to interpolate the variables from the donor grid (as it is done for the current time level information). The result obtained by this procedure is plotted by the dashed line in Fig. 5.42. This shows that very large amplitude errors occur in the computations.

The source of these fluctuations can be found in the interpolating procedure. The previous time-level data for the new fringe points is obtained by using the interpolated database at the current time level. However, the grid point locations from the previous time-level should have been used. Using the current time-level database for these points results in incorrect interpolation coefficients and incorrect donor points. When this error was corrected the computed Cp value shown by the solid line in Fig. 5.42 was obtained. When the previous time-level data is not available for the newly created boundary points, the time integration for these points is changed to first-order. The time differencing in the next time step will be second-order backward differencing because the previous time-level information has been established from the current time-step calculation.

The present flapping foil example illustrates the issues encountered in actual simulations—especially when bodies of relative motion are involved. The fidelity of computed results depends not only on algorithm and geometry modeling but also on computational procedures like interpolation schemes between grids. These are realistic issues users of CFD tools need to resolve when dealing with real-world simulations. In addition, physical modeling plays an equally, if not more important role in computing a wide range of flow regimes. Some of these features are discussed further in later chapters, in conjunction with examples.

5.5 External and Juncture Flow

The flow around a cylinder plate or a wing-body juncture produces interesting viscous phenomena due to the interaction between the boundary layer from the plate and viscous layer from the cylinder. The 3-D separation of the boundary layer and subsequent formation of the so-called horseshoe vortex and its development is very challenging to analyze both experimentally and numerically. This juncture flow can occur in many practical engineering problems. Flow around a wing-fuselage junction and around an appendage submarine body are just two examples, and the flow near the end-wall of turbomachinery blades might be one of the most complicated juncture flow problems in engineering. One major motivation for studying this type of juncture flow is related to the flow analysis of the Space Shuttle main engine (SSME), which will be presented in detail in the next chapter. In the SSME, liquid oxygen (LOX) posts are densely packed in the main injector region. Even though a single cylinder-plate flow is an extreme idealization of the flow in the actual oxygen-post region in the SSME, validating the computational procedure in this simplified model problem is of considerable value in extending the simulation procedure to realistic cases where detailed experimental measurements are very difficult and scarce.

5.5.1 Cylinder on a Flat Plate

Most of the earlier studies on cylinder/flat plate juncture flow have been experimental. Baker (1979) shows that laminar juncture flow is confined to a very limited region. A similar result was obtained later by Thomas (1987). Eckerle and Langston (1986) reported a single primary vortex and saddle point contrary to multiple vortex systems observed earlier by other researchers. Interpretation of the phenomena also varies (Thomas, 1987; Peake and Tobak, 1980).

Computational simulation of these flows involves distinctively different features from those of external aerodynamics. For instance, the thickness of the viscous layer for these types of flows is of the same order as the characteristic flow-field dimension, while the viscous region tends to be confined in a thin layer near the body for external flows. Realistic juncture flows under an internal flow environment are likely to have a large amount of deflection, as in the case of LOX post regions in the SSME. Several numerical studies on this flow have been attempted. Kaul et al. (1985) reported a numerical study on a single cylinder-plate flow using the INS3D code. Highlights of this and other studies were reported by Kwak et al. (1986). Independently, Kiehm et al. (1986) reported a numerical study of flow around a single post in a channel. These computational results show qualitatively similar phenomena. Several representative results using INS3D are summarized below.

In Fig. 5.43a, the computational domain for a single post on a flat plate is illustrated. The upstream boundary layer thickness is varied by using partially and fully developed channel flow profiles. The convergence characteristics of the flow solver

Fig. 5.43 Flow around a
single post on flat plate at
Re = 1,000 using the artificial
compressibility method
(INS3D code): (**a**) grid; (**b**)
convergence history

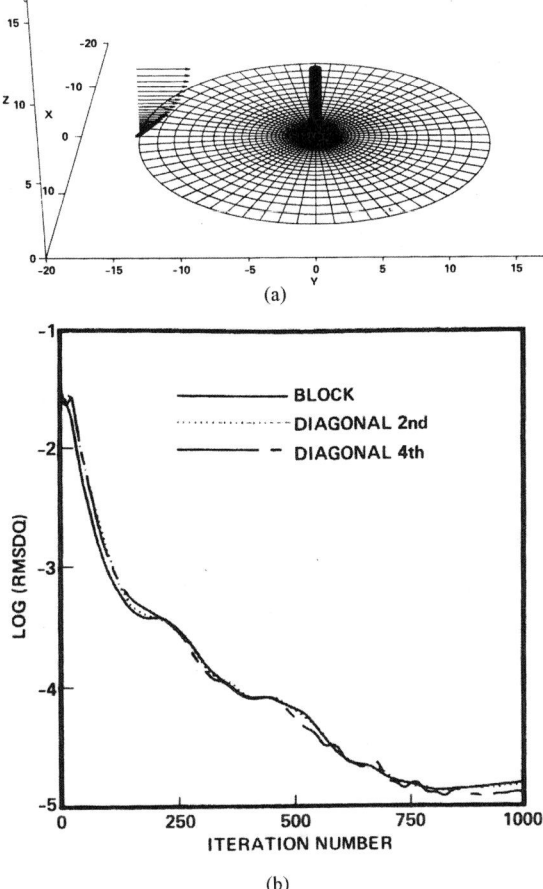

are shown in Fig. 5.43b by the history of RMSDQ, which denotes the root-mean-square value of the change per iteration in the pressure and velocities. The three curves in the figure show three variations of the INS3D code; namely a block tri-diagonal, a diagonal version with second-order implicit smoothing terms, and a diagonal version with fourth-order implicit smoothing terms, as explained in Chapter 4. The flow solver converges quickly to about four orders of magnitude reduction in RMSDQ. The computing time per iteration per grid point is 91 μs for the block tri-diagonal version and 32 μs for the diagonal version of the code. These timings were based on computations performed on the Cray 2 supercomputer in the early days of high-end computing.

In Fig. 5.44, particle traces for a single post at Re = 1,000 are shown. A saddle point separation and a horseshoe vortex can be seen from the traces near the flat plate. The secondary flow in front of the cylinder wraps around toward the

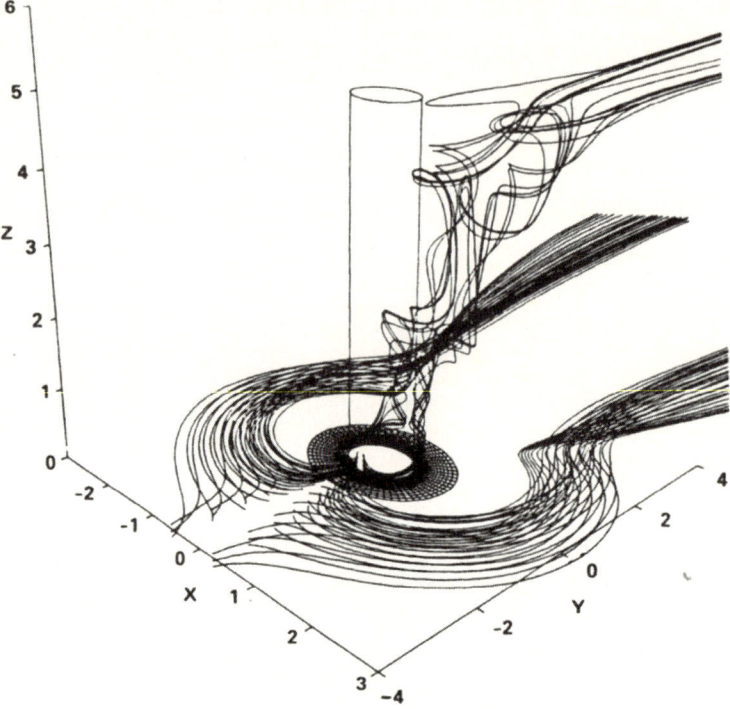

Fig. 5.44   Particle traces for a single post on a flat plate at Re $= 1,000$

wake region and forms a counter-rotating pair of vortex filaments. These spiraling twin vortices demonstrate a striking difference between this type of juncture flow and a 2-D cylinder. The vortex filaments are washed upward and attenuate as they interact and move downstream. In reality, vortex shedding and possible unsteady motion take place at this stage. These tornado-shaped vortices are very difficult to observe experimentally, and validation of this phenomenon was very much needed in the 1980s. G. Schewe (1985, private communication, DFVLR, West Germany) produced oil flow visualization around a single post that shows clear evidence of the twin vortex behind the cylinder, as shown in Fig. 5.45 (*special thanks to Dr. G. Schewe for providing this picture*). This experimental observation is qualitatively similar to the computed results, shown by the particle traces in Fig. 5.44. This juncture flow structure will lead to a strong variation in skin friction and pressure along the cylinder, and hence significantly affects the overall loading on the post.

5.5.2 Wing-Body Junction

Wing-body flow has been of interest for resolving juncture flow and tip vortex roll-up and propagation. At the junction region, the flow has similar characteristics to the

Fig. 5.45 Oil flow
visualization around a single
post at Re $= 1.85 \times 10^5$
(G. Schewe, DFVLR, 1985)

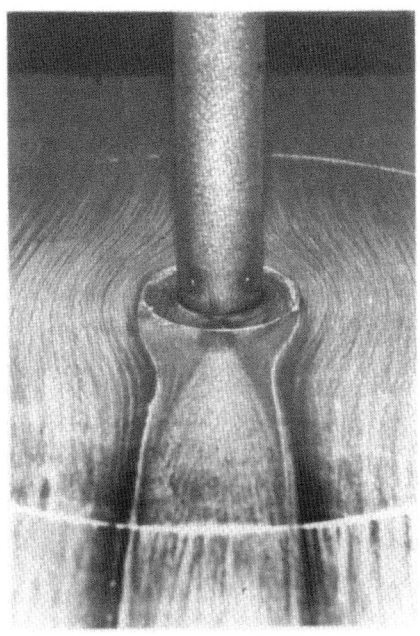

cylinder on a flat plate shown above. Wing-body juncture flow has been of interest
to airplane designers in resolving the wing-fuselage area that is close to high-lift
devices. For naval hydrodynamics, wing-body is a generic case for a submarine
hull-appendage flow, and thus provides a good validation case for CFD procedures
to be used for the entire submerged vehicle.

5.5.2.1 Wing-Body Juncture Flow

An example case of a wing on a flat plate is discussed next. The results presented were obtained by Burke (1989). Numerical results were obtained by applying
INS3D and comparing them with the experimental data of Dickinson (1986). The
wing is a hybrid shape consisting of a 1.5:1 elliptic nose and a NACA 0020 tail
joined at the location of maximum thickness. This is a generic configuration characterizing a wing-fuselage juncture of an aircraft or a hull-appendage juncture of
ships or submarines. The Reynolds number of the experiments, based on the chord
length of the wing, is 5×10^5. The coordinate system is shown in Fig. 5.46.

Turbulence modeling for this type of complex flow is very challenging. A high-
level turbulence model may be necessary for detailed study on the juncture flow
itself. Since our study was done in preparation for extending the flow solver to more
realistic applications, computing efficiency was of major importance. Therefore, the
computational simulation was done by devising a simple algebraic turbulence model
derived from Patankar et al. (1979). Considering the simplicity of the turbulence

Fig. 5.46 Coordinate system
for the wing-flat plate study

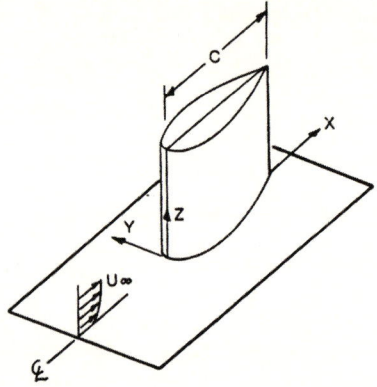

model, the results compare quite well with experimental data capturing important
features of the flow. Some of the results are reproduced here.

In Fig. 5.47, surface pressure on a flat plate near the wing is compared with
experiments while, in Figs. 5.48 and 5.49, velocity contours at two different vertical planes are shown. Overall, the computed results compare favorably with the
experimental data. However, in this numerical experiment, some flow details close
to the juncture region are not studied. Also, the Reynolds number for realistic cases
is orders of magnitude higher than the current laboratory experiment, requiring further validation. Since the wing has a finite height, another aspect to consider is the
wing tip vortex effect, especially in the wake region. A wing tip vortex roll up and
propagation is presented next.

5.5.3 Wingtip Vortex Flow

The wingtip vortex flow has been of interest in many areas of fluid engineering, and
its significance has been seen in many practical problems. For example, tip vortices
generated by wings of large aircraft have been known to affect other aircraft following closely behind. In rotorcraft aerodynamics, interaction of the tip vortex and
blade can directly affect the aerodynamic performance of the vehicle and can cause
substantial vibration under some flight conditions. On ship propellers or submarine
propulsors, the tip vortex is of great concern in conjunction with cavitation inception
and wake propagation (see, for example, Arndt and Maines, 1994). In liquid rocket
propulsion, tip vortex and cavitation generated from turbopump blades can be very
damaging to the pump, as well create system-level vibration.

Although there has been a great deal of tip vortex work in the form of theoretical,
experimental and computational studies, the current understanding of the intricacies
of this flow is still not very comprehensive. General characteristics of tip vortex
formation are well known; however, the details of their formation, initial roll-up,
and downstream development is still a subject of research.

Fig. 5.47 Pressure on flat
plate along y = constant at
Re = 5. × 10^5

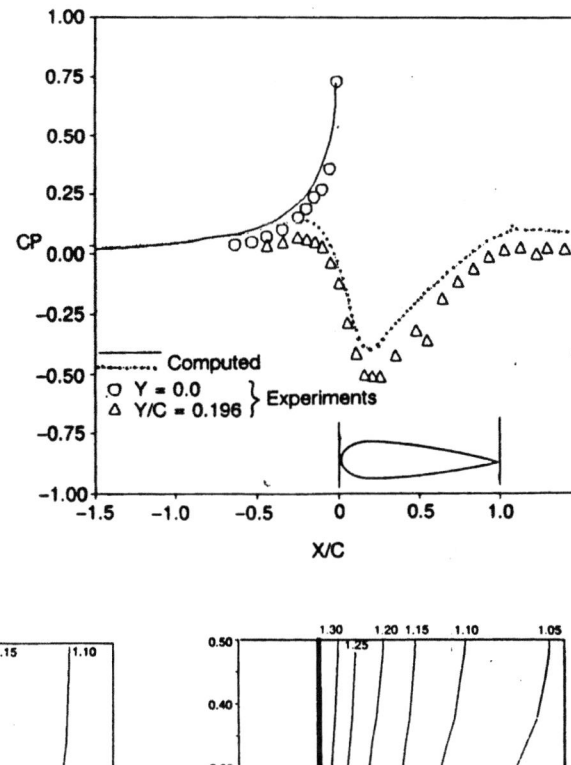

Fig. 5.48 Velocity contour for wing on flat plate at x/c = 0.18 and Re = 5. × 10^5: (**a**) experiment
(Dickinson, 1986); (**b**) computation (Burke, 1989: INS3D)

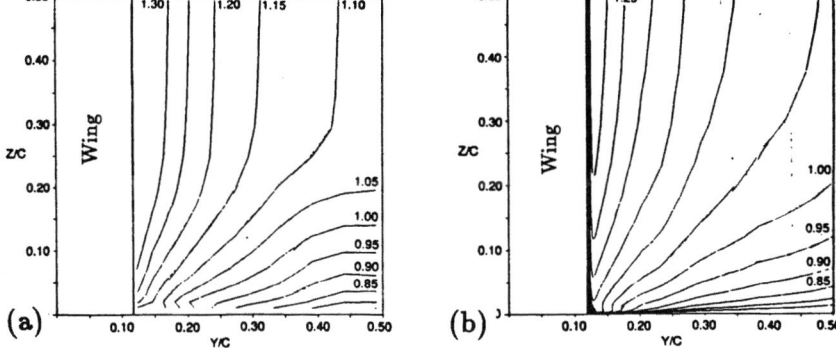

5.5.3.1 Experimental-Computational Validation Approach

Computational studies on tip vortex formation and propagation have been performed perhaps most extensively in conjunction with rotorcraft aerodynamics.
Inaccuracies in computational studies can be attributed to computational procedures
and physical modeling. Accurate modeling of tip vortices requires resolution not
only of the viscous boundary layer region, but also certain areas with high flow gradients such as the core of the vortex region. Sufficient grid density and appropriate
distribution are essential. The inaccuracy caused by poor grid resolution manifests
itself in the form of excessive numerical dissipation. Thus, use of a high-accuracy

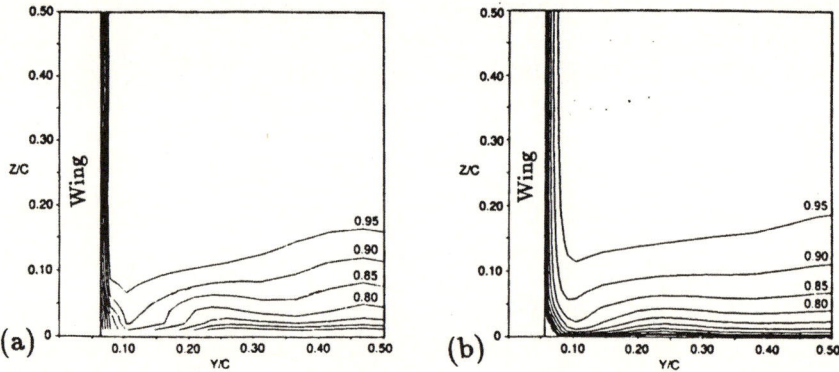

Fig. 5.49   Velocity contours for wing on flat plate at x/c $= 0.75$ and Re $= 5. \times 10^5$: (**a**) experiment (Dickinson, 1986); (**b**) computation (Burke, 1989: INS3D)

scheme can be very helpful. The flow field generated by the tip vortex is highly three-dimensional and can be highly turbulent, or can be dominated by inviscid dynamics, as in the wake region.

A computational capability for predicting a detailed flow field, especially the rollup of the tip vortex, is of interest for validating computational procedures, as well as for assessing turbulence models being used in conjunction with production codes. For the purpose of studying tip vortex details, a low-speed experimental study was performed by Chow et al. (1991) and Zilliac et al. (1993) at NASA Ames Research Center. The experimental setup is shown in Fig. 5.50. Even though the test was performed with air, the flow speed is low in the incompressible regime. Therefore, the experimental results from this study have been used to validate incompressible flow solvers we discussed earlier.

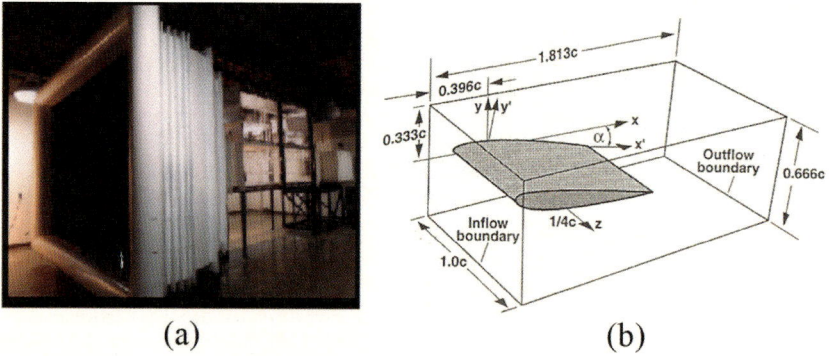

Fig. 5.50   Experimental model: (**a**) 32-inch $\times$ 48-inch low-speed wind tunnel at NASA Ames Fluid Mechanics Laboratory; (**b**) NACA 0012 wing with round tip, Re $= 4.6 \times 10^6$, $\alpha = 10°$

The flow domain includes a rectangular wing with a NACA 0012 airfoil section and a rounded wing tip. The wing has an aspect ratio of 0.75 and is mounted inside a wind tunnel at 10° angle of attack. The flow is turbulent with a Reynolds number of 4.6 million based on the chord length. Both the artificial compressibility code, INS3D-UP, and the pressure projection code, INS3D-FS, are used to study the roll-up of a vortex in the tip region and the near-field propagation behind the trailing edge.

5.5.3.2 Geometry

The full computational geometry is a close approximation to the 32 inch × 48 inch low-speed wind tunnel and wing setup, as discussed above. Two major factors affecting the accuracy of the computation are numerical errors primarily due to approximation from discretization and lack of grid resolution, and turbulence and transition modeling. Adequate grid resolution is especially important not only for the viscous boundary layer region but also for the region with high flow gradients, such as the vortex core and near-wake region. Sufficient grid density and appropriate distribution are essential. Our initial calculations indicated that an extremely fine mesh is required to resolve the tip vortex flow field. Consequently, a subset of the full geometry problem is devised that only includes the wake region. This enables extensive study of several contributing factors to the accuracy of computed results, such as grid refinement, discretization effects, and sensitivity to turbulence modeling.

(i) Wake-only Problem

The computational domain for this problem includes the region from the trailing edge to 0.69c downstream of the trailing edge. The experimental velocity profile in a cross-flow plane at the trailing edge of the wing was imposed at the inflow of this wake model. The inflow boundary condition for the pressure was computed based on the method of characteristics using one-dimensional Riemann invariants. An experimental pressure distribution is prescribed at the exit boundary. The velocity components at the outflow were calculated by using one-dimensional Riemann invariants. This information is not what one usually has available in numerical simulations. For this study, however, the boundary conditions are set up to resolve vortex roll-up on the wing surface and subsequent propagation in the near field with minimum influence of boundary procedures.

(ii) Complete Geometry Problem

The computational domain for the complete geometry case includes a rectangular half-wing with a NACA 0012 airfoil section and rounded wing tip, as shown in Fig. 5.50. No slip boundary condition is imposed at the solid surface, and the normal pressure gradient is set to be zero. The inflow and outflow boundary conditions are prescribed in the same manner as for the wake-only problem.

5.5.3.3 Grid

Several different grid generation strategies can be selected, depending on the flow solver to be used. In the present validation, a single grid topology is chosen for the entire wing region with an optional embedded grid for the vortex core region. This will minimize grid-related error. However, in more realistic problems involving complex geometry, generating single-zone grids may not be straightforward. Multiple zones, possibly with overlapping regions, are common in structured grid setups—otherwise unstructured grids can be employed. Equally important is the grid distribution to resolve flow features of interest. The automatic adaptive grid method is not discussed here. The base grid for the wing is shown in Fig. 5.51a.

For the wake-only problems, a single grid is generated such that a uniformly clustered grid covers the vortex region. This is then stretched out to uniform spacing away from the vortex core region. As one approaches the side and top walls, spacing is reduced to resolve viscous layers. The grid for the wake-only problem is shown in Fig. 5.51b.

The generation of a single-block grid for the entire geometry of the experimental setup is not straightforward. Specifically, the restrictions of the outer wind tunnel walls and corner regions can cause problems because of their closeness to the wing surface and wingtip. For this reason, a smoothly varying grid is difficult to obtain. Also, the inflow location of the domain is situated very closely to the wing.

For the computed results presented here, a single-zone grid of C-O type is chosen without introducing grid singularity at the tip region. A two-dimensional base grid is first generated around the airfoil section on the wind tunnel wall using an elliptic grid generator. The volume grid is generated by stacking the 2-D grid along the straight section of the wing. Then, the grid is wrapped around the rounded wingtip. Viscous spacing of 1×10^{-5} c is imposed on the body and root wall, while viscous spacing of 1×10^{-4} c is imposed on the wind tunnel wall. Local mesh refinement based on the knowledge of the tip vortex location from a previous

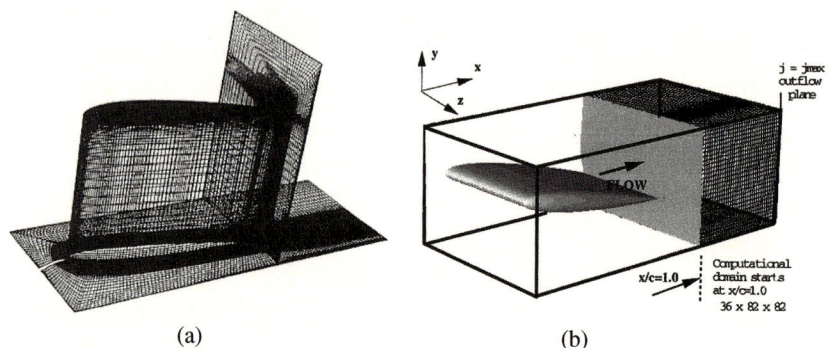

(a)   (b)

Fig. 5.51 Grid topology: (**a**) wing region grid 130 × 145 × 73; (**b**) wake region grid 36 × 82 × 82

solution is also incorporated. Even though this grid feature adaption was done manually at the time this computation was performed, more sophisticated grid adaptation methods are now available that can be coupled to the solution procedures in general applications.

5.5.3.4 Turbulence Modeling

Varying levels of turbulence models have been used to study vortical flows and wake vortex problems. Algebraic models such as the Baldwin–Lomax (1978) model are primarily designed for boundary layer type problems and are not suitable for complex vortical flow problems. Models of this type do not take into account the transport and diffusion of turbulence, and history effects are not captured. These effects are important for wingtip vortex flows, recirculating flows, separated flows, and interacting shear layers. Higher-level models, such as one- or two-equation models, have similar deficiencies. However, higher-level models offer a possibility for adding ad hoc revisions to characterize tip vortex flow.

To study the sensitivity of turbulence models to solution accuracy, a modified form of the Baldwin–Barth one-equation turbulence model (1991) was experimented with in the current study. The model can be implemented in a straightforward manner, since the turbulence length scale is automatically handled in the model equation. This type of model over-predicts the eddy viscosity level in the core of a vortex. It attempts to fix this issue by modifying the production term in the model equation. In the standard Baldwin–Barth one-equation model, the production term, P, for νR_τ is approximated by:

$$P = C_1 \nu R_\tau X \tag{5.1}$$

where C_1 is a constant, ν is the laminar kinematic viscosity, R_τ is the turbulent Reynolds number, and X is a scalar measure of the deformation tensor. There are several choices of X. For example, X can be based on the magnitude of vorticity, $|\omega| = \left(2\Omega_{ij}\Omega_{ij}\right)^{1/2}$ where Ω_{ij} is the vorticity tensor, on the strain rate $|s| = \left(2S_{ij}S_{ij}\right)^{1/2}$ or on the norm of the entire tensor, as discussed by Spalart and Allmaras (1992). The option of basing X on $|\omega|$ is the simplest to implement, and has a theoretical motivation, in that turbulence can be related to vorticity. A similar idea can be found in the $k - \omega$ two-equation modeling. A modified form of the production term was used by Dacles et al. (1993) and Kiris et al. (2001) for the wake vortex study presented here. This ad hoc production term combines $|\omega|$ and $|s|$ as follows:

$$P = C_1 \nu R_\tau \left(|\omega| + 2\min\left(0, |s| - |\omega|\right)\right) \tag{5.2}$$

The modification is devised to reduce the eddy viscosity in the regions where the vorticity exceeds the strain rate, such as in a vortex core where the flow is nearly pure rotation. This modification represents an attempt to empirically adjust the production term for vortex-dominated flows. Note that the factor 2 in this equation is an arbitrary constant that can be adjusted depending on the amount of diffusion the turbulence model gives.

In addition to this modification, other forms of the production term are also experimented with using INS3D-FS, such as:

$$X = |\omega|$$
$$X = \min(|\omega|, |s|)$$

This numerical experiment provides information on the sensitivity of the turbulence model on the solution for highly vortical flow. For more general applications, the turbulence model has to be designed to automatically reflect this change in turbulence production in the vortex core region. Existing one-equation models produce excessive dissipation regardless of the grid resolution.

Transition to turbulence is dictated by the experimental setup and is set at $s/c = 0.0417$ from the leading edge. This transition strip is located on the pressure and suction side of the wing and wraps around the tip. Early computations reveal that the result, which includes this transition location, is not much different from that of a fully turbulent assumption for the entire region. Therefore, all computed results presented here do not include the transition location.

5.5.3.5 Near Wake Computation Using the Artificial Compressibility Method

Computed results by Dacles–Mariani et al. (1993, 1995a, b, c) using the artificial compressibility code INS3D-UP, are presented next for wake vortex propagation in the near field. For this wake-only problem, a $35 \times 103 \times 103$ grid (371,000 gird points) is used. We found that the third-order accurate differencing scheme for the convective terms in the momentum equations was too diffusive. In the numerical experiment, both third-order and fifth-order schemes are compared, combined with production model modifications in the Baldwin–Barth turbulence model using Equation (5.2).

As shown in Fig. 5.52a, b, the turbulence model modification has the greatest impact on velocity at the core during the near-field propagation in the wake. An additional 5% improvement in accuracy was observed by increasing the differencing scheme to fifth order. In Fig. 5.52b, the core center velocity is under-predicted by approximately 25% for the cases with no turbulence model modification. The reason for this difference can be seen in Fig. 5.52c: the magnitude of eddy viscosity in the vortex core region between the modified and unmodified models is quite different, indicating that the turbulence model in use needs improvement to correctly represent flow physics in highly vortical flow.

Capturing the flow features of the propagation of the tip vortex flow depends primarily on two factors: numerical procedures including grid resolution and turbulence modeling. The static pressure coefficient in the core region is very sensitive to numerical accuracy. It also depends on how well the turbulence model represents the flow in the core region, as briefly illustrated above. More detail can be found in the references cited.

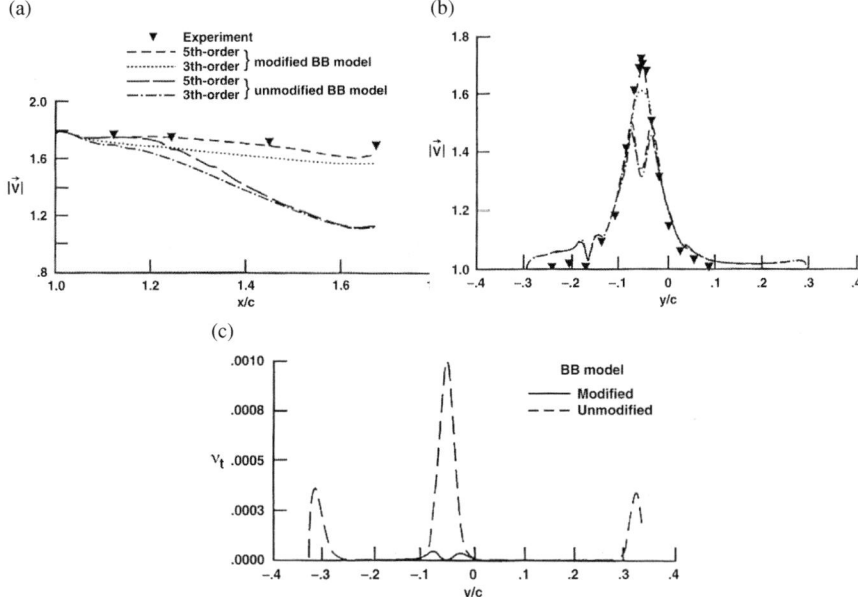

Fig. 5.52 (**a**) Peak velocity magnitude at vortex core; (**b**) Total velocity magnitude across vortex core at x/c = 1.241; (**c**) Eddy viscosity profile across vortex core at x/c = 1.241

5.5.3.6 Near Wake Computation Using the Pressure Projection Method

We computed the same case using the pressure projection code, INS3D-FS. To illustrate a validation process and to characterize the near wake computations using a pressure projection approach with the fractional step procedure, results presented by Kiris and Kwak (2001) are summarized next.

The computational domain includes the region from the trailing edge of the wing (x/c = 1.0) to 0.673 of the chord-length, c, downstream of the wing using the H-H grid topology. Extensive experimental data by Chow et al. (1991) are available at x/c = 1.0, 1.12, 1.24, 1.447 and 1.673. The experimental velocity profile at the trailing edge (x/c = 1.0) is imposed as an inflow boundary condition so that we can focus on wake propagation aspects. Pressure distributions at boundaries are calculated from the compatibility condition.

As a first step, we performed the computations on a coarse grid with a 36 × 42 × 42 mesh in x, y, z directions (i-, j-, k-directions). The velocity peak at the vortex core was under-predicted using this grid; therefore, we next increased the grid dimensions to 36 × 82 × 82—essentially doubling the grid in cross-flow directions. The peak velocity at the core was improved, but not significantly. This is consistent with the same computations using the artificial compressibility method by Dacles–Mariani (1993). The primary reason for this is due to excessive dissipation in the core region, even with increased grid resolution. Physically, the vortex core is dominated by dynamics and not much viscous effect exists. To account for this physics, the production term in the turbulence model is modified as explained in Equations (5.1) and (5.2). The sensitivity of this term affecting the solution accuracy, especially in the core region, is shown in the series of plots presented next.

We have experimented with several combinations of convective term differencing, grid resolution, and modification of the production term in turbulence modeling. A list of these test cases and legends used in the figures presenting computed results are summarized in Table 5.1. First, in Fig. 5.53, axial progression of velocity magnitude and static pressure coefficient C_p are shown along the vortex core line in the wake. As can be seen, even with the best choice of differencing scheme and modified turbulence model, the accuracy of the solution deteriorates when the grid is coarse. Using increased resolution and a higher order differencing scheme, the error in the core region can be reduced to less than 2%.

The amount of dissipation is automatically computed in upwind differencing. It also depends on grid spacing. Computationally, higher-order schemes are less expensive to implement compared with increasing the grid resolution to produce comparable results. These cases are compared in Fig. 5.54 by plotting velocity magnitude across the wake vortex at three interior stations (x/c = 1.12, 1.24 and 1.47) and at the exit boundary (x/c = 1.673). The most sensitive quantity in these plots is the peak velocity at the vortex core. For our computations, the grid resolution near the wind tunnel wall boundary layer was not sufficiently high, especially considering that the Reynolds number is 4.6 million. Also, the turbulent Reynolds number at the inflow boundary for the computation is set to 1 in the Baldwin–Barth model. Therefore the computations and the experimental results do not match very well near the wind tunnel wall, and the discrepancy is largest at the exit plane.

Similarly, the cross-flow velocity across the wake vortex at four different locations is plotted in Fig. 5.55. Using a different production term in the turbulence model equation does not have a measurable influence on cross-flow velocity compared to the velocity magnitude.

Since the cross-flow velocity is zero at the vortex core, the dissipation introduced by the turbulence model, the grid resolution, and the order of accuracy in the differencing scheme primarily affect the axial flow velocity components.

In Fig. 5.56, the comparison of C_p across the wake vortex is shown at three different locations. Since the pressure has not been prescribed at the inflow

Table 5.1 Legend of computed results in Fig. 5.53 through 5.56

Computed results	Experiment, Chow et al. (1991)		
	Convective terms diff.	Grid size $j \times k \times l$	Baldwin–Bart model production term
	Third-upwind	$36 \times 82 \times 82$	$\min(|s|, |w|)$
	Fifth-upwind	$36 \times 42 \times 82$	$\min(|s|, |w|)$
	Fifth-upwind	$36 \times 82 \times 82$	$|w| + 2.0 * \min(0.0, |s| - |w|)$
	Fifth-upwind	$36 \times 82 \times 82$	$|w|$
	Fifth-upwind	$36 \times 82 \times 82$	$\min(|s|, |w|)$

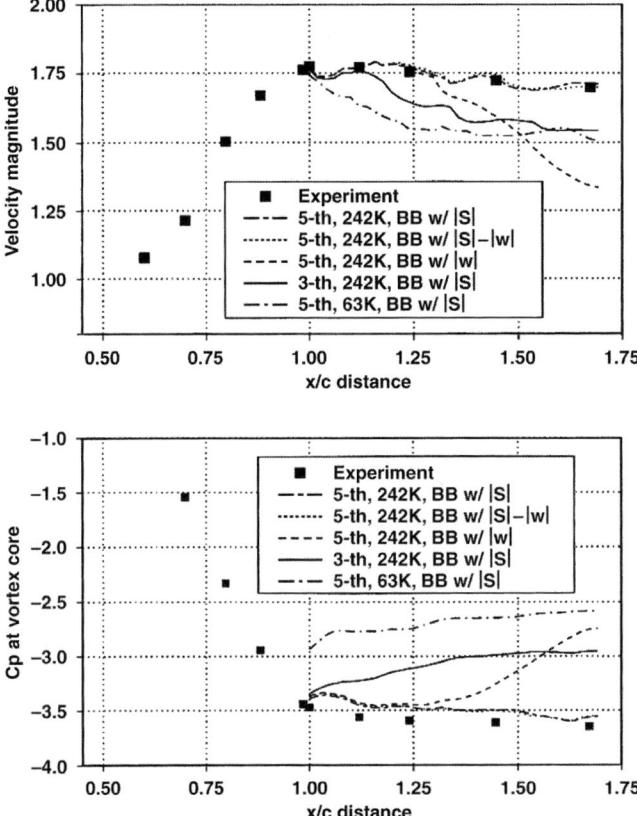

Fig. 5.53 Axial progression of flow quantities along vortex core line

boundary, computed C_p values are compared at the inflow and outflow boundaries, as well. The comparison is quite satisfactory.

In the current pressure projection method, the incompressibility condition is satisfied automatically at each physical time step. This feature makes it possible to use this method in time-accurate computations.

5.5.3.7 Initial Rollup of Round Wingtip Vortex

Initial rollup of the wingtip vortex flow has been of significant importance because of its relevance to many practical problems including formation of aircraft wake, interaction of tip vortex and rotorcraft blades, tip and wake vortex from propellers of naval vehicles, and tip vortex from turbopump blades and its impact on cavitation. Tip shape, as well as flow conditions, impact the rollup process. In our experimental-computational study, only a round tip shape was considered. However, by presenting a detailed study, we hope to shed some light on how numerical methods and physics

Fig. 5.54 Comparison of velocity magnitude across wake vortex

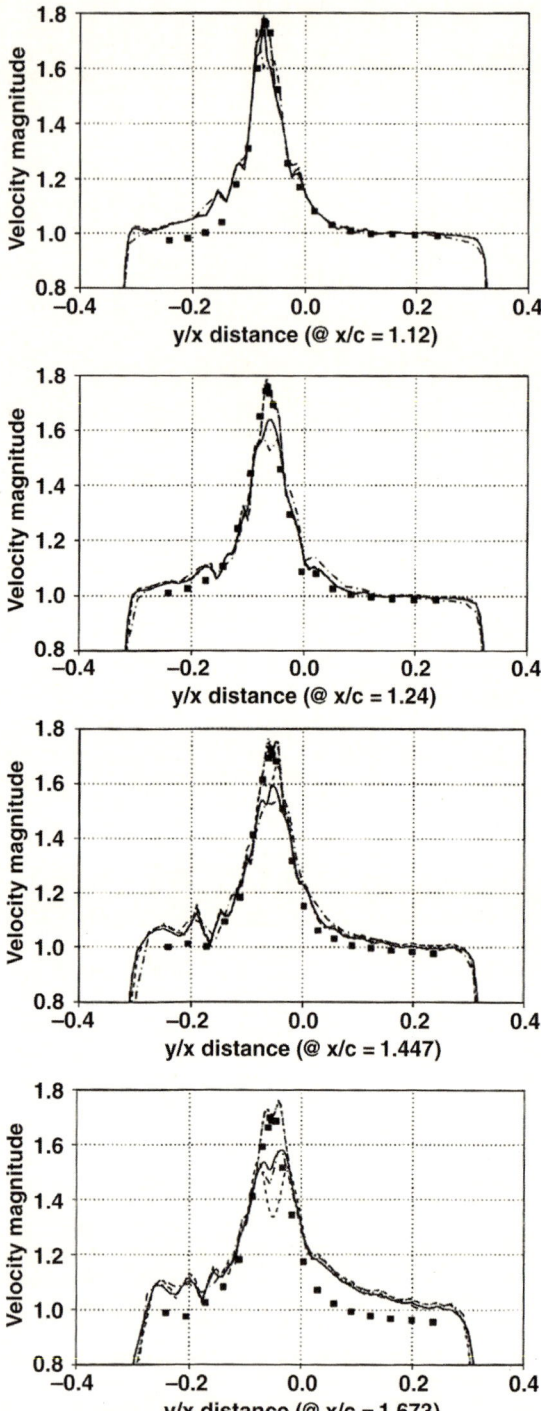

Fig. 5.55 Comparison of cross-flow velocity across wake vortex

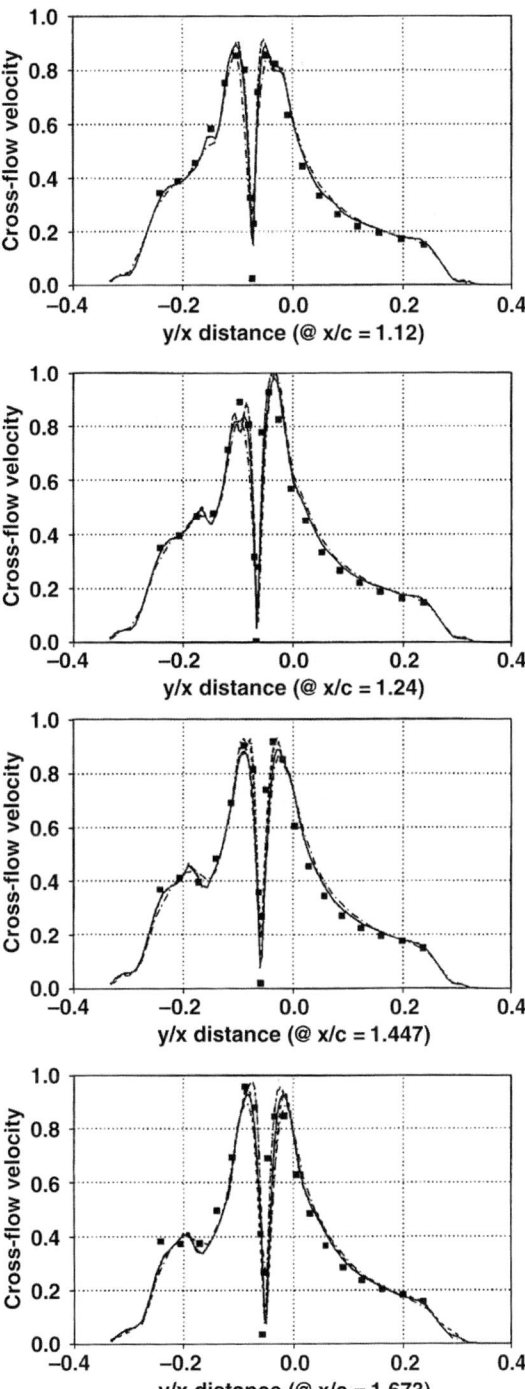

Fig. 5.56 Comparison of C_p across wake vortex

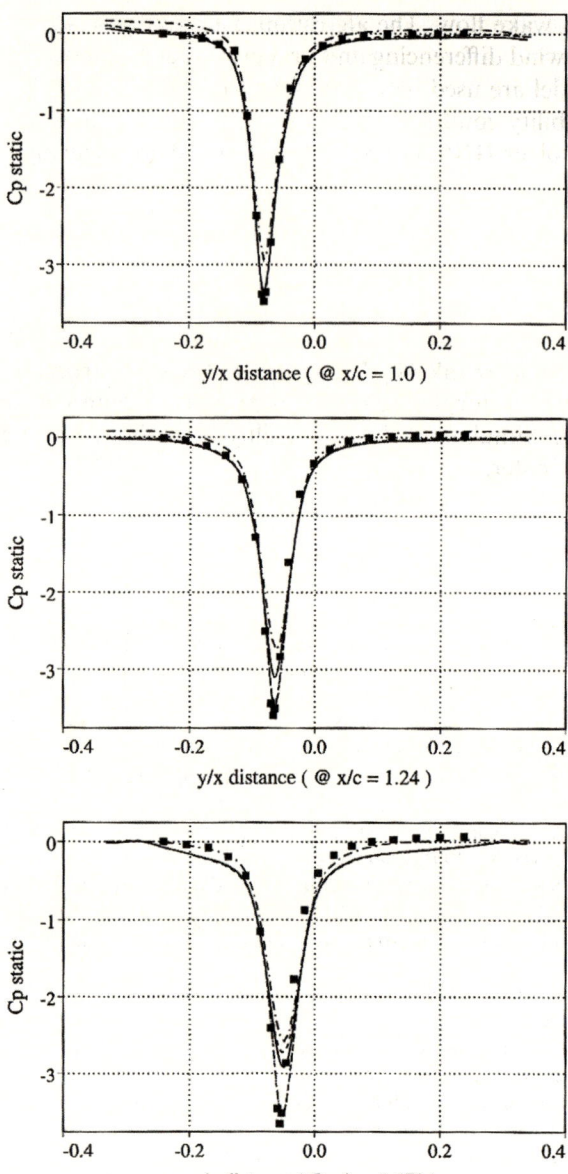

affect the tip vortex rollup and propagation. In the previous sections, we discussed how the numerical procedure and turbulence model affect propagation of the wake vortex in the near field. In this section, the rollup process is studied using the same wing with a round tip.

The tip vortex formation process for a wing with nearly constant loading is schematically shown in Fig. 5.57. A discrete vortex forms at the tip fed by vorticity from the boundary layer near the tip. As the vortex moves downstream, it rolls up more and more of the wing wake, until its circulation is nominally equal to that of the wing. The rollup distance is small compared to the distance between interacting lifting surfaces, such as the strake or foreplane, and the main wing on a close-coupled fighter or consecutive blades on a helicopter rotor. The flow in the near-field rollup region is therefore important in its own right, as well as in providing a possible means of controlling the far-wake vortex.

As depicted in Fig. 5.50, the experimental inflow is at 10° of angle of attack. Near the tip high, cross-flow velocity whips around the wing tip from the pressure side to the suction side, as sketched in Fig. 5.57. Here, we assume the entire flow field is turbulent, so a transition model is not considered. For implementing a turbulence model—either Baldwin–Barth (1991) or Spalart–Allmaras (1992)—the grid spacing on the boundary layer must be small enough to capture the rollup. In many practical problems, it is common to use $y^+ = 1.0$ for the first grid spacing in the boundary layer. To resolve the turbulent boundary layer during the initial rollup, we found it is necessary to use much finer grid spacing, $0.16 < y^+ < 0.36$, for the one-equation turbulence model used.

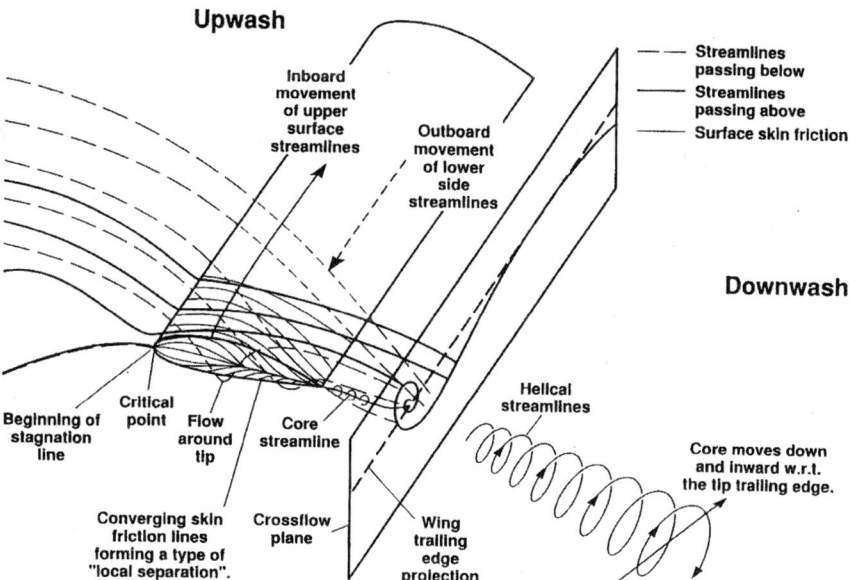

Fig. 5.57 Schematic of initial rollup of a wingtip vortex for a round tip configuration with an angle of attack $\alpha = 10°$

The entire rollup is visualized both by experiment (measured by seven-hole pressure probe) and by computation. The computed results presented here are obtained using the artificial compressibility code and the Baldwin-Barth turbulence model, unless otherwise specified. In Fig. 5.58, the velocity magnitude is compared for the rollup and the near-wake region. The black outline in both figures represents the outer boundary of the measured planes. The first sign of the vortex rollup can be seen in the farthest upstream plane in the figure. At the trailing edge, the circulation of the vortex is 87% of its final level (Zilliac et al., 1993). Notice also that the root vortex (horseshoe vortex region) was captured by the computation designated here by the blue patch on the root wall. In the experimental image (Fig. 5.58), the density of measured points near the wall was not enough to accurately map the low momentum region.

Since the flow is rich in flow physics, visualization of the entire field to capture all details is a challenge even in this simple geometry problem. In Fig. 5.59, initial rollup of the tip vortex is visualized using particle traces colored by velocity magnitude.

To validate the quality of the numerical solution on the wing surface, the surface pressure contour is compared with experimental data, in Fig. 5.60. The general trend and pressure levels of the computed and measured results compare quite well. Of particular interest is the pressure peak induced by the presence of the vortex above the suction side of the wing (green patch on the outboard portion of the wing near

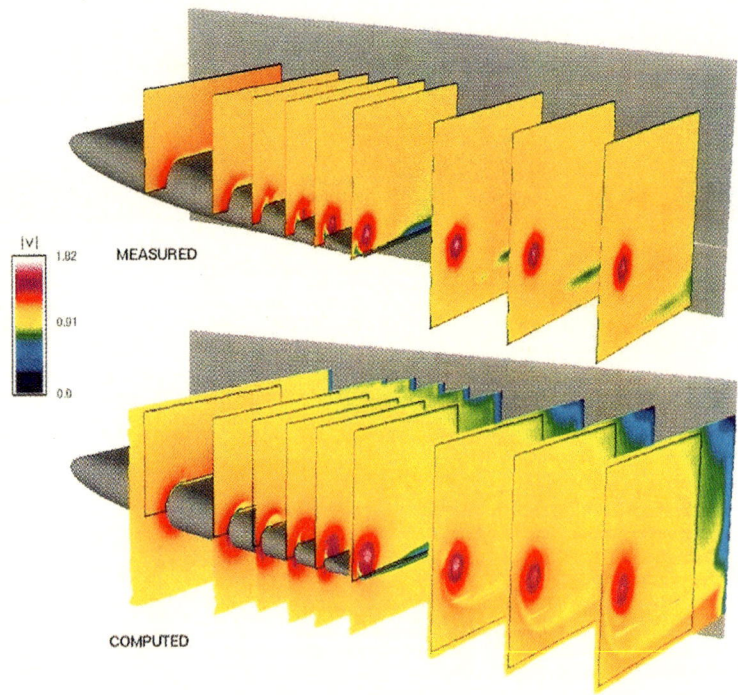

Fig. 5.58 Comparison of velocity magnitude contours

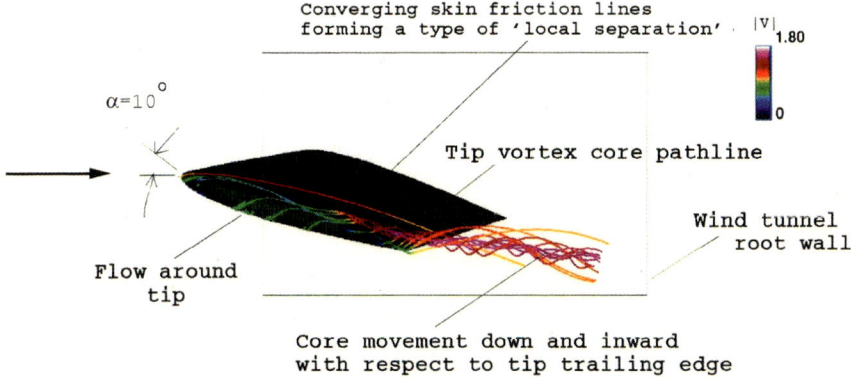

Fig. 5.59 Visualization of computed particle path lines colored by velocity magnitude

the trailing edge). The magnitude and extent of this pressure region are sensitive to the grid resolution and the turbulence modeling. A chord-wise line plot of the pressure coefficient, C_p, is shown in Fig. 5.61 at the location $z/c = 0.667$, which is approximately under the tip vortex. Although some differences occur, the vortex-induced peak is well captured by this numerical procedure.

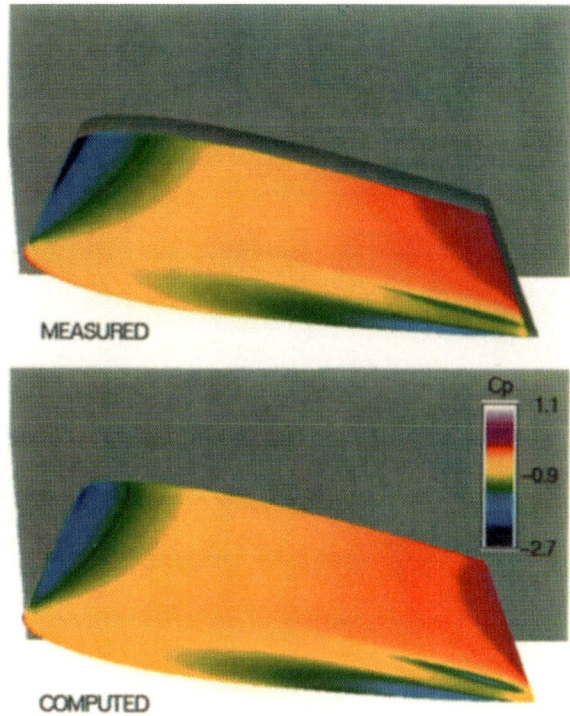

Fig. 5.60 Comparison of surface static pressure for awing tip vortex rollup problem

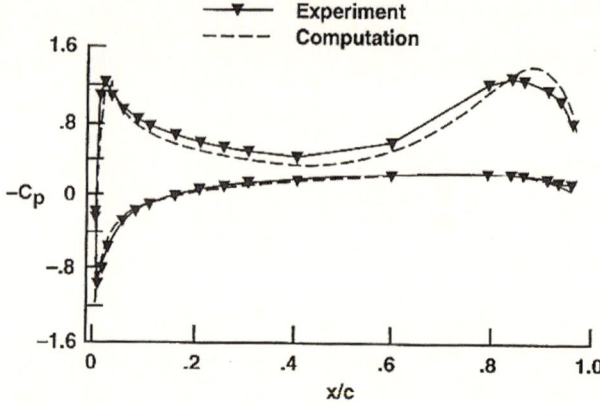

Fig. 5.61 Static pressure profile at $z/c = 0.667$ (roughly below the tip vortex)

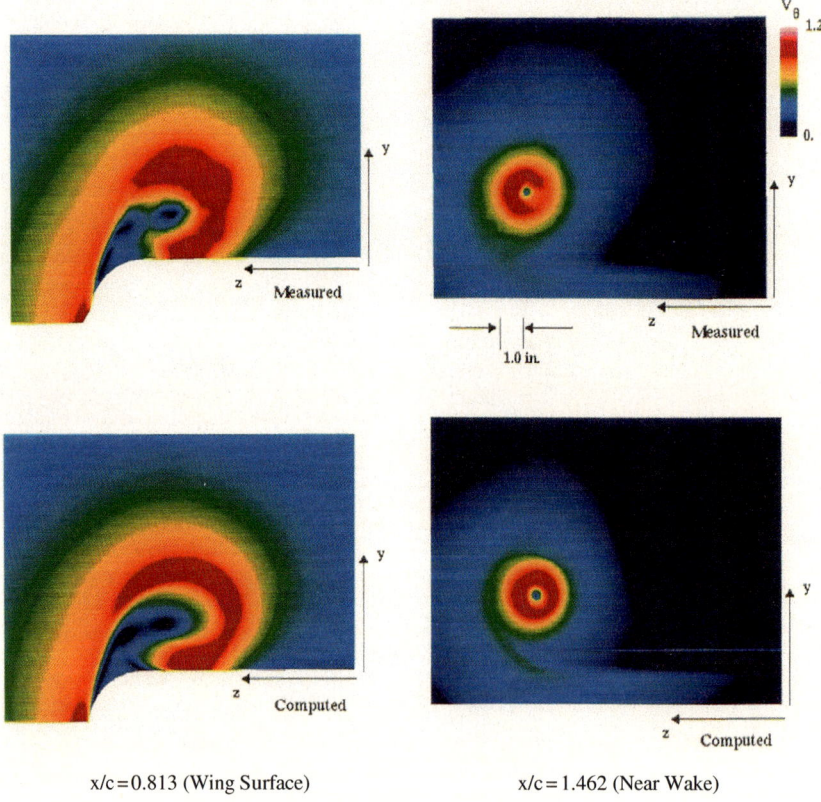

x/c = 0.813 (Wing Surface)   x/c = 1.462 (Near Wake)

Fig. 5.62 Cross-flow velocity comparison at two streamwise locations

Figure 5.62 shows a close-up of the cross-flow velocity, depicting details of how the vortex rolls up, detaches from the surface, and forms free vortex flows. The general trends agree well although some of the computed details do not exactly match the measured contours. In both the computed and measured results, the shear layer detachment point is on the suction side of the wing tip. This result was expected because the turbulent boundary layer can withstand a limited amount of adverse pressure gradients. This illustrates that the choice of turbulence models plays an important role in accurately predicting a vortex rollup procedure. In the literature cited, we have experimented extensively using different turbulence models and grid resolutions, and found that turbulence modeling significantly affects the results. Since we only present the numerical procedure, we simply emphasize the significance of the physical modeling aspects in conjunction with this highly vortical flow simulation.

An important measure of the validity of the Navier–Stokes computation is whether the surface skin-friction topology is computed correctly. Figure 5.63 shows a comparison between the surface skin-friction magnitude and the results of an oil-flow wind tunnel experiment. As seen in the figure, the location and extent of the primary and secondary convergence lines agree well with those of the experiment. The lower surface boundary layer flows around the tip to a line of surface streamline confluence or convergence, as shown in the particle traces. The convergence line indicates the departure of the shear layer from the surface, and occurs as the fluid moving in a cross-flow direction encounters an adverse pressure gradient. Near the rear of the wingtip a second convergence line is shown in both the experiment and computation, indicating the presence of a secondary vortex. On the suction-side mid-span region of the wing, the computation shows a small area of separation near the trailing edge. As shown by the skin-friction magnitude level, the shear is

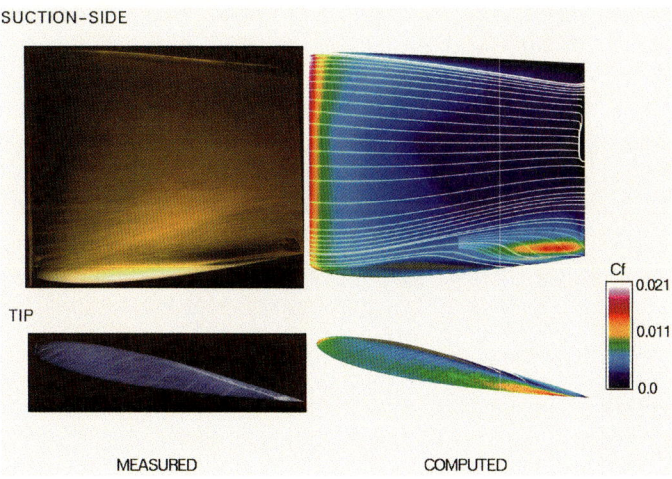

Fig. 5.63 Comparison of skin friction coefficient and particle path lines: $115 \times 157 \times 83$ grid

low upstream of this region. Although not readily visible, this behavior is similarly reflected in the oil flow pattern.

In this section, we presented a wingtip vortex flow validation to illustrate the procedures and modeling involved. When this case was first studied in the 1990s— when computing capability was measured in megaflops—1.5 to 2.5 million grid points for a small region around a wingtip were unrealistic for real-world problem solving. Now (in 2010) the case is measured by multiple teraflops and is expected to go toward petaflops or higher—so even if only a fraction of any computing facility's resources are available, it is feasible to attain the grid resolution at the level we discuss for a CFD simulation of a realistic configuration. Of course, researchers can also achieve this high-accuracy through a combination of high-order numerical schemes, grid refinement, and higher-level turbulence models as necessary.

In this chapter we have presented validation cases selected to cover various different types of flow features of fundamental nature. We tried to cover both steady and time-dependent procedures using internal, external and juncture flows. Some of the algorithmic and physical characteristics presented in conjunction with these cases will appear in following chapters when we solve real-world engineering problems.

Chapter 6
Simulation of a Liquid-Propellant Rocket Engine Subsystem

From an engineering point of view, CFD is a tool for preliminary design, design improvement, risk analysis, mission planning and operations. In this chapter, we will present engineering aspects of CFD through a task where CFD has played a significant role in accomplishing the goal of a real mission.

In this chapter, we focus on the engine subsystem related to complex internal flow. In applications involving real-world problems, CFD simulation is often confined to a truncated geometry, since modeling the entire configuration could be unrealistic or may not be necessary. Therefore, in addition to the approximations due to numerical algorithms, uncertainties can come from geometric simplification, approximate boundary conditions, and assumed initial conditions. Physical modeling often involves transition, turbulence, and multi-phase phenomena. Engineering problems require varying degrees of rigor and accuracy in obtaining flow solutions. Some of these features can best be discussed through real examples.

Previously, we presented some details of algorithms and computational procedures, including issues related to grid generation and designing boundary conditions, and validation using fundamental fluid dynamics problems. Even when all these fundamental issues are clarified and usable tools have been developed, it takes another step to make all these technologies useful to accomplish goals required for specific missions.

In this and subsequent chapters, some of the issues often encountered in CFD applications are presented. (Turbopump flow is discussed separately in the next chapter.) The level of rigor can only be determined depending on the requirements of particular simulations, whether it be dictated by economy, as in the commercial world, or by enabling computations, as in mission support arenas. This chapter illustrates the steps required for "mission computing," a term used for those classes of problems that require enabling analysis in designing and operating a specific mission, such as space exploration vehicle development and operation. This is in contrast to those problems dictated by economy in the commercial world where similar tasks are repeated with some variations to meet customer requirements.

Even though computing procedures and priorities may vary, this chapter is intended to illustrate the "next challenges" faced by those who go into real-world practices after formal training in CFD. The technical issues discussed in this

D. Kwak, C.C. Kiris, *Computation of Viscous Incompressible Flows*, Scientific Computation, DOI 10.1007/978-94-007-0193-9_6, © US Government 2011

chapter are limited to flow simulation aspects. However, for a complete engineering analysis, fluid dynamic features must be coupled to multidisciplinary aspects.

Typical 3-D CFD applications require a large number of mesh points to solve complex flow involving multiple zones, which include skewed and stretched grids, as well as physics modeling. Therefore, for flexibility of setting up computational models, it is natural to choose a primitive variable approach, which has been the primary focus of this monograph. We used our flow solvers developed at NASA Ames Research Center. However, there are a wide variety of algorithms, procedures, and associated codes available for incompressible flows. These codes and procedures can be utilized for obtaining results similar to those presented here. Users of off-the-shelf codes, however, need to be fully aware of the capabilities and limitations of the tools they choose.

In applying these computational tools (even though computer speed and memory have increased substantially in the recent past) the turnaround time still dictates the problem size one can choose for modeling and simulation—a constraint that will remain for the foreseeable future. In many engineering applications, it is very important to generate solutions in a timely fashion to have any impact on design and analysis. However, although varying degrees of solution fidelity, numerical simulations, when combined with engineering and physical understanding of the problem at hand, can provide valuable complementary information to measured data—thus reducing the number of experimental trials required for developing advanced flow devices. And, even when component-level optimization has been performed, a more sophisticated approach, such as computer optimization of a reasonably complete design, can be attempted with the increased computer speed through massive parallelism. A specific example related to the Space Shuttle is presented in this chapter.

6.1 Historical Background

The development process of the Space Shuttle as the nation's primary space transportation system (STS) began in early 1969. The design process, starting from concept to selection of the major design configuration, was much affected by political, budgetary, and technical considerations. Budgetary constraints impacted the selection of the vehicle configuration. The budgetary considerations included limiting development cost as well as lowering operational cost. The shuttle development program was subjected to formal economic constraints, perhaps for the first time in space exploration. Thus, the terminology "mission computing" implicitly includes cost and effectiveness as well as functionality.

The shuttle design also involved huge challenges of integrating various subsystems as well as coordinating discipline experts in different organizations and tasks. Eventually, in March 1972, the orbiter configuration comprising solid rocket boosters and external tanks with three engines was adopted, which is similar to that flown in shuttle missions we have observed. The three-engine configuration was selected to enhance various abort capability.

The Space Shuttle main engine (SSME) development work began as a part of the two-stage-to-orbit vehicle development. The SSME was the first reusable liquid booster engine requiring performance substantially beyond the engines available at that time. Successfully developed by Rocketdyne, the SSME has been used in all shuttle missions starting from the first mission, STS-1, flown on April 12, 1981.

A wide spectrum of fluid dynamics issues has been associated with the SSME where CFD has been very valuable for analysis and design. However, in the beginning, CFD technology and computing power were at a fundamental development stage and were not available for any large-scale engineering applications such as shuttle development. Right after the first flight, several upgrade projects began for enhancing the safety, reliability, and performance of the shuttle. The SSME powerhead upgrade, called the Phase II+ redesign, was under way in the early 1980s. At the same time, flow solvers were being developed at NASA, and Cray-class computers became available. In this chapter, we will discuss how CFD began making impacts on SSME hardware development in conjunction with integrated design, experiments, and full-scale tests.

While the CFD application work began with this historical background, the approaches and application processes discussed in this chapter are relevant to CFD application processes for "mission computing" in general. It is hoped that the examples given in this chapter will provide CFD users with some valuable insights into the potential issues and how CFD can influence real-world applications.

6.2 Flow Analysis in the Space Shuttle Main Engine (SSME)

Rocket propulsion systems using liquid propellant have been used for boosters and spacecraft starting in the early 1950s. The Space Shuttle became NASA's workhorse for the Space Transportation System (STS) in 1981, and has been the primary vehicle for the agency's human space program ever since. For shuttle launches, three powerful main engines operate in addition to two solid rocket boosters. The SSME is 14 feet long, weighs approximately 7,000 pounds, and is 7.5 feet in diameter at the end of the nozzle. It is propelled by liquid hydrogen and liquid oxygen (LOX), producing a specific impulse of 453 s and 512,264 pounds of thrust in a vacuum. In this staged-combustion cycle engine, the fuel and oxidizer are fed by a turbopump. The engines operate for 8.5 min during liftoff and ascent and burn more than 500,000 gal of cryogenic fuel and oxidizer.

The SSME was originally developed without the help of CFD technology. However, it has been the only operational liquid booster engine designed for human space flight to date (as of 2009). Since its initial design, NASA has continued to increase the reliability and safety of shuttle flights through a series of enhancements. Modifications include new high-pressure fuel and oxidizer turbopumps, a redesigned powerhead, and a new heat exchanger and large-throat main combustion chamber. These modifications support the increasingly extended role of the shuttle for scientific and commercial applications. As a part of these activities, the upgrade of the SSME powerhead was initiated in the early 1980s to substantially increase the operating margin and engine durability. To achieve this goal without increasing the weight and size of the existing engine, it became essential to understand the dynamics of the hot-gas flow in the powerhead.

Generally, there are three categories of fluid dynamics related sub-elements in liquid-propellant engine systems: (1) combustion devices, (2) turbopumps, and (3) complex internal flow subsystems. Analyses of these components offer three corresponding categories of simulation challenges: Mixing and chemistry; flow with cavitation; and complex internal flow in subsystems, respectively.

Because of the complexity of the geometry, an experimental approach is extremely difficult, time consuming, and expensive. Therefore, computational simulations have offered an economical alternative to complement experimental work in analyzing the original SSME powerhead configuration, and to suggest new, improved design possibilities. During this redesign work, major milestones were established for the computational effort. Highlights of this task are presented here to explain features encountered in this mission support application from a CFD perspective.

The late Werner von Braun said:

> ... Behind each apparent miracle, however, stands the flawless performance of numerous highly complex systems. All are important. The failure of only one portion of a launch vehicle or spacecraft may cause failure of an entire mission. But the first to feel this awesome imperative for perfection are the propulsion systems, especially the engines ...

If we call the Space Shuttle a great engineering masterpiece, the SSME is certainly an engineering marvel. We'll discuss next how CFD can be applied to meet some of the fluid dynamics challenges encountered in the SSME powerhead upgrade.

6.3 Flow Analysis Task and Computational Model for the SSME Powerhead

The SSME hot gas manifold (HGM) is the structural backbone of the engine and contains two preburners, a main injector, and various propellant and oxidizer ducts and lines (see Fig. 6.1). It also includes two high-pressure turbopumps, the main combustion chamber, and a gimbal bearing that attaches the engine to the shuttle orbiter. Together, all components are called the "powerhead."

In the SSME staged combustion cycle, the fuel is partially burned at very high pressure and at relatively high temperature in the preburners. The resulting hot gas is used to run the two-stage turbines in the high-pressure turbopumps. Hot gas discharged from the gas turbine enters the annular turnaround duct (TAD) and makes a 180° turn to flow back along the outer surface of the pump through the annular fuel bowl. The fuel-rich hot gas is directed into transfer tubes that link the bowl with the injector. The hot gas enters the injector and flows across a forest of LOX posts. It

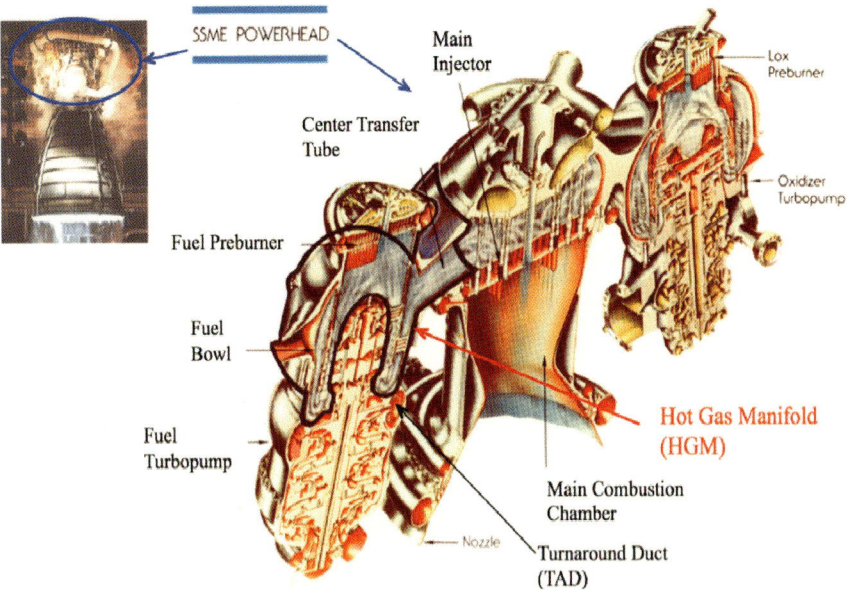

Fig. 6.1   Sketch of the SSME powerhead indicating various components (courtesy of Rocketdyne, Canoga Park, California, USA)

enters concentric tubes around the LOX posts and flows into the main combustion chamber where it burns with the oxygen.

In Fig. 6.1, the bold outlined area from the turbine outlet to the transfer tube connected to main injector assembly is called the "fuel side" of the hot gas manifold, and is the focus of the redesign effort discussed in this chapter.

The Reynolds number of the primary flow in the manifold is on the order of 10^6 per inch. Because of the high gas temperature, the Mach number is less than 0.12. The flow is turbulent and is incompressible for all practical purposes. At the time of the HGM redesign work, our preconditioning technique, as discussed in Chapter 4, was not mature, so it was not possible to use the compressible Navier-Stokes flow solver at this low Mach number range. Therefore, we utilized the INS3D incompressible flow solver that was then being developed. Similarly, other low Mach number flow problems not involving local high-speed compressible flow regions can be simulated using incompressible flow solvers available today.

6.3.1 Computational Model Description

In the original design of the SSME powerhead components, the gas flows into the main injector through three transfer ducts on the left side of the powerhead, as sketched in Fig. 6.1 (fuel preburner side), and enters into the top portion of the

Fig. 6.2 Schematic of liquid
oxygen (LOX) post
arrangement in the main
injector: top view of a
horizontal cross section of the
main injector assembly

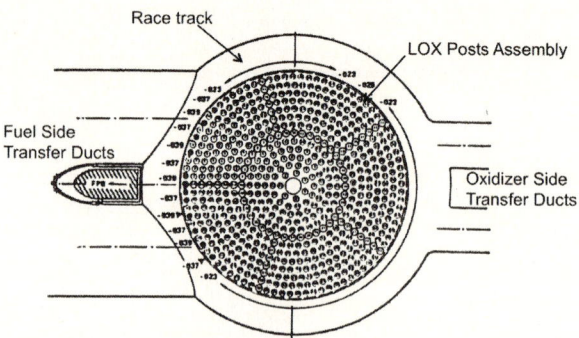

main injector assembly where many liquid oxygen (LOX) posts are bundled (see
Fig. 6.2) on top of the main combustion chamber. On the right side of the power
head (oxidizer preburner side), two transfer ducts are connected to the right side
of the main injector assembly. Around the bottom portion of each LOX post in the
main injector assembly are a number of small holes through which the hot gas flows
into the main combustion chamber. There, it mixes with the oxidizer, which comes
through a circular passage along the centerline of each LOX post.

As a part of the HGM redesign effort, the CFD study began with an analysis of
the hot gas flow in the original powerhead configuration. When this effort began,
only the first version of the INS3D code was available, so the computed results
presented in this section have been obtained using the original version of INS3D.

We have chosen a computational model of the powerhead to analyze critical areas
where the dynamics of the hot gas flow are expected to have a significant effect on
the overall performance of the SSME. As shown in Fig. 6.3, the model starts from
the gas turbine exit on the fuel preburner side, and extends to the main injector
assembly. The main injector consists primarily of a bundle of LOX posts, which
is physically modeled by a porous media. The engine was in operation when the
upgrade effort was initiated, and then it was identified that the modification was
needed on the fuel side of the HGM. Since the hot gas from the oxidizer side and
fuel side meet in the racetrack region in the main injector assembly, the computational model for the fuel side was truncated at the racetrack where outflow boundary
conditions are imposed.

The fuel-side HGM model geometry and grid topology are shown in Fig. 6.3.
The grids in the horizontal and vertical cross section of the HGM are shown in
Fig. 6.3a, b. The model shows only one half of the fuel-side HGM geometry, as
shown by the B-B cut in Fig. 6.3a. There are three transfer ducts connecting the fuel
bowl and main injector assembly. Shown in the figure are one side duct and half of
the center duct.

The H-grids for these ducts are generated using an algebraic process starting from
a unit circle. Near the duct boundary the grid lines are concentric circles except in
the vicinity of the four singular points. Using the nearly orthogonal grid in this unit

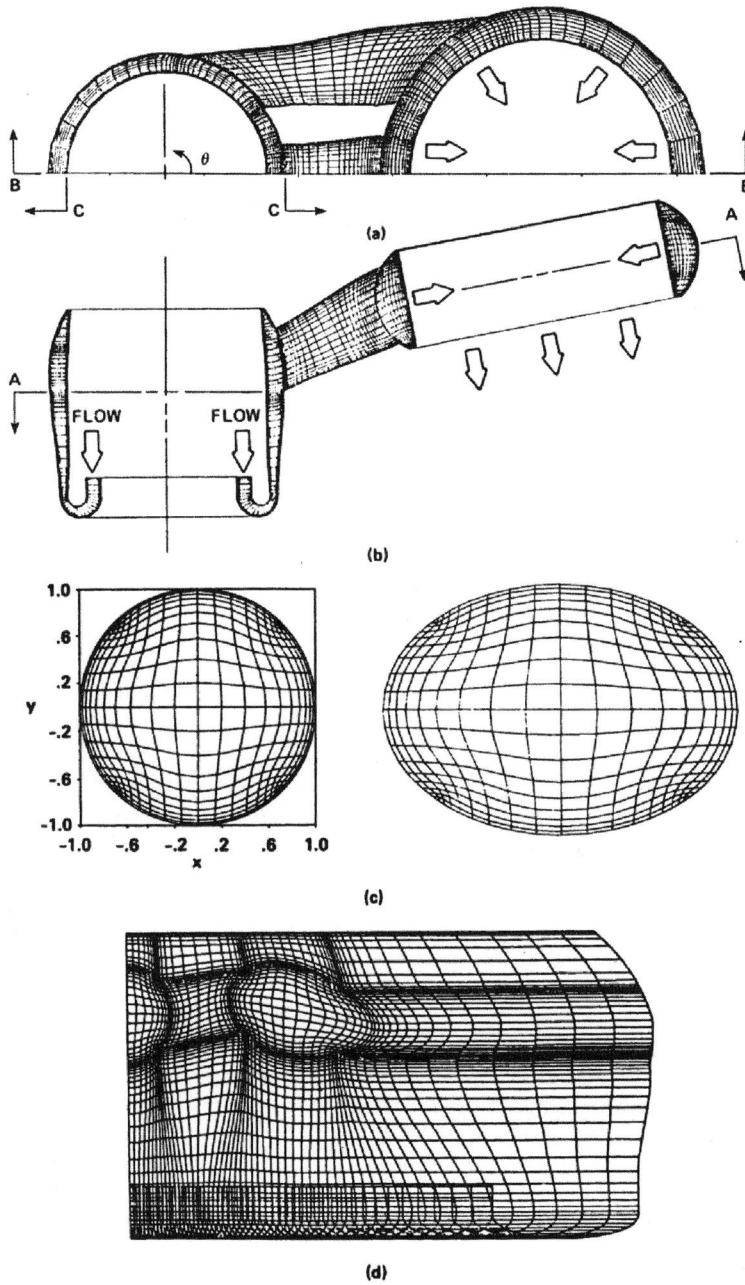

Fig. 6.3 Geometry and grid of the original SSME three-duct hot gas manifold model:(**a**) horizontal view (cross section A-A); (**b**) vertical view (cross section B-B); (**c**) H-grid for circular and elliptic cross section of transfer duct; (**d**) unwrapped surface of annular fuel bowl (cross section C-C)

circle, one can obtain H-grids for tubes or ducts of any given shape and dimension by a simple linear transformation. These duct-grids are connected to the fuel bowl and circular racetrack-shaped region (see Fig. 6.2) of the main injector. As shown on left side of Fig. 6.3c, the original duct design used a circular cross section. On the right side of Fig. 6.3c, the grid for an elliptic duct is also shown, which allows a redesign option if deemed beneficial. Note that any geometry modification must include structural analysis as well as fluid dynamics.

Two approaches can be considered in coupling two computational regions, namely, the fuel bowl and the transfer ducts. Overlapping grids will require interpolation, while perfectly matching grids simplify this at the expense of adding complexity in grid generation. In our computation, matching grids were employed. Figure 6.3d shows an unwrapped surface of the annular fuel bowl with openings. The H-type grid topology for the transfer duct then allows smooth transition from the axisymmetric TAD to the transfer ducts. This arrangement is also convenient in clustering the grid near the duct wall. For general development work, overlapping a grid arrangement may be more convenient for parametric comparison of various geometric options.

The grid for the entire HGM system is generated by using algebraic functions, and is written with a high degree of flexibility for changing geometric configurations. By specifying the shape, dimension, and preferred number of transfer ducts, a grid for a variety of new HGM configurations can be obtained in a short time. The ducts described in this section are connected directly to the fuel bowl without any fairings, while in the original engine the three transfer ducts were connected smoothly to the annular fuel bowl with fairings. This configuration, with an abrupt change in geometry, is more demanding computationally than smooth configurations. At the time this task was performed, the particular grid generation routine was customized for this particular application, partly because automatic grid generators were not versatile enough to accommodate this configuration. Now, commercial packages and software in the public domain are capable of handling complex geometry gridding and are readily available.

6.3.2 Multiple-Zone Computation

A large number of mesh points are required to solve the 3-D turbulent flow in the SSME. To facilitate numerical simulations, the domain of interest is divided into several zones. This requires a special treatment at the zonal interface for a smooth continuation of the solution between zones.

Figure 6.4 shows a five-zone arrangement for the HGM flow field. Zone 1 includes the TAD and fuel bowl. Zones 2, 3, and 4 comprise the three transfer ducts for the original engine configuration. The annular region, denoted as the racetrack of the main injector, is represented by Zone 5. Also shown in the figure are some overlapping grids in the various zonal interfaces. The grid is chosen to be continuous and smooth across zonal boundaries. In the first attempt at simulating this model, the racetrack (Zone 5) was not included in the computation. In a later computation

Fig. 6.4 Multiple-zone
arrangement for the SSME
powerhead simulation

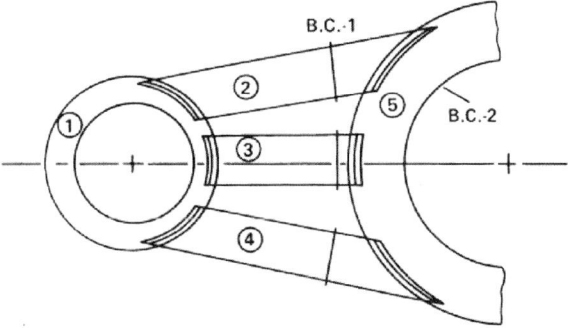

ZONE
① TAD AND FUEL BOWL
② ③ AND ④ TRANSFER DUCTS
⑤ RACE TRACK (FUTURE WORK)
OVERLAPPING GRID IN ZONE INTERFACES.

involving a new design configuration, the racetrack region was included, along with the main injector assembly that was modeled by a porous medium.

More detailed study of the LOX post in the main injector assembly was also performed using the results of the HGM computation as the boundary condition, but is not be given here. Since the vertical plane through the center of the fuel bowl and the main injector is taken to be a plane of symmetry, results for the HGM presented in this chapter were obtained using only half of the HGM. When the swirling flow from the gas turbine exit is included, the symmetry assumption should not be used.

In the artificial compressibility formulation, waves are propagating in both up- and downstream directions while the solution approaches a steady state. In the present problem, the interfaces between zones are locations where the geometry changes abruptly. Therefore, in the neighborhood of those interfaces, the flow is expected to experience a rapid change. To maintain a smooth continuation of the solutions across these zones and, consequently, to achieve a stable and fast-converging computation, means for providing adequate communication for the traveling waves must be established. Overlapping regions and a proper zonal interpolation scheme are required for this purpose.

A forward or backward differencing operation, if applied to the interfaces of multiple zones, would distort the geometric representation. To maintain a smooth transition of the flow field across a zonal boundary, the Jacobian and the metrics at the interfaces are computed using grid points in neighboring zones. Then the pressure and velocities, Q, are updated explicitly at each iteration.

Let values at $n + 1/2$ denote conditions to be used to advance the computation to $n + 1$. The values of $Q^{n+1/2}$ Zone 1 at the exit plane are obtained from the values of the corresponding plane of Zone 2 at n, that is:

$$\left[Q^{n+1/2}_{B.C.}\right]_{Zone1} = \left[Q^n_{interior}\right]_{Zone2}$$

Values of $Q^{n+1/2}$ for Zone 2 at the zonal interfaces are taken from the latest computed result of Zone 1 as:

$$\left[Q_{B.C.}^{n+1/2}\right]_{Zone2} = \left[Q_{interior}^{n}\right]_{Zone1}$$

When more than two points overlap, the latest values in the interior of this overlapping region must be properly transmitted to the next zone. A number of ways are available to treat this problem. The simplest one is to take an average of the two values computed in Zones 1 and 2, as below:

$$\left[Q^{n+1/2}\right] = \frac{1}{2}\left(\left[Q^{n}\right]_{Zone2} + \left[Q^{n+1}\right]_{Zone1}\right)$$

A scheme using updated zonal boundary values, but original interior values, has also been tested. Either way, converged steady-state solutions have been obtained. However, the scheme with interior updates converges at a much faster rate.

6.3.3 Grid and Geometry Effects

The INS3D flow solver has been validated by computing fundamental fluid dynamics problems, as illustrated in previous chapters. However, many other aspects need to be clarified in real-world applications, such as grid-induced error due to skewness and 3-D stretching, and grid clustering and its relation to numerical dissipation. In real geometry, the flow is likely to go through strongly curved sections. This introduces yet another subject—that of strong curvature effect on turbulence structure. The grid selected for the current simulation could introduce errors such as grid singularities in H-grid, as well as skewness while trying to enforce one-to-one matching between grid interfaces, and stretching. Some of these aspects are examined here.

The approximate factorization (AF) algorithm implemented in the original version of the INS3D flow solver integrates the difference equations along the transformed coordinates. At the junction of the two H-grid directions, flow particles in the two coordinate directions could communicate only indirectly via the interior mesh points. This produces a corner-effect error. Even though this error is not as severe as the one caused by external flows, an ad hoc method of eliminating this corner effect is devised based on a finite-element concept.

Let j and k denote the indices for the grid points along the increasing ξ and η directions, respectively. Then let $j = k = 1$ be the corner point, which is singular. First, the pressure at this point, p_{11}, is determined by an extrapolation along the diagonal direction $j = k$. Second, p_{13} and p_{31} are obtained in the usual manner. Then p_{12} and p_{21} are established by an interpolation along the circular surface (see the H-grid singularity in Fig. 6.3c).

The full viscous term given by Equation (2.13a) can be simplified to Equation (2.13b) when the grid is orthogonal. Even though full viscous terms can be used, it is convenient and economical to keep only the orthogonal part. It is, therefore, of practical interest to estimate the overall error caused by using the orthogonal formulation when a non-orthogonal grid system is used.

As a quick measure of an overall error caused by grid skewness, a 2-D channel flow is computed using two grids, namely, (1) a stretched Cartesian grid (orthogonal) and (2) a non-orthogonal grid where the skewness is controlled on the upper half of the channel, as sketched in Fig. 6.5. In the computation only orthogonal terms are kept. Converged solutions on the lower half, where the two grids are identical, are then compared. Total error depends additionally on the Reynolds number and the third-directional skewness. However, this quick experiment indicates that

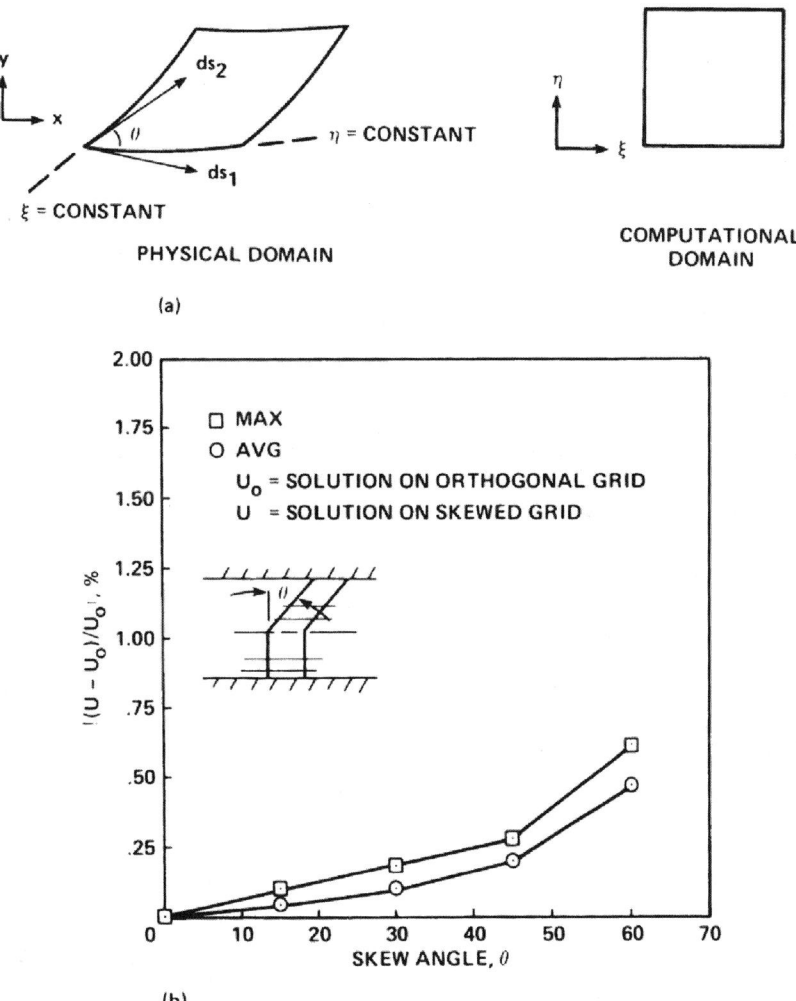

Fig. 6.5 Grid effect on channel flow solution at Re $=$ 1,000: (**a**) definition of grid skewness, (**b**) relative error due to skewness

the orthogonal assumption can be used without significantly impacting the overall solutions if the grid is basically orthogonal in most of the computational domain.

This is a qualitative study only, and a full investigation should include more severe cases involving separation and recirculation regions, as well. When computational resources are not severely limited, the full viscous terms can be used on the right-hand side of the momentum equations, which will remove the uncertainty associated with non-orthogonal viscous terms.

6.4 Turbulence Modeling Issues

Turbulence models for RANS calculations of aerospace vehicles have been developed mostly for external flows such as on airfoils or wings. The primary focus for external flow has been predicting the boundary layer development and separation. Since these models lack universality, turbulence models for RANS calculations are adjusted to match characteristics of different types of external flow such as boundary layer or free shear flow.

Historically, various levels of turbulence models have been developed, such as algebraic, one-equation, two-equation and Reynolds stress models. Review articles on these turbulence modeling are numerous (for an early review, see, for example, Reynolds, 1976). These modeling approaches based on Reynolds-averaged Navier-Stokes equations are lacking in prediction capability. More physics-based approaches such as large eddy and direct simulation emerged most notably in the 1970s.

The three-dimensional time-dependent computations of large-scale structures of turbulence were of primary interest to meteorological flow simulation (see for example, Deardorff, 1973; Fox and Lilly, 1972; Smagorinsky, 1963). Later, the turbulence research group at Stanford University started using the term "Large-Eddy Simulation (LES)" (see Leonard, 1973; Kwak et al., 1975). The original hope was that the sub-grid scale (SGS) model would provide the necessary dissipation for unresolved scale turbulence independently from numerical algorithm. However, it was found that the numerical scheme acts as spatial filter simultaneously. Moreover, handling of the boundary layer region turned out to be very challenging.

A hybrid LES-RANS approach was then developed to circumvent the problem related to the boundary layer region (e.g. Spalart et al., 1997, 2009). A large number of research reports have been published over the years, and readers are referred to existing literatures on this subject. A comprehensive review of this subject is beyond the scope of the present monograph.

Since the task we discuss in this chapter involves internal flow, straight applications of the usual RANS models developed for external flows will not produce a reliable basis for analysis and redesign involving HGM flow. Even though one- and two-equation models are incorporated into the INS3D family of codes, in lieu of developing a new turbulence model for internal flow, we discuss how to handle the turbulent flow encountered using a simple algebraic expression. Even though an algebraic model lacks prediction capability and does not represent turbulent flow physics in a strict sense, when tuned for a particular type of turbulent flow the model can still produce engineering solutions in an economical way.

The primary goal of our task was to analyze the current configuration and then suggest a modified design configuration of the HGM in time to be used for the flight hardware development. As will be explained in the latter part of this chapter, the entire process of developing an upgraded engine with flight certification takes a very long time. Like many mission support tasks, the preliminary design phase had to be done in a relatively short time to make any impact on flight hardware development and testing. Of course, this meant the turbulence model implementation and ad hoc adjustment had to be done with expediency.

6.4.1 Selection of Turbulence Model for Internal Flow

The simplest turbulence model for a RANS computation is an algebraic model. The basic assumption of algebraic models (or of any eddy viscosity models) is that the turbulence is in local equilibrium; that is, the production and dissipation of turbulence are the same. For this modeling approach, turbulence length scale distribution needs to be prescribed based on empirical input. The empiricism has been mostly based on boundary-layer type flow. Even though this modeling approach cannot handle transport and history effects, simplicity of the expression has been the major advantage, and many external flow problems have been successfully computed using this approach. However, for flows with streamline curvature, separation and/or rotational effects, as well as unsteady flow, it becomes very difficult to define a generally applicable length scale.

In external flow problems, boundary layer thickness or displacement thickness is often used as a length scale for the turbulence. However, boundary layer thickness is difficult to define in a numerical approach. In addition, computation of displacement thickness requires very fine grid resolution to maintain accuracy. In one of the most frequently used algebraic models, Baldwin-Lomax model (1978), the length scale of the turbulent eddies is determined by the location of the maximum moment of vorticity. This length scale is well defined in external flows when the flow is not separated. However, the maximum moment of vorticity is not as well defined for fully turbulent internal flows or juncture flows. In particular, the moment of vorticity is almost constant for a fully developed pipe or channel flow except in the thin sublayer region and near the centerline.

For fully developed internal flows in a duct with mild streamwise curvature, the location of the maximum velocity may be used to determine the eddy sizes. However, in the case of the present turnaround duct, as illustrated in Fig. 6.3b, the location of the maximum velocity is not a good measure of the boundary layers associated with the two opposite walls. In this case, the flow consists of a pair opposite vorticities. Therefore, there must be a location where both of them vanish, which essentially divides the two boundary layers from opposite walls. In practice, however, it is difficult to locate the position where vorticity vanishes. One alternative is to use vorticity thickness as the distance from the wall to the position where vorticity becomes minimum, as will be explained next.

6.4.1.1 An Extended Prandtl-Karman Mixing Length Model for Internal Flow

An algebraic model is based on the idea that the turbulence level is represented by a length scale. To represent internal flow turbulence by an algebraic eddy viscosity model, we attempted to modify the Prandtl-Karman mixing length expression that matches internal flow characteristics.

An eddy viscosity model for the wall region is typically given by:

$$v_t = l^2 \frac{du}{dy}$$

$$l = ky$$

where

v_t = eddy viscosity
l = mixing length
k = Karman constant

Away from the wall, the mixing length approaches a constant value representing the turbulent eddy size. For fully turbulent internal flow in a duct, the two boundary layers associated with the opposite walls merge in the flow field. In the middle of the duct, these opposite vortices cancel each other. Following Chang et al. (1985b, 1988a), a simple algebraic model can be devised as follows, based on the Prandtl-Karman mixing length theory combined with the strength of vorticity:

$$v_t = l^2 \, |\omega| \tag{6.1}$$

where

$$l = \kappa^2 \left[1 - \exp\left(-\frac{y}{\kappa \delta_\omega} \right) \right] \delta_\omega$$

$$\kappa = 0.4$$

Here, $|\omega|$ is the strength of the 3-D vorticity and y is the distance from the wall. The vorticity length, δ_ω, is defined to be the distance between the maximum vorticity and the point where vorticity first vanishes. This length can be interpreted as a measure of turbulent eddy sizes. The location of the maximum vorticity for an attached flow is at the wall. When the boundary layer separates, typically under adverse pressure gradient, this location starts to move away from the wall. The points of vanishing vorticity can then be replaced by the location of minimum vorticity in merged layers for the purpose of numerical study.

Near the wall, corrections for the viscous sublayer effects are required. For attached layers, one of the first damping functions by Van Driest (1956) can be applied. In the neighborhood of boundary layer separation, as well as for the separated reverse flow region, this damping function has to be modified. In view of the fact that viscous effects in the shear layer are concentrated in the vicinity of the maximum vorticity, an effective vorticity length can be defined as:

$$\Lambda = \frac{u_{ref}}{|\omega|_{max}} \tag{6.2}$$

where u_{ref} is the average velocity at the inlet. Then, following the spreading rate of the laminar viscous region, an effective thickness of the viscous region can be defined as:

$$\delta^+ = \sqrt{\frac{\nu \Lambda}{u_{ref}}} = \sqrt{\frac{\nu}{|\omega|_{max}}} \tag{6.3}$$

The inner sub-layer scale is thus given by:

$$y^+ = \frac{y}{\delta^+} \tag{6.4}$$

For an attached turbulent boundary layer, the maximum vorticity coincides with the maximum rate of strain at the wall. Then Equation (6.4) reduces to the usual form for y^+. Near the inception of separation and the separated flow region, however, the point of maximum vorticity will move away from the wall.

In the outer layer, the intermittency factor, γ, given by the Klebanoff's turbulent boundary layer, is applied with a slight modification, as follows:

$$\gamma = \frac{1}{2} \left\{ 1 - erf \left[5 \left(\frac{y - y_{\omega \, max}}{\delta_\omega} - 0.8 \right) \right] \right\} \tag{6.5}$$

where $y_{\omega \, max}$ is the distance between the wall and the maximum vorticity. The maximum point in the Gaussian distribution is located at $0.8\delta_\omega$, corresponding to about 0.76δ in Klebanoff's original expression. To sum up, the composite form of the eddy viscosity can be written as:

$$\nu_t = \gamma \left[l \left(1 - e^{-y^+/A^+} \right) \right]^2 |\omega| \tag{6.6}$$

In the reverse flow region of the separated layer, there exists a small-size vorticity with an opposite direction to its adjacent outer mixing or wake-type shear layer. The eddy size of the reverse flow is determined by the distance from the wall to the point of minimum vorticity, somewhere in the vicinity of the largest negative velocity. Since reverse flow regions are usually very thin and close to the wall, the Prandtl mixing length with wall damping is adequate for this region.

6.4.1.2 Application to Pipe and Channel Flow

As a precursor to turnaround duct calculations, the above algebraic model is first applied to the fully developed pipe and channel flow. Here, $\delta_\omega = R$, where R is the radius of a pipe or the channel half-height. The mixing length given by Equation (6.1) is compared with classical experimental data by Nikuradse (1932), as shown in Fig. 6.6.

In these flows:

$$|\omega| = \left|\frac{du}{dy}\right|$$

and the shear stress varies linearly from the wall:

$$\tau = \tau_w(1 - \hat{y}), \quad \hat{y} = \frac{y}{\delta_\omega} \tag{6.7}$$

By using the relation for the turbulent shear stresses and the eddy viscosity defined by:

$$\tau = \rho v_t \frac{du}{dy} \tag{6.8}$$

The eddy viscosity for the fully developed channel or pipe flow can then be obtained by:

$$\frac{v_t}{u_\tau R} = \kappa^2 \left(1 - e^{-\hat{y}/\kappa}\right) (1 - \hat{y})^{1/2} \tag{6.9}$$

where

$$u_\tau = \left(\frac{\tau_w}{\rho}\right)^{1/2}$$

The result of Equation (6.9) is plotted as the broken line in Fig. 6.6b. It compares quite well with the experiments until the distance from the wall reaches about 0.7. The solid line in the same figure is the computed result for the inlet of the turnaround duct where a logarithmic velocity profile is specified. In the center region the velocity profile is almost flat. It is difficult to obtain accurate velocity gradients from the old measured velocity data. The calculated eddy viscosities in this region therefore have some uncertainty. It is commonly believed that the eddy viscosity approaches some constant value in this region. If a constant value for this region is acceptable, v_t for $y/R \geq 0.7$ can be set to its value at $y/R = 0.7$ as determined by the above equation.

The results shown above from the ad hoc modification of an algebraic model based on mixing-length concept is to verify whether the mixing length approach proposed here is adequate for application to the TAD analysis, and is not intended for developing a new turbulence model generally applicable to internal flows. When

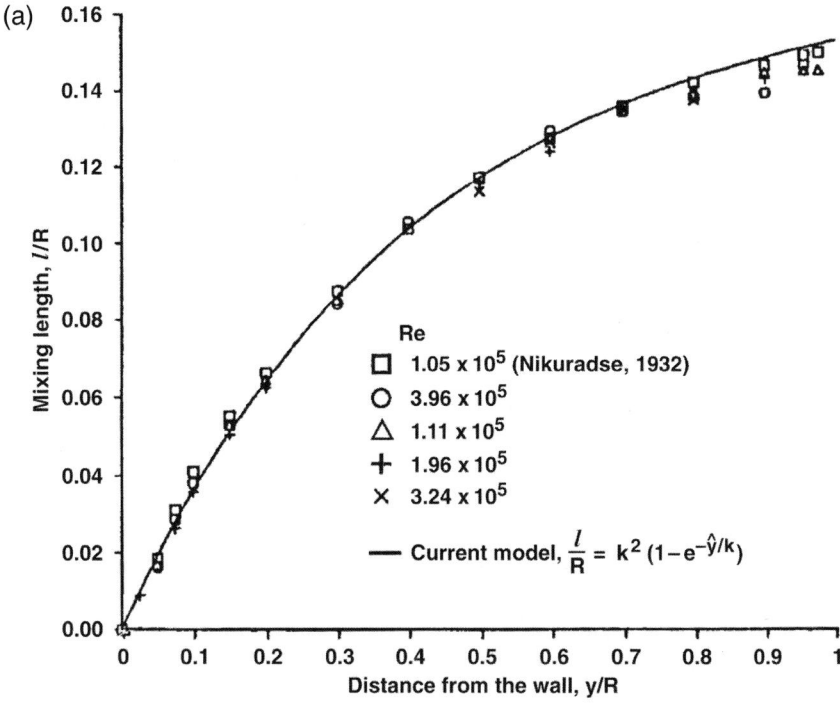

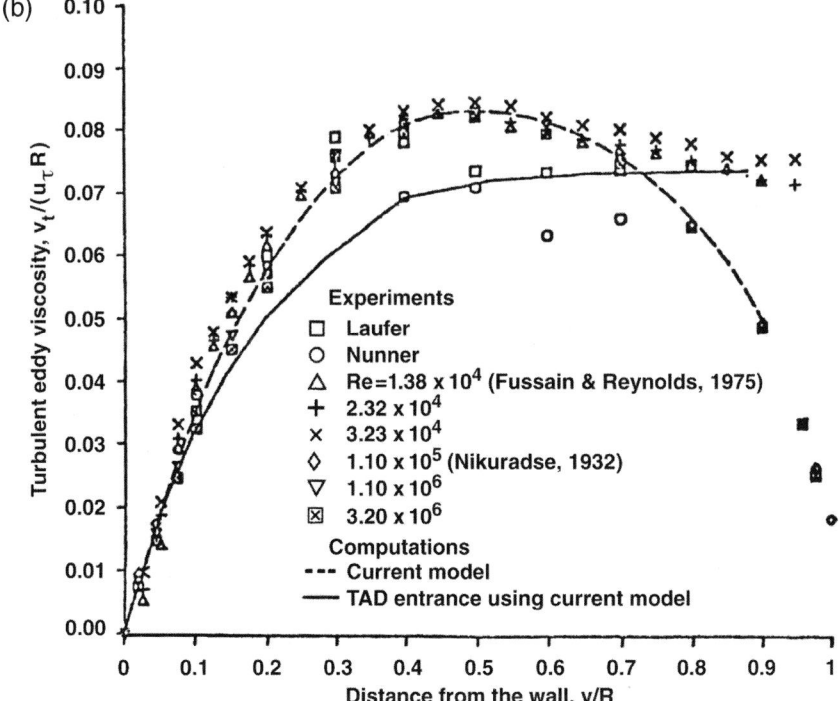

Fig. 6.6 A mixing length model for a channel flow: (a) mixing length, (b) turbulent eddy viscosity

the TAD analysis was in progress, available turbulence models were not designed for a direct application to this type task. In addition to the internal flow aspects where the viscous effects are not confined to the boundary layer, TAD geometry poses additional challenges for defining strong curvature effects in a diverging three-dimensional channel. In many mission applications, similar situations may arise where theoretical or empirical studies are not available. In those cases, judicious implementation of available physical models, modified as needed, may be required for timely impact of the CFD approach on mission schedules.

6.4.2 Turbulence Modeling Issues Involving Strong Streamwise Curvature

The flow through the SSME powerhead offers a variety of rich internal flow phenomena. Among others, the curvature effect of turbulence was most uncertain at the time the current simulation was started. Therefore, we raised the question on the accuracy of the computed results when the turbulence model being used did not account for the strong curvature effects. In this section, our process for addressing this issue is described.

In the current configuration of the SSME hot gas manifold, the partially combusted hot gas from the pre-burner operates the turbopump (see Fig. 6.1), then enters into the axisymmetric annular 180° TAD, as shown in Fig. 6.3b. In the sharp U-turn region of the TAD, the ratio of the internal shear layer thickness to the radius of curvature is of order one. On the convex side (that is, the inner wall), the turbulence is expected to be greatly reduced, while it is expected to be substantially enhanced on the concave side. In addition, strong adverse and favorable pressure gradients coexist and interact with each other. Because of the sharp turn, the streamwise variations and the normal gradient of the streamwise velocity are of the same order of magnitude. Furthermore, the flow rapidly changes direction. Understanding the structure of turbulence in this flow and quantifying the effects on the flow are crucial for a successful computation of the TAD.

Since the original work by Prandtl in 1929, the effect of mild streamline curvature on the structure of a turbulent boundary layer has been studied by many investigators (for a comprehensive review, see, for example, Bradshaw, 1973). It is well known that a mild curvature produces a profound effect on the turbulent flow structure. For example, if the ratio of the boundary layer thickness to the radius of the wall curvature is merely 1%, there will be approximately 10% or more change in the turbulent quantities. In most external flows such as the flow over an airfoil, the streamline curvature is designed to be very mild *except* near the nose region where flow remains laminar. For problems of this kind, the literature on the curvature effect is quite extensive (see, for example, Wattendorf, 1935; Eskinazi and Yeh, 1954; Gillis and Johnston, 1983; Moser and Moin, 1984).

However, study on a strong curvature effect is very limited. The question was how the strong curvature in the TAD affects the turbulence quantity in a very short duration while the flow passes through the 180° turn region. To analyze the fluid dynamic performance of the SSME powerhead, it is thus essential to quantify the strong curvature effect under the flow condition in the TAD.

Since virtually no data was available at the time, experimental investigations were started to study the fundamental structure of turbulence associated with sharp U-turn internal flows. Two different experimental configurations were designed: (1) two-dimensional 180° turn with a constant cross-sectional area, at NASA Ames by Monson et al. (1989, 1992), and (2) axisymmetric U-duct to simulate the TAD, at Rocketdyne by Sharma et al. (1987). Even though the Ames experiment was motivated by the SSME flow analysis, is was designed to study the fundamental aspects of turbulent flow with strong curvature; while the Rocketdyne experiment was designed to provide more specific information relevant to TAD flow. The Ames experiment and related turbulence modeling issues will be discussed next, followed by the axisymmetric U-duct modeling case.

6.4.2.1 Two-Dimensional U-Duct Study

Many internal flows in engineering encounter curvature of various degrees. Experimental and computational studies have been done on pipes and ducts with bends, in conjunction with liquid rocket engine and hydroelectric power generation, among many engineering practices. Two-dimensional U-duct experiments at Ames do not provide directly applicable data to SSME simulations. For instance, this geometry does not have the large expansion of flow passage as in the SSME TAD, and the Reynolds number is much lower than the actual flow. To obtain 2-D results, a three-dimensional duct with a constant cross sectional area was constructed where two end-walls are separated by 10 channel height. The data taken along the centerline of the channel are to be used for 2-D computations. Therefore, end-wall effects along the centerline need to be minimized such that the data can be used for model validation, and then to extend to future model development.

The experimental configuration conducted at NASA Ames High Reynolds Channel I (HRC I) is depicted in Fig. 6.7. HRC I is a blow down facility using unheated dry air at ambient temperature. After an entrance nozzle, a straight rect-angular upstream duct is located with a dimension of 3.8-cm high and 38-cm wide (that is, an aspect ratio of 10), and 83-cm long. It is followed by a 180° bend with constant channel gap spacing equal to the centerline radius of 3.8 cm (that is, a radius to gap ratio of 1). Following the bend, another 54-cm long straight down-stream section is placed. A throttle plate at the exit of a bottom-settling chamber controls the flow rate. This was adjusted so that the Mach number of the flow was 0.1, which is the same condition as in the SSME TAD.

Tests were conducted with p_t at 1.2 and at 12 atmospheric pressure to achieve Reynolds numbers of 10^5 and 10^6, respectively. Large Plexiglas side windows allowed optical access to the entire bend, and to 12H up- or downstream regions. Inner windows incorporated suction slots spaced H apart upstream of the bend to remove the side-wall boundary layers. The suction (combined with the high AR) was intended to keep the flow as two-dimensional as possible. The full detail of this experiment is documented by Monson and Seegniller (1992).

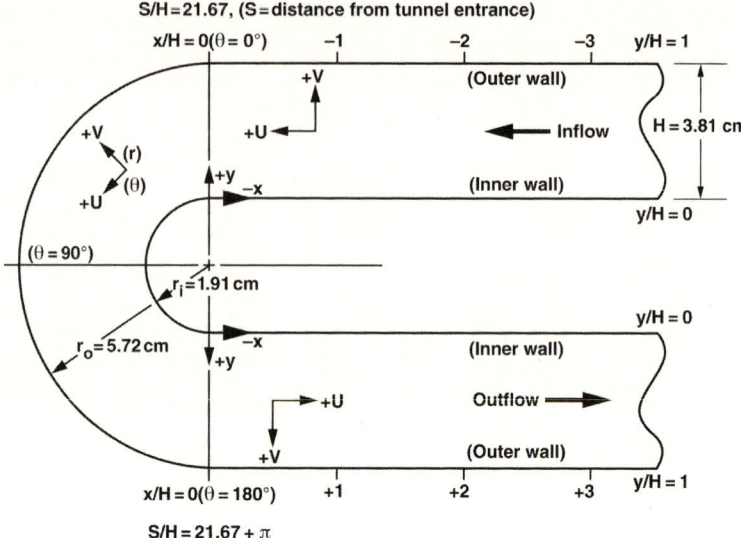

Fig. 6.7 Geometry of 2-D turnaround duct experimental model at NASA Ames High Reynolds Number Channel I (HRC I)

A numerical study performed by Monson et al. (1989) assessed the ability of a simple algebraic modeling approach for computing this type of internal flow with strong curvature. An algebraic model given by Equation (6.1) was used for this exercise. The length scale is calculated following Burke (1989), who used the ad hoc formula below to account for n walls or surfaces:

$$\frac{1}{l} = \sum_{i=1}^{n} \frac{1}{\kappa y_i \left[1 - \exp\left(-y_i^+ / A^+\right)\right]} \tag{6.10}$$

This simple model was chosen because Burke successfully simulated wing-body juncture flow at a Mach number of approximately 0.1 and $Re = 10^6$, and it is simple enough for a quick implementation for 3-D applications. It is natural to expect some limited prediction capability in this approach. However, this can give a quick overall picture to users seeking first-order results before embarking on high-fidelity computations. A few selected results, measured and computed using a modified mixing length model by Monson et al. (1989, 1992), are presented below.

In Fig. 6.8, velocity profiles measured at 4H upstream of the bend are shown. The overlapping data in the figure are taken from separate runs from each wall. The curve fits, shown as a solid and a dotted line in the figure, are used as upstream boundary conditions for computation. The profiles show that about the middle third of the channel contains an inviscid core indicating that the flow is turbulent but not fully developed at this location. The intensity of this core is measured by the LDV at about 1%, which is the resolution of the instrument. The boundary layers on the inner wall

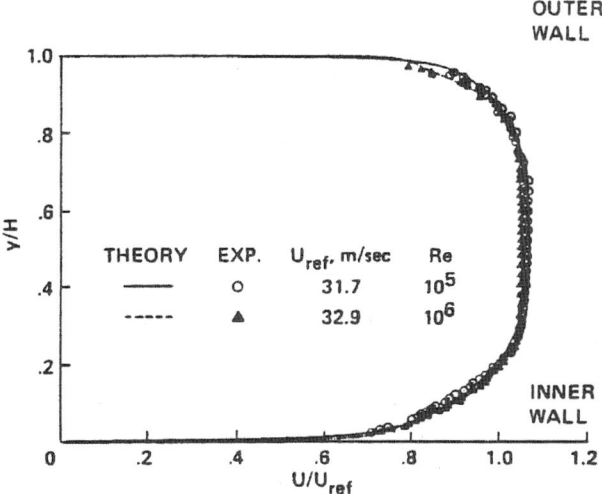

Fig. 6.8   Longitudinal velocity in 2-D U-duct at x/H $= -24.0$, M $= 0.1$

are somewhat thicker due to an asymmetric test section of the inlet nozzle. The flow angle was measured as zero across the channel. Additional velocity and flow angles taken at several transverse positions in the duct show excellent two-dimensionality of the flow at this inflow location before entering into the bend region.

In Fig. 6.9, computed pressure coefficients and data are compared. The reference pressure, p_r, was chosen such that computational results agree with the measured data at the start of the calculation. The experiment shows a moderate pressure rise on the outer wall along the bend and a strong suction peak on the inner wall. The maximum difference in pressure coefficients between the two walls is about 2.6, which is about the difference one may expect from centrifugal forces. Computed results and experimental data agree very well along the outer wall. However, along the inner wall, the differences widen farther around the bend, probably due to differences in the separated region. Downstream of the bend computed results show higher pressure, indicating that the skin friction losses along the bend are underpredicted by the computation. The slopes of the computed and measured pressure are about the same downstream of the bend, which indicates that skin friction agrees better in that region.

In Fig. 6.10, computed and experimental velocity profiles are compared at the start of the bend ($\theta = 0°$). The measured profile shows that the flow has begun to accelerate near the inner wall due to a favorable pressure gradient, and the peak velocity is much closer to the inner wall. On the other hand, the flow near the outer wall is decelerated. Generally, the computed and experimental results agree well at this point. This is probably because the flow is mainly driven by the pressure field, and turbulence structure has very little effect, so the algebraic model works reasonably well.

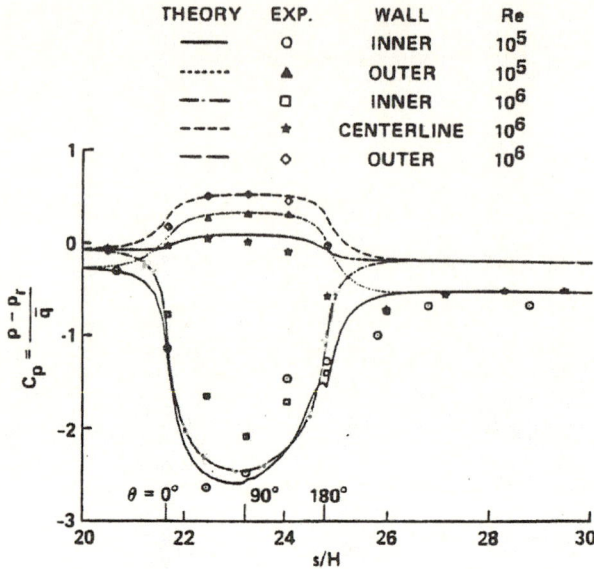

Fig. 6.9 Comparison of static pressure coefficients for 2-D turnaround duct at $M = 0.1$

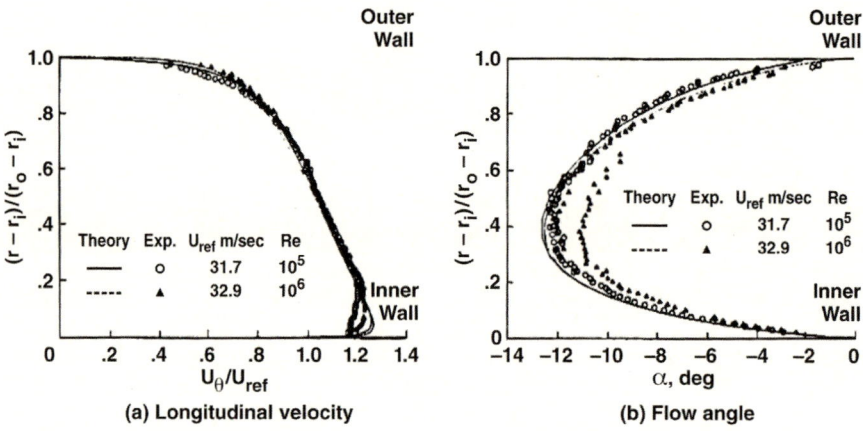

(a) Longitudinal velocity   (b) Flow angle

Fig. 6.10 Longitudinal velocity profile in 2-D turnaround duct at $M = 0.1$ and $\theta = 0°$

In Fig. 6.11, the velocity profile is shown halfway through the bend ($\theta = 90°$). The flow continues acceleration near the inner wall and deceleration near the outer wall. Very few differences are observed between the flow at the two Reynolds numbers.

The boundary layer on the inner wall is extremely thin and may actually be relaminarized. The computed results over-predict the boundary layer thickness on the outer wall. A similar trend is observed in an axisymmetric case that uses an

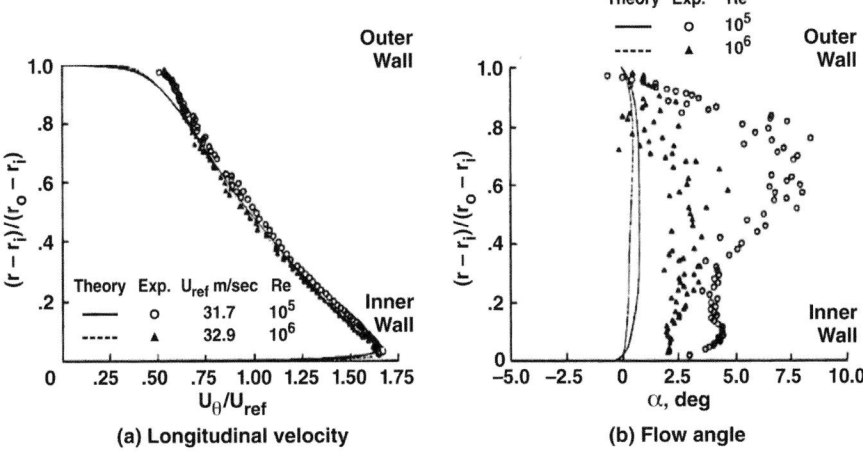

Fig. 6.11 Longitudinal velocity profile in 2-D turnaround duct at $M = 0.1$ and $\theta = 90°$

algebraic model by Chang and Kwak (1988a). Possible factors contributing to this discrepancy come from the high level of turbulence and unsteadiness in this region, observed during the LDV tests. Sandborn (1988) presented evidence for the existence of highly time-dependent instabilities in this region of a turnaround duct. The computed mixing length along the outer wall is nearly constant, indicating that an algebraic modeling approach is likely to over-predict boundary layer thickness in the outer wall region. The accuracy of an algebraic model suffers in this region. Using other models will result in different boundary layer thickness. For example, Chen and Sandborn (1986) reported thinner boundary layer thickness using a $k - \varepsilon$ model. The difficulty may have started from the basic assumption of equilibrium turbulence in using an eddy viscosity model. This type of flow may require higher-level modeling. However, considering that the flow is highly unsteady at this point and will separate after this point, the simple algebraic recipe for eddy viscosity produces remarkably good results.

In Fig. 6.11b, flow angles are measured. The measured data indicate that flow turns with a much larger positive turning angle. The scatter in the data is due to the unsteadiness in this region. The computed results deviate widely from the measured flow angle, and this may be due to the extent of separation along the inner wall downstream of the turn.

In Fig. 6.12, the velocity profile is shown at the end of the bend ($\theta = 180°$). The velocity profiles indicate that separation bubbles exist on the inner wall just downstream of 150° in the bend—a small one for $Re = 10^5$ (height of 3.1% of H) and a much larger one for $Re = 10^6$ (height of 8.5% of H). The reattachment occurs at x/H between 1 and 2 for $Re = 10^5$, and at x/H = 2 for $Re = 10^6$. The separation bubbles observed from the real-time Doppler signals on an oscilloscope appear to be very unsteady and unstable at both Reynolds numbers. At times, the bubbles are completely swept away—then instantly reestablish at aperiodic frequency. The

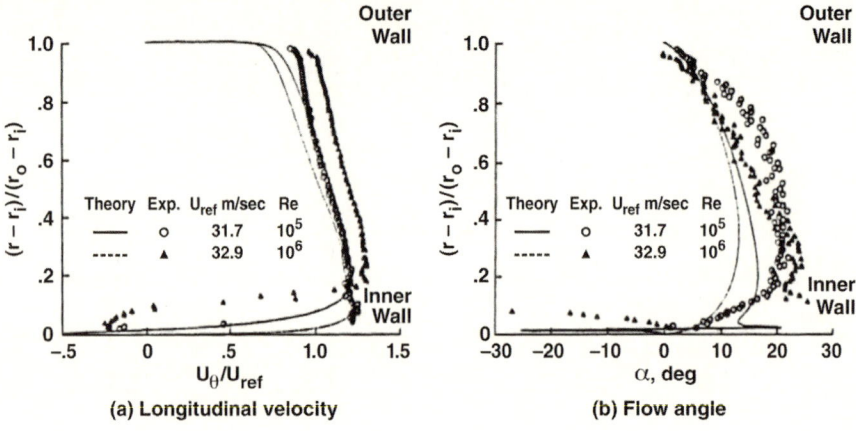

Fig. 6.12 Longitudinal velocity profile in 2-D turnaround duct at M = 0.1 and $\theta = 180°$

instability of separation bubbles could be a major reason for highly scattered measurement in this flow. Outside the separated region, the flow is accelerated around the bubbles. Sandborn (1988) observed a similar phenomenon in his water flow test at Re = 10^5.

Using an algebraic model, one might expect that the accuracy of computed results is very much lost. However, our results show varying degrees of success in matching the experimental data. For Re = 10^5, separation is reasonably well predicted with the separation angle at 153° into the bend, reattachment at x/H = 1.1, and a bubble height of 5.7% of the channel height, H. The velocity profile matches quite well except that, near the outer wall, the mixing length model overpredicted the boundary layer thickness. For Re = 10^6, the mixing length model predicted smaller separation bubbles than it did at Re = 10^5. Once the separation is not predicted correctly, the computed result past the separation point becomes worse in matching with the experiments. The bubble size becomes smaller as the Reynolds number increases—which is opposite to experimental observation. As a reference point, the Navier-Stokes prediction of Sandborn (1988) using a multiple-scale $k - \varepsilon$ model with low-Reynolds number $k - \varepsilon$ model predicts larger separation zones than the mixing length model, as described here. Our mixing length model predicts smaller separation bubbles for Re = 106, which is expected. Because of the poor separation prediction, results for the downstream side of this flow are expected to be poor. This shows that the modeling approach needs to incorporate more physics than a simple mixing length description.

Many attempts have been made to develop a higher accuracy, more predictable model, and many review articles can be found (for example, Reynolds, 1976). From a CFD applications viewpoint, trade-offs between accuracy versus uncertainty bounds or consistency need to be assessed for a timely impact on engineering.

A typical velocity profile after reattachment is shown in Fig. 6.13 at x/H = 2 and 12, respectively. The computed results show a gradual approach toward a fully

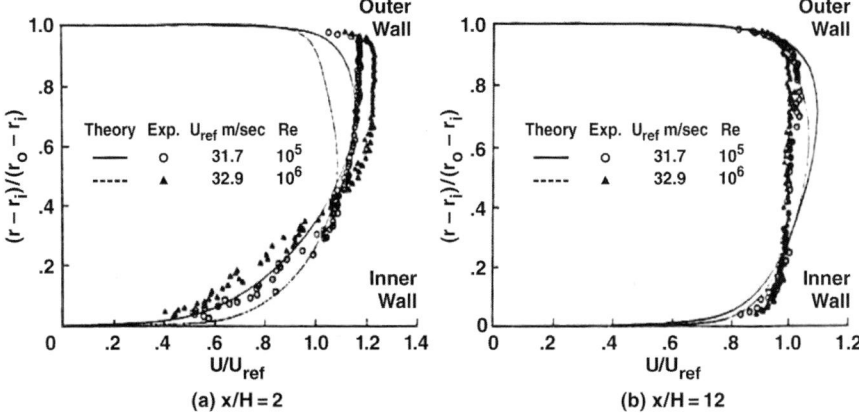

Fig. 6.13 Longitudinal velocity profile in 2-D turnaround duct at M = 0.1

developed channel profile, while the experiments show more plug flow type profiles. At this stage, differences existed between the experiments and computation—such as unsteadiness and presence of a weak 3-D structure in the experiment, among other things. For full validation in this region, three-dimensionality needs to be included as well as a more physically correct turbulence model.

6.4.2.2 Axisymmetric U-Duct

The two-dimensional U-duct discussed above was intended to produce turbulence data for model development and validation. However, the cross sectional area of the 2-D duct remains constant, while in the SSME TAD case, the cross-sectional area for flow passage diverges starting from the beginning of the bend. The flow in the actual TAD is under much higher adverse pressure gradient due to expansion of this channel configuration. Therefore, another experiment was designed to account for the divergent channel effect in addition to the strong curvature. That experiment was performed at Rocketdyne by Sharma et al. (1987).

A schematic of this axisymmetric experiment is shown in Fig. 6.14. The duct width is 2 inches, and the radius of the center-plane of the annular duct is 10 inches upstream of the turn and 14 inches downstream of the 180° bend. The radius of curvature is 1 inch for the convex inner wall, and 3 inches for the concave outer wall. From the beginning through completion of the turn, the annular cross-sectional area undergoes a 40% increase. The boundary layer on the inner wall side is therefore expected to separate in the neighborhood of the 180° turn. The experiment is conducted at atmospheric pressure. The Reynolds number is approximately 10^5, and the Mach number is about 0.1. Hot wire and hot film probes are used for data acquisition.

For such complicated turbulent flow as encountered in the SSME, high-level turbulence models such as a two-equation model or Reynolds-stress model may

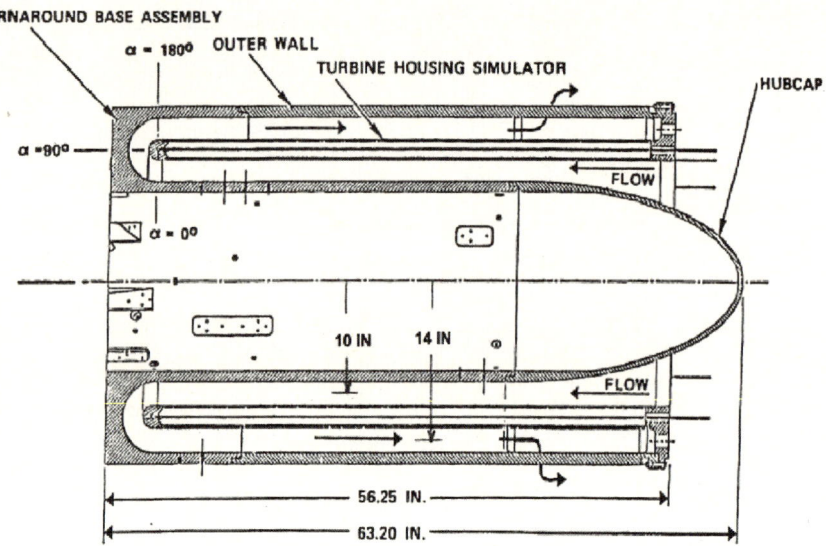

Fig. 6.14 Schematic of experimental axisymmetric turnaround duct model

be needed. However, as discussed in the previous section, for many engineering applications, a simple yet adequately tuned model as proposed by Equation (6.1) will be of considerable value for design purposes. The previous study with a 2-D U-turn was intended to determine the sensitivity of the strong curvature effect on the existing turbulence models. Since the existing models are primarily developed for external flow, a modified algebraic model was proposed and tested with the experimental data.

It is to be noted that the purpose of this study was to assess whether the proposed algebraic model is adequate for the SSME HGM redesign task on hand. One primary concern was the strong curvature effect where no empirical data was available for the evaluation of the model adequacy. From the study, it was observed that strong unsteadiness, separation, and adverse pressure gradient have more dominant impacts on the computed results. The SSME TAD, however, has annular axisymmetric geometry in the 180° turn region, followed by a non-symmetric transfer duct connected form the fuel bowl to the main injector assembly. Therefore, the actual TAD to be redesigned includes the rapidly diverging area in the flow direction, as well as the strong curvature effect.

The axisymmetric experiment was designed to study these features in a laboratory setting. The related computations presented here are natural extensions to the two-dimensional U-duct study, and the computations are used to assess the sensitivity of the proposed algebraic turbulence model for the analysis of the actual SSME geometry.

For computations, only three azimuthal planes are required because of the axisymmetry. The center plane and the other two planes are set on either side of the center plane offset by a small angle. Across the duct width, 81 grid points are distributed with an expansion factor of about 1.2. The smallest grid size next to the wall is 0.02% of the duct width that corresponds to y^+ of less than 5. In order to resolve the detail of the separated flow profile, the largest grid size in the core region is set to be 0.025 times the width. The 180° U-turn region is divided by 42 equal intervals. An orthogonal cylindrical grid system is used here to eliminate any inaccuracies associated with non-orthogonal grids. The grids upstream and downstream of the bend are then fixed by smooth expansions from the bend region. These grids are orthogonal close to both walls. Therefore, the flow path is represented by a $3 \times 81 \times 95$ grid in the circumferential, radial, and flow directions, respectively. This grid in the U-turn region is shown in Fig. 6.15a.

The measured streamwise velocities at 10 inches upstream of the bend (i.e. x/H = −5) are used as the inlet conditions. Due to hot-wire probe infusion, reliability of the measured velocity normal to the wall is uncertain. Since the normal velocities are small, they are set equal to zero.

Figure 6.15b shows the comparison of the static pressure distributions between the measured data and the computed results. The measured data were obtained in an earlier experiment in which no turbulent tripping mechanism was used. The resulting boundary layer on the outer wall at 8 inches upstream of the bend, that is, x/h = −4, is very thin. By using these measured velocities as an inlet condition at that location, the computed result shown with a solid line agrees very well with the experiment. In the neighborhood of the 180° turn, pressure on the inner wall shows a dip followed by a long recovery from the trough.

In external flows, boundary layer separations usually interrupt the pressure-recovering processes. The resulting pressure downstream of the separation will

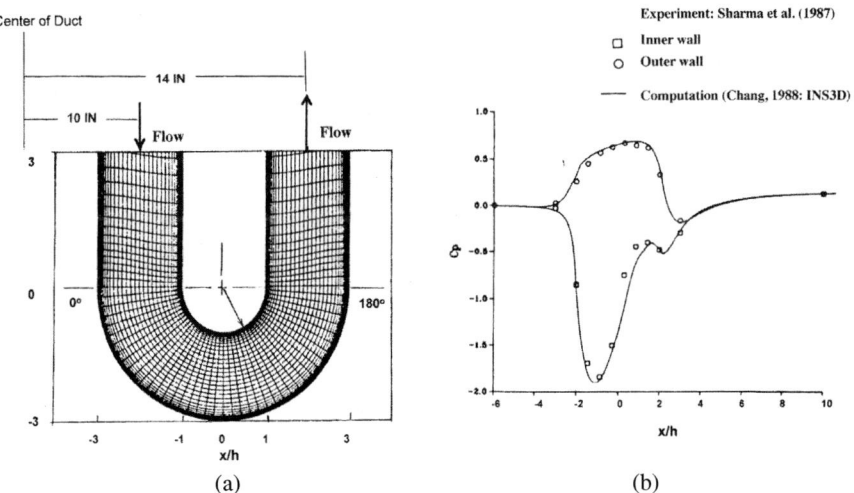

Fig. 6.15 A computational model for an axisymmetric turnaround duct and computed pressure coefficient: (**a**) grid; (**b**) pressure coefficients on the inner and outer wall

be almost constant. In the present confined duct, the flow that is pushed out-ward due to inner layer separations is squeezed back by the flow on the opposite side. As a result, the effective flow area is reduced, creating a region of local acceleration that results in a local pressure dip. In a later experiment, the layer is tripped and artificially thickened to about 0.5 inches at the position 10 inches upstream. Because of the thickened layer at the outer wall, a larger portion of kinetic energy is carried by the inner-wall region, inducing a larger acceleration in the bend.

Comparisons of the measured and computed streamwise velocity profiles are shown in Fig. 6.16. Agreement is good, in general, except near the outer con-cave wall in the bend region. The numerical results show a slight separation from the beginning of the turn through about 103°, while the experimental data shows attached layers all the way through the turn.

Accuracy of the measured velocities in this region is still uncertain due to two factors. First, because of the sharp concave wall effect, the enhanced turbulent intensities obtained are as high as 45% of the measured mean velocities. Since the hot-wire probe measures only the rate of heat transfer, it does not detect the direction of the flow. In a region where heat transfer is significantly affected by the turbulent fluctuations, it is extremely difficult to obtain the true mean flow velocities, as evi-denced by experiments in the low-speed region of a mixing layer. Second, the mass flow obtained by integration of the measured velocities over the cross-sectional sur-face of these few planes is substantially larger than the amount of mass coming from the inlet. The computed velocity profiles conserve the mass flow to an order of 10^{-5}.

Because of these uncertainties, we have drawn no conclusion for the discrepan-cies. On the convex inner-wall side, the layer begins to separate at about 125° into the turn. The data obtained by the hot-wire probe registered a high value due to large turbulent fluctuations in the separated region, and the data do not reveal the direction of the flow. Just as the computations need assessment of uncertainties coming from numerical procedures and physical modeling, experimental data need to be carefully evaluated to determine validity and error bound.

In Fig. 6.17, the Reynolds stress, $\overline{-u'v'}/u_{ref}^2$, is shown. The computed results agree qualitatively with the experiments in the U-turn region. Again, in the outer concave region the turbulence level is much higher than the inner convex part of the turn region.

To capture the curvature and diverging channel effects, a higher-level turbu-lence model will be necessary, possibly with curvature effect terms included. From our study, however, the accuracy of the solution using the proposed alge-braic turbulence model is regarded as acceptable for studying preliminary design changes—providing the final design configuration is verified by higher accuracy modeling and testing. Therefore, this model has been used extensively for simu-lating the SSME powerhead for the analysis and redesign tasks. In the followings sections we discuss how this CFD approach has been utilized in the redesign task, and present computed flow solutions in a variety of different HGM configurations at various Reynolds numbers. Here, the Reynolds number is based on the mean velocity and the duct width at the entrance of the TAD.

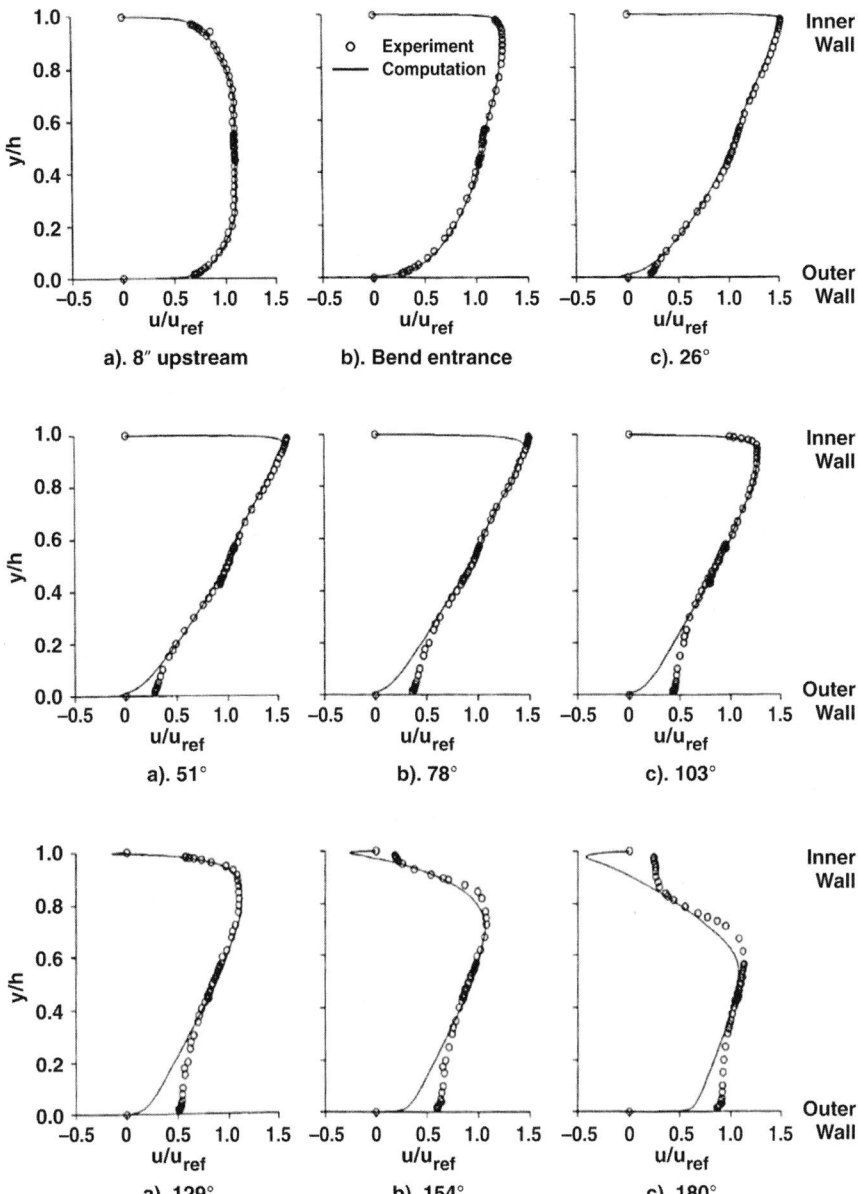

Fig. 6.16 Comparison of normalized streamwise velocities for axisymmetric U-duct

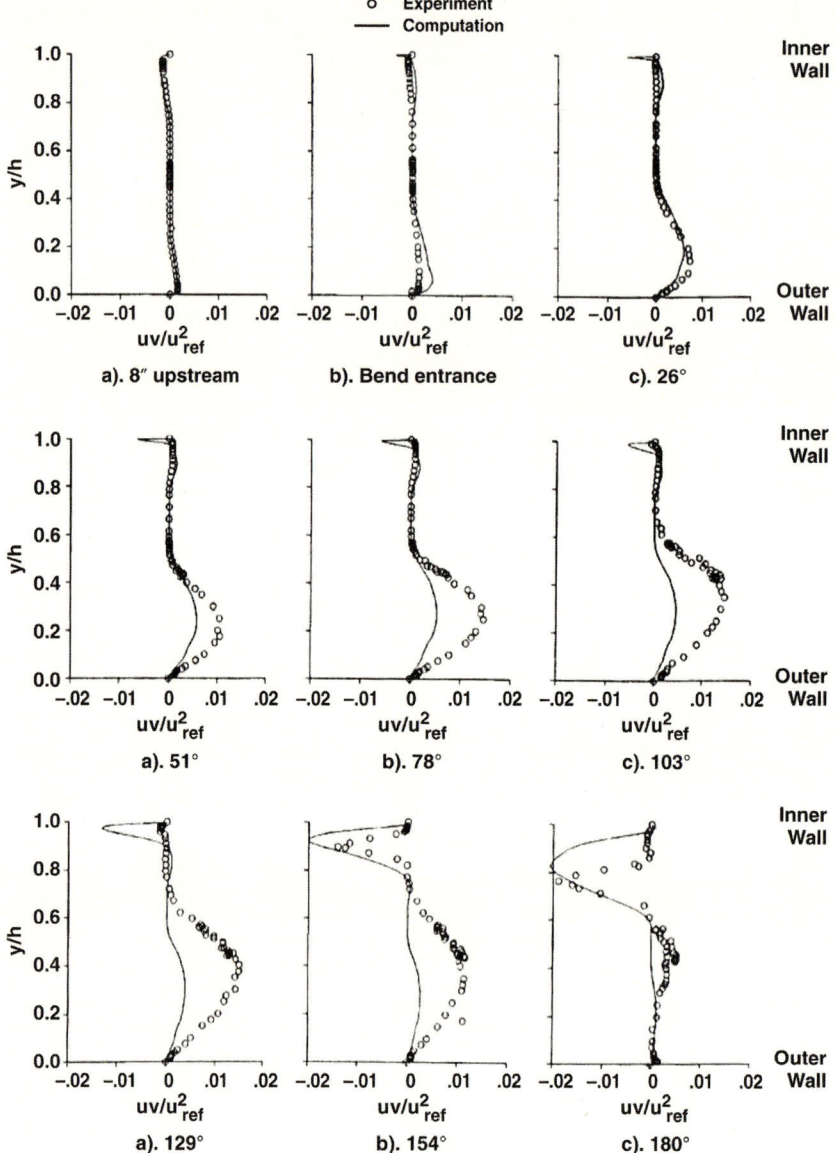

Fig. 6.17 Comparison of normalized Reynolds stress for axisymmetric U-duct

6.5 Analysis of the Original Three-Circular-Duct HGM Configuration

One of the main purposes of the SSME redesign was to increase the rated thrust. To proceed with this task, called the Phase II+ redesign, the first step was to analyze the flow in the powerhead. Turbulence modeling issues with quantifying this complicated flow have been discussed in previous section. Here, we present a step-by-step process for analyzing the current design configuration, followed by an analysis of the new design in the next section.

The entire rocket engine development process, including design of the basic configuration, experimental validation of component level functions, and final certification testing, requires enormous time and cost in terms of computational resources. In order to make any impact on the configuration design, the CFD contribution has to be timely as well as reliable. Therefore, geometry modeling and boundary condition procedures are often truncated and approximated, and physical models chosen for economy—as long as one can obtain reasonably consistent results at the expense of some inaccuracies such as in the separated flow regions. We present the engineering process used in conjunction with the Phase II+ redesign of the SSME for the purpose of illustrating a CFD application procedure for supporting a mission-driven task.

In the design of rocket engines and vehicles, size and weight are two major constraints to work within. The original SSME configuration was designed with an axisymmetric annular turnaround duct. The duct width is on the order of 1 inch and the radii of curvature are approximately 0.6 and 1.4 inches for the convex inner and concave outer walls, respectively. The Reynolds number, based on the width of the duct entrance, is on the order of 10^7. The Mach number of the hot gas flow is less than 0.1, which makes the incompressible assumption mathematically valid.

For the first step, steady-state solutions are obtained for a truncated model of the current three-circular-duct HGM. For this initial computation, the intention is to define the function of the three transfer ducts sketched in Fig. 6.4. These ducts connect the fuel-side fuel bowl (left-hand side in Fig. 6.1) and the main injector assembly at the center. On the right-hand side of the schematic shown in Fig. 6.1, a similar configuration is located with the oxidizer turbopump. The two sides are not symmetric and thus a full analysis requires coupling of these ducts and the racetrack of the main injector. For the fuel side, the two computational domains, namely, the fuel bowl with TAD and main injector assembly with the racetrack, are connected by three ducts. Therefore, for the analysis of the fuel-side TAD, a computational model truncated at the junction of the transfer ducts and the racetrack, is generated. A three-dimensional view in Fig. 6.18 shows surface grids for this model.

To investigate the function of the three ducts, a laminar flow at Re = 1,000 was first computed. Even though the boundary layer is thicker than a turbulent case and the subsequent separation region would be large, the resulting flow will give a conservative estimate for engineering purposes. In this model, the three transfer ducts are assumed to discharge the flow separately, which results in no communication of the pressure between the center and outer ducts at their exit planes. Therefore, the downstream condition is not a good approximation of the real case. For this reason, small residual waves have remained in the computed results. However, the root-mean-square value of the change in the flow variables, RMSDQ, has dropped below 10^{-5}, and an essentially steady-state solution has been obtained.

Interpretation of the computed results is another challenge for this problem. The overall energy required for the flow to reach the main injector assembly can be one measure. A locally non-uniform flow field that can cause high structural load on

Fig. 6.18 Inner and outer
surface grids for a three-duct
Hot Gas Manifold model

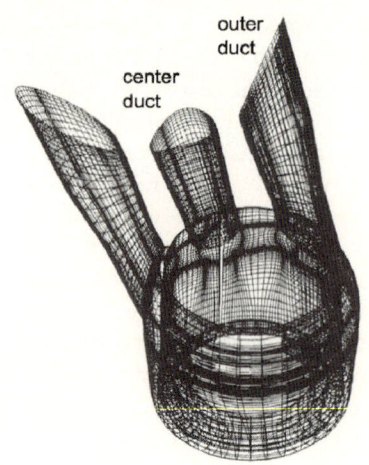

some of the LOX posts along the racetrack needs to be identified. Three-dimensional
flow visualization can shed some light on this aspect.

In Fig. 6.19a, b, velocity vectors are shown in the horizontal and vertical cross-
sections corresponding to the cut view of the grid. The flow in the center transfer
duct is highly non-uniform, and a large separation region is formed just downstream of the entrance to the transfer ducts. By comparison with the vector length
in the figure, the flow in the center duct is much slower than in the outer ones. The

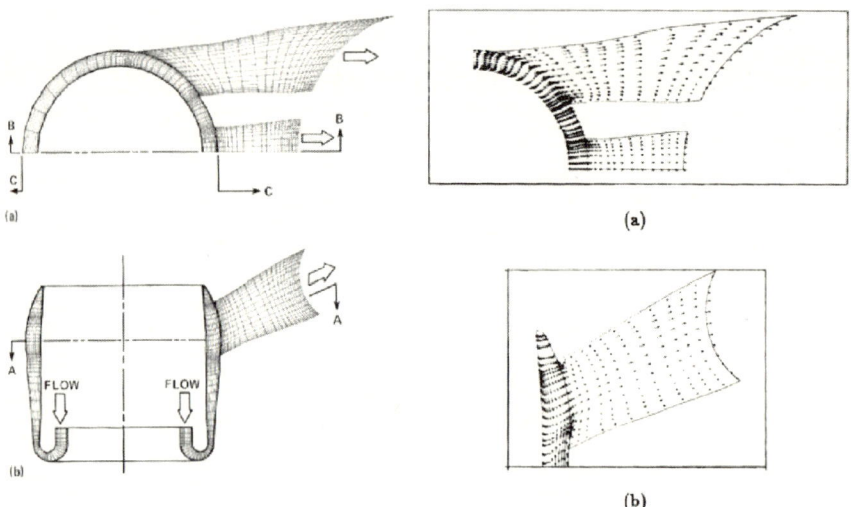

Fig. 6.19 Computed velocity distribution at Re = 1,000: (**a**) top view, (**b**) vertical cross-section of
center transfer duct

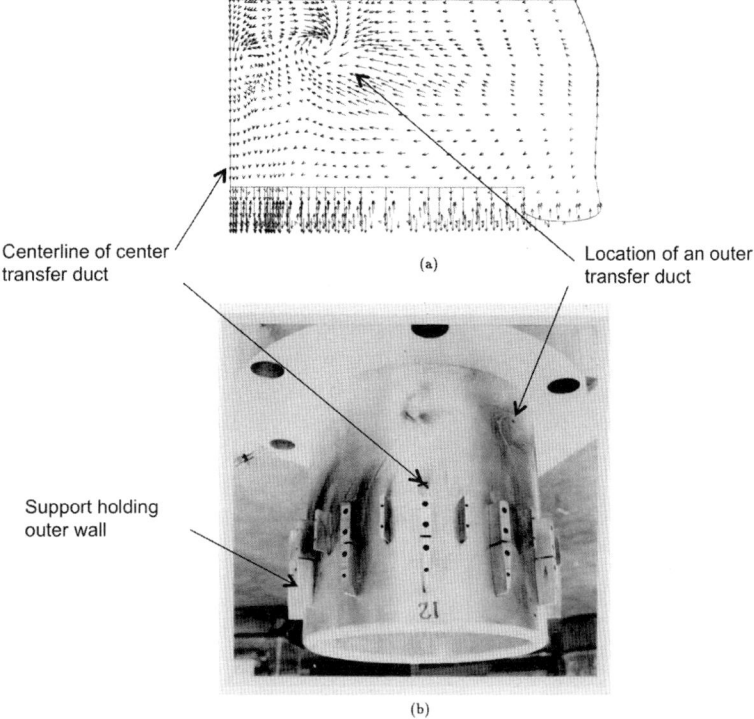

Centerline of center transfer duct

(a)

Location of an outer transfer duct

Support holding outer wall

(b)

Fig. 6.20 Particle traces on unwrapped inner wall surface: (**a**) computed vectors indicating particle traces, (**b**) experiment (at Rocketdyne) showing particle traces on inner wall of the Hot Gas Manifold

results shown in this figure agree qualitatively with the airflow test data conducted at Rocketdyne with a Reynolds number of about 10^6. The predicted mass flow through the center duct is 9.8% of the total mass flow, which agrees with the test data.

Figure 6.20a illustrates the 3-D velocity vectors at an unwrapped plane near the inner wall of the fuel bowl. A reverse-flow pattern is clearly visible near this wall. Three-dimensional swirl patterns are predicted in the vicinity of the entrance to the transfer ducts. Figure 6.20b is a photograph taken after removing the outer wall following flow measurements at Rocketdyne, which indicates, by means of surface-streak (shear-pattern) visualization, the similar swirls at the corresponding locations in the airflow test.

The existence of the swirls can be explained as follows. The flow coming from below the SSME powerhead has a large momentum due to the relatively small width of the annular duct. Among the streamlines of this flow in between the two ducts, there exists a dividing streamline. This streamline has a stagnation point at the top of the fuel bowl, as shown in Fig. 6.20a. On the left side of this dividing streamline, the flow is bent leftward to the center duct. Due to symmetry, a rightward flow is also approaching the center from the other side. When these opposite currents approach each other, another dividing streamline is formed with a stagnation point, again at the top wall. The stagnation pressure forces the streams to bend downward, and at the same time, the streams make a right-angle turn into the circular duct. Conservation of momentum consequently results in the formation of swirls.

The pattern of the swirl and its center depends on the relative strength of the approaching currents. Near the center duct, double swirls of equal strength are formed due to symmetry. In the vicinity of the entrance to the outer duct, the current approaching leftward from the rear part of the bowl is larger than the one approaching rightward. A stronger swirl is thus formed, located sideways toward the weaker stream.

Figure 6.21 shows perpendicular cross-sectional views of three different sections of the transfer ducts; namely, near the entrance, at the midsection, and near the exit plane. Near the entrance, the velocity vectors in the center duct have symmetric double swirls, while the outer duct has a strong swirl accompanied by a much weaker one. The swirling velocities are largely reduced at the midsection and are physically dissipated before entering the main injector regions.

In this original HGM design, the three transfer ducts are to carry evenly divided loads for transferring hot gas from the fuel bowl side to the main injector. However,

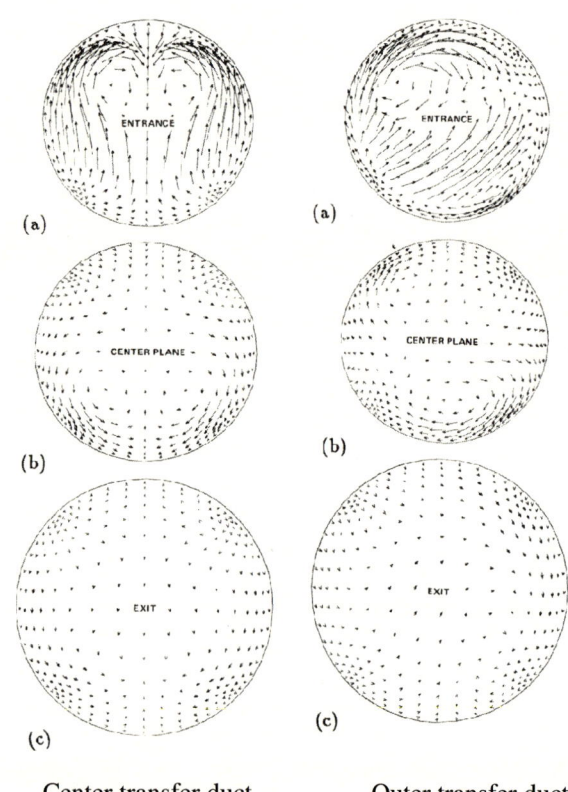

Fig. 6.21 Computed velocity vectors at three different vertical cross sections of three-duct Hot Gas Manifold

Center transfer duct   Outer transfer duct

from our analysis, it was found that only about 10% of the total mass flow goes through the center duct. The large separated region at the entrance of the center duct does not help alleviate this uneven distribution of the flow. Subsequently, a new design idea was developed using two larger area transfer ducts.

The original design used a circular cross section based on structural reason. For the new two-duct design, an elliptic shape was chosen. This maximizes the total cross-sectional area, given the limited space in the powerhead. The new two-duct design is analyzed using the same numerical procedure and then compared with the original design. This provides relative change in performance and flow quality. Then, to quantify the flow more accurately, the Reynolds number and turbulence modeling were chosen closer to the real case. This and other features are discussed next.

6.6 Development of New Two Elliptic-Duct HGM Configuration

From the computational flow analysis and experiments, the center duct of the original three-duct HGM is found to transfer a limited amount of mass flow (about 10% of the total flow). In addition, the transverse pressure gradient remains large, with a large bubble of separation after the 180° turn. To improve the quality of the flow, a large-area, two-duct design concept has been developed (see Yang et al., 1987; Lin et al., 1987). In addition, the ducts are chosen to have an elliptical shape in order to distribute the mass flow evenly to the main injector region. A cross-sectional view of the computational grids for the new two-duct power head model is shown in Fig. 6.22.

The main injector assembly consists of several hundred LOX posts (Fig. 6.2), and is located at the center section in both Figs. 6.22 and 6.23. The flow goes through

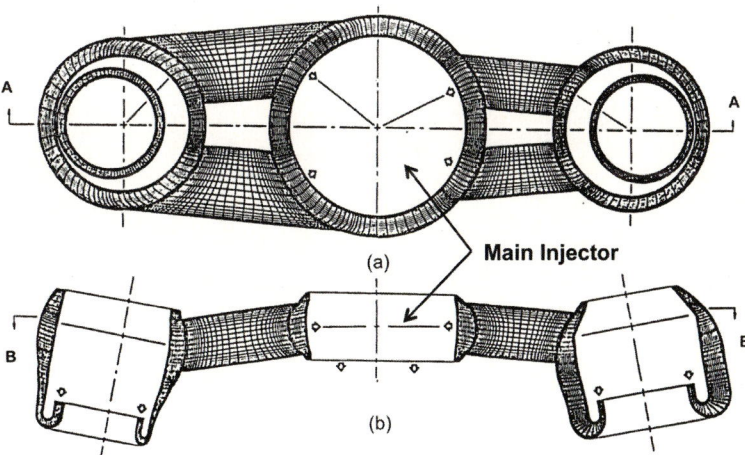

Fig. 6.22 Computer model of the new two-duct SSME power head: (**a**) horizontal cross section (B-B), (**b**) vertical cross section (A-A)

the gap between the posts and small holes between shields supporting these posts. We therefore assumed that the flow enters into the main injector assembly from the racetrack in the radial direction.

The fuel side of the HGM delivers 70% of the total flow. Therefore it is assumed that approximately 70% of the main injector area along the racetrack receives 70% of the mass flow. Airflow tests confirm this approximation. The primary task of this redesign is to modify the fuel side of the powerhead (left-hand side of the cross-sectional view in Fig. 6.1). Thus, for the analysis of the fuel-side HGM, a truncated geometry was created from the entire powerhead model. The computational model for the fuel-side HGM is shown in Fig. 6.18. Multi-zone computations were performed, as sketched in Fig. 6.4 with the extended turbulence model discussed in Equation (6.1). Both laminar and turbulent flow solutions are obtained for comparison. The Reynolds number is based on the width of the TAD entrance, and the inlet velocity is 10^3 for the laminar flow and 1.9×10^6 for the turbulent flow case. The same boundary conditions are used for both flow simulations.

One important improvement goal of the new design was to reduce the size of the separation bubble after the 180° turn. The computational approach for this task is achievable using laminar flow analysis. For the purpose of comparing relative change in flow quality under a geometric modification of this type, laminar flow analysis can be used for a conservative estimate and to minimize uncertainties stemming from turbulence modeling. Once the new configuration is outlined, a parametric study and optimization can be performed to arrive at a preliminary redesign configuration. As shown in Fig. 6.24, the separation bubble size is substantially reduced after the geometry has been modified.

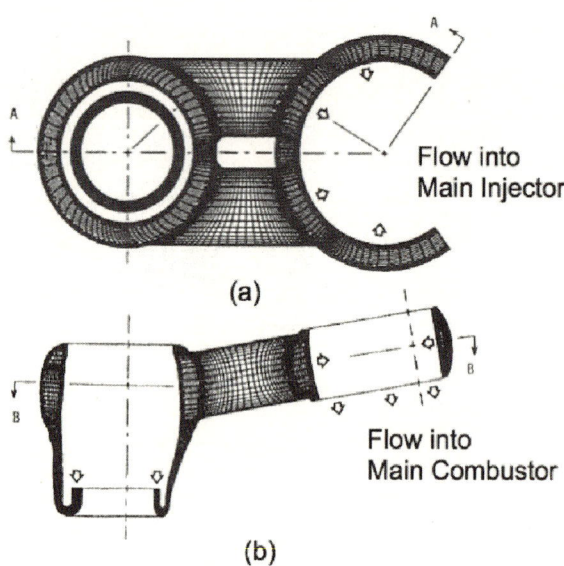

Fig. 6.23 Computational grids for the fuel-side HGM analysis: (**a**) horizontal cross section, B-B, and (**b**) vertical cross section A-A

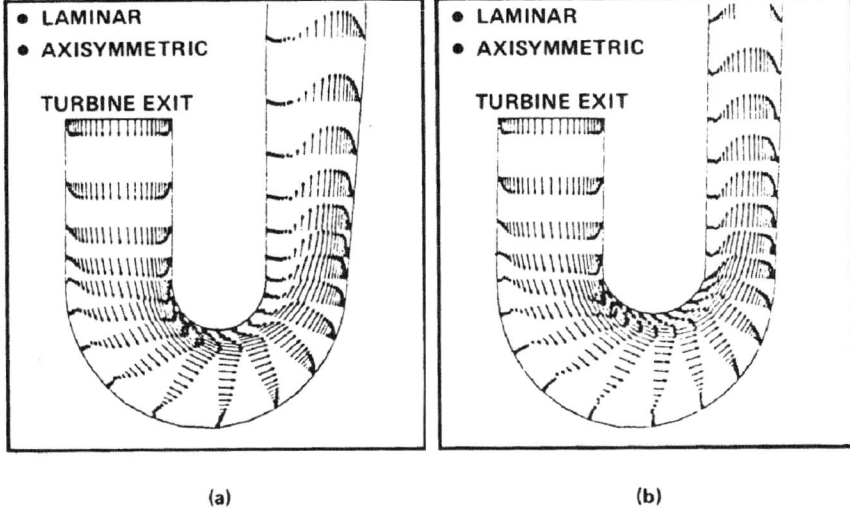

(a)   (b)

Fig. 6.24 Comparison of velocity vectors on vertical cross-section of HGM to observe the separation bubble after the 180° turn: (**a**) original three-duct design, (**b**) new two-duct design

Because laminar flow computations were utilized for the design change, it is of interest to compare computed results for laminar and turbulent flow conditions. Therefore, comparison of computed solutions are made among three different cases: Laminar solutions for the original three-duct design, and laminar and turbulent solutions for the new two-duct design.

For the new two-duct design, Fig. 6.25 shows the velocity vectors at three sectional planes of the TAD with laminar and turbulent flow conditions. For the laminar computation, a fairly large separated zone exists starting right after the 180° turn, while a much smaller separated region is predicted for turbulent case. For the turbulent case, the boundary layer is much thinner than in the laminar flow case. Since the turbulent kinetic energy is much higher in the boundary layer, relative to the corresponding laminar case, a smaller separation bubble develops under the adverse pressure gradient after the turn.

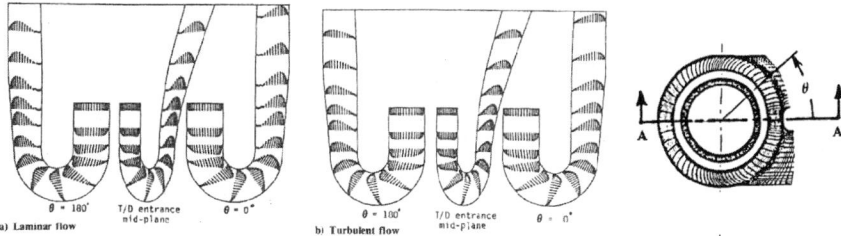

Fig. 6.25 Comparison of cross-sectional velocity profile between laminar and turbulent flow for the two-duct design

When the flow goes into the transfer duct, a swirling flow pattern develops, as shown in Figs. 6.19 and 6.20. The intensity of the swirl depends on the upward velocity from the TAD. Since the laminar flow has a larger blockage than the turbulent case due to larger separation after the turn, the magnitude of the swirling is expected to be larger in the laminar case than in the turbulent flow case. The computed maximum magnitude of the swirling velocity near the duct entrance is 0.44 of the inlet velocity for the laminar case, and 0.33 for the turbulent case. The swirl gradually dissipates as the flow goes farther into the duct. Also it is observed that a large separation bubble existing at the entrance of the transfer duct in the original design is practically removed in the new configuration.

One important objective of the redesign was to reduce the transverse pressure gradient at the exit of the gas turbine underneath the preburner (see Fig. 6.1). Therefore, the laminar and turbulent solutions for the new two-duct HGM is compared to the computed results of the original configuration. In addition, the experiments performed using both the three-duct HGM and the new two-duct configuration are compared with the computed results. In Fig. 6.26, pressures around the fuel bowl of the HGM are compared among these cases. In the original configuration, there exists a large pressure gradient from the back side ($\theta = 180°$) to the transfer duct side ($\theta = 0°$). In the new two-duct configuration, both the experiment and the computation show a much lower pressure gradient.

As sketched in Fig. 6.1, the hot gas flowing into the computational model of the HGM shown in Fig. 6.26 is from the exit of the gas turbine run by the hot pre-burned gas. Therefore, the pressure at the exit of the gas turbine blades has a large pressure gradient in the case of the original design, while this gradient is substantially

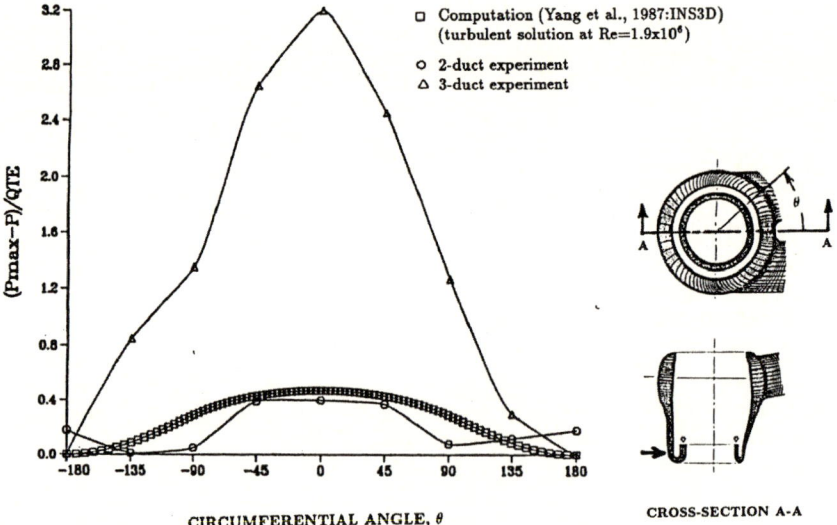

Fig. 6.26 Comparison of transverse pressure coefficients after 180° turn along circumference of the turnaround duct

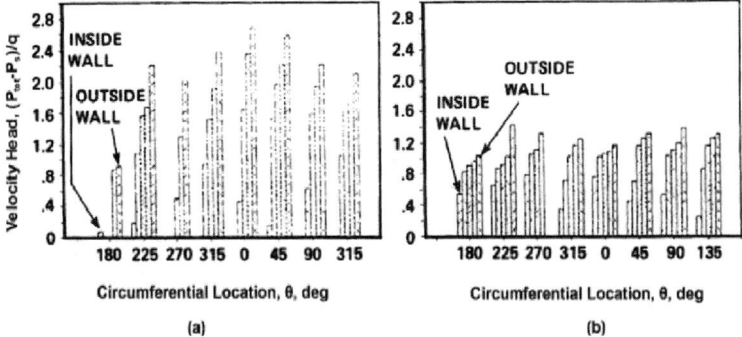

Fig. 6.27 Velocity head measurements inside the turnaround duct after 180° turn: (**a**) original 3-duct design, (**b**) new 2-duct design

reduced in the new design. From a hardware perspective, the new design results in a substantially reduced load on the bearings that hold the gas turbine and the turbopump through the center of the HGM (see Fig. 6.1).

A large number of parametric computations were performed to optimize the HGM configuration. Because of the limited computer speed and capacity in the 1980s, only manual optimization was performed to find the best TAD, fuel bowl, and transfer duct geometry. Various laboratory experiments were performed to fine-tune the HGM design, initially using cold air. Some water tests with bubbles injected for qualitative visualization were also tried at the beginning of this redesign effort. Because of the three-dimensional nature of the flow, visualization was also a challenge.

To compare the differences between the two designs, the velocity head is measured within the HGM. The measured velocity head at five different locations across the channel between the inner and the outer wall is shown in Fig. 6.27a, b for the original and new designs, respectively. To find the most favorable flow conditions, over 20 different two-duct configurations were studied computationally, potentially providing the optimum geometry to designers. Ideas from researchers were incorporated into the manual optimization process. Fully automated numerical optimization may be implemented utilizing faster codes and computers compared to the 1980s.

One of the objectives of the computational analysis was to pinpoint the locations where flow experiences the greatest energy losses. An important measure of the energy losses is the mass-weighted average total pressure along the flow. The total pressure coefficient C_{po} is defined as:

$$C_{po} = \frac{\bar{p}_o - \bar{p}_{o1}}{\bar{p}_{o1}}$$

where

$$\bar{p}_o = \frac{1}{M} \int \left[P + \frac{1}{2} \left(u^2 + v^2 + w^2 \right) \right] dm$$

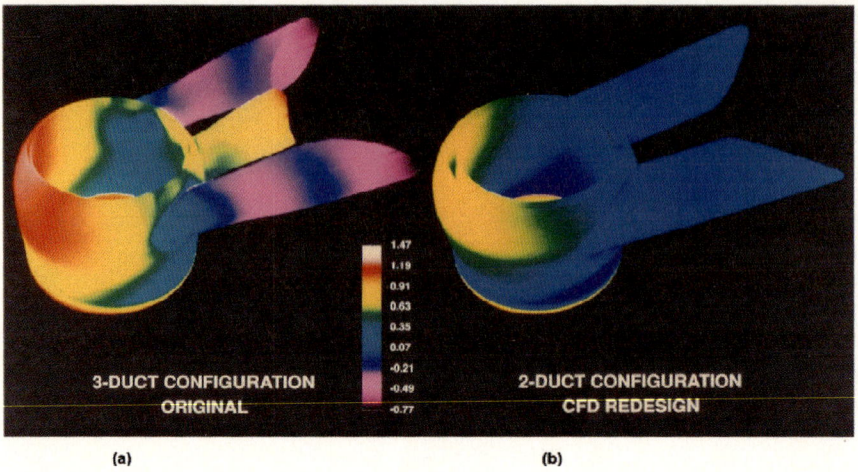

(a)   (b)

Fig. 6.28 Comparison of surface pressure between the original and the new Hot Gas Manifold design

To visualize the pressure distribution, surface pressure maps for the original and newly designed HGM are shown in Fig. 6.28. The flow is much more uniform in the new configuration and the variation between the maximum and minimum pressures is much lower. Visualization of various quantities is presented by Belie (1985).

Figure 6.29 illustrates the decreasing coefficient of the mass-weighted total pressure along the centerline of the TAD, the fuel bowl, and the transfer duct. The discontinuities shown in the figure correspond to the entrance of the transfer duct, where energy fluxes are computed over different planes. In the figure, three different HGM configurations are compared. The initial two-duct design shows 28% less

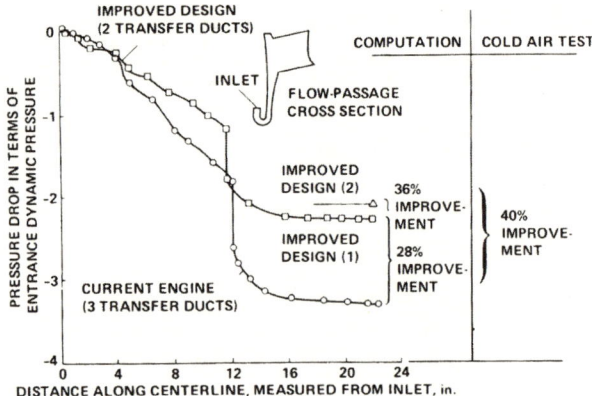

Fig. 6.29 Comparison of measured pressure losses between the original and new, computationally improved Hot Gas Manifold design

total pressure drop compared to the original three-duct version. After fine-tuning the two-duct configuration computationally, the pressure drop decreased even further to 36% less than the original configuration. This final optimized configuration was then tested using cold-air flow, which shows 40% reduction in pressure loss under cold air test condition.

6.6.1 From Redesign to Flight

After laboratory experiments and CFD validation, the new two-duct hot gas manifold became part of a new powerhead design. In the course of integrating this new configuration, Rocketdyne incorporated a new and improved design process. This design lowers the temperatures in the engine during operation, as well. In addition, this design reduces stress on the main injector and requires fewer welds, eliminating potential weak spots in the powerhead, and had 52 fewer parts, leading to a 40% cost reduction.

The new two-duct engine made its first flight on Space Shuttle Discovery's 20th mission (STS-70) in July 1995, and has been used in all subsequent shuttle missions.

Chapter 7
Turbopumps

Rotating machinery that involves liquid has been in use for many centuries. The history of various forms of pumps goes back to the early days of human civilization, just as that of hydraulic turbines or marine propellers. Hydraulic turbines, for example, have used water for hydroelectric power generation. Another example is the Francis turbine developed in the nineteenth century—still one of the most popular water turbines in use.

Highly sophisticated pumps driven by high-speed turbines have emerged as liquid-propellant rocket engine technology has advanced, prompting use of the term "turbopump," particularly in conjunction with rocket engines for space exploration. Requirements of a turbopump vary depending on the specific impulse required and the associated engine design approach. For example, the inlet and exit pressures and flow rate for a gas-generator cycle engine are different from that of a staged-combustion cycle engine. In general, liquid-propellant rocket engine turbopumps operate under very severe conditions and are a challenge for numerical simulation.

Major advances in turbo pump technology have been made for more than three decades, just as CFD became an engineering tool during the same period in parallel with computer hardware advances. State-of-the-art of turbopumps depend on many factors such as material, the manufacturing process, bearing and seals, inlet and outlet arrangement, and the dynamics of the unsteady flow environment. Although all these factors vary among designs—sometimes significantly—CFD simulation procedures can be discussed in a unified manner. In this chapter, we discuss the CFD issues in simulating turbopump flow, based on our experience with the most advanced staged-combustion cycle engine, the Space Shuttle Main Engine (SSME). The discussion is focused on the computational procedures rather than engine specifics.

7.1 Historical Background

As briefly introduced in Section 6.1, in the early 1980s several major upgrades were made to the first flight engine used as the First Manned Orbital Flight (FMOF) SSME. One of the major modifications made on the FMOF engine was a new

D. Kwak, C.C. Kiris, *Computation of Viscous Incompressible Flows*, Scientific Computation, DOI 10.1007/978-94-007-0193-9_7, © US Government 2011

two-duct powerhead as well as an upgraded high-pressure oxygen turbopump. These modifications were incorporated to produce the Block I engine. The upgraded Block I engine was first flown successfully on STS-70 in July 1995, after which, a series of upgrades were undertaken to develop the Block II engine. Modifications in the Block II engine include the addition of a large throat combustion chamber, a new high-pressure fuel turbopump, and new material and manufacturing processes. The first model of this engine flew in 2001 and the first flight mission with all three Block II engines was STS-110 in April 2002.

From a CFD perspective, these upgrades require high-fidelity simulation capabilities, especially for combustion devices, turbine stages, pump inducer and impeller, and complex internal flows. Incompressible flow simulation capabilities are of critical importance for the analysis of pump-related flow and internal flow of cryogenic liquid. In this chapter, computational steps are discussed in detail, using the shuttle and shuttle-derived examples.

7.2 Turbopumps in Liquid-Propellant Rocket Engines

In designing a liquid-propellant rocket engine, several different approaches can be used in selecting the power cycle, depending on the mission requirements and the cost. To obtain high thrust, the chamber pressure has to be high, so it is necessary to increase the thrust-to-weight ratio. A turbopump in this environment needs to be compact to minimize weight. The most complicated and sophisticated design to date is the staged combustion power cycle used in the SSME. Historically, a gas-generator power cycle was developed and has been used very successfully in many rockets. More recent engines use expander or hybrid power cycles. From a pump design viewpoint, each design approach will result in different turbopump arrangements. Since the scope of our discussion is limited to CFD procedure development, any turbopump from these engines would satisfy our purpose. In this chapter, therefore, the examples are selected from our computational activities related to the shuttle engine and shuttle-derived configurations

In a staged-combustion engine, liquid propellant (typically cryogenic) is pumped from the fuel tank at low pressure into the combustion chamber. A turbine powered by partially burned combustion product is sketched in Fig. 7.1. The fuel turbopump is shown on the left-hand side of the figure. A typical turbopump assembly for a liquid-propellant rocket engine is shown in Fig. 7.1. Issues related to fluid dynamics are also indicated in the figure.

Fuel at the inlet of the turbopump is at low pressure, and is delivered to the combustion chamber at high pressure. Since a rocket engine needs to achieve a high thrust-to-weight ratio, turbopumps typically operate at an order-of-magnitude higher speed than other turbomachinery, such as turbojets or conventional pumps. Even at highly pressurized operating conditions in a rocket engine, cavitation is still a major issue because cavitation-induced vibration can cause serious damage to the pump structure, such as the impeller and diffuser blades. From a simulation point

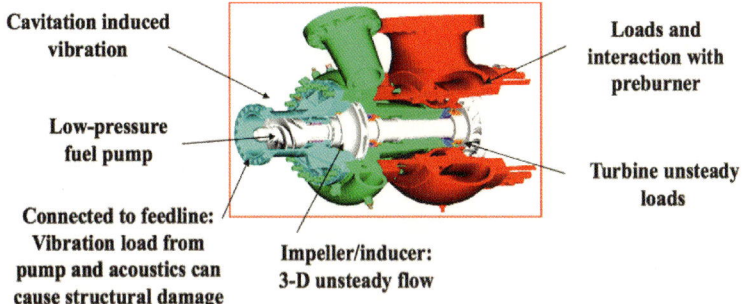

Cavitation induced
vibration

Low-pressure
fuel pump

Connected to feedline:
Vibration load from
pump and acoustics can
cause structural damage

Impeller/inducer:
3-D unsteady flow

Loads and
interaction with
preburner

Turbine unsteady
loads

Fig. 7.1   Sketch of a generic fuel pump arrangement for a liquid-propellant rocket engine

of view, the major advantage of accurately simulating the unsteady environment in
a turbopump is to design cavitation-free or minimally cavitating impeller configurations. Considering that the turbopump has to operate in a relatively wide range,
finding an optimum configuration including off-design conditions is not straightforward. In this chapter, we discuss a general CFD simulation procedure using a
generic pump similar to those used in the SSME.

The turbopump subsystem is the most crucial and expensive element in a
liquid-propellant rocket engine. However, turbopumps have been developed semi-
empirically for many decades and the unsteady three-dimensional viscous flow
phenomena have not been fully accounted for in the design process. Even though
CFD applications for turbomachinery have been reported in the literature, applications for turbopump design have been very limited. This may be due to the
difficulties associated with quantifying the unsteady three-dimensional flow in turbopumps that include inducers, impellers, and diffusers (stationary). In addition, it
takes many years from design to flight to develop a new or improved turbopump
system. To predict pump performance, which is directly tied to the engine performance, salient features of 3-D viscous flow phenomena must be resolved, including
wakes, boundary layers in the hub, shroud and blades, blade-hub juncture flows, and
tips clearance flows (see, for example, Hah et al., 1995). Another important feature
related to 3-D unsteady flow in turbopumps that affects the safety and reliability
of a rocket engine, is the fatigue on the structure due to flow-induced vibration.
Quantifying damaging frequencies and amplitude of this flow-induced vibration is
an important challenge to CFD simulation of turbopumps.

One of the early efforts for improving turbopump subsystem designs in liquid-
propellant rocket engines began in the early 1990s at NASA's Marshall Space
Flight Center, where a pump CFD consortium was established involving univer-
sities, industry, and NASA (see Garcia et al., 1992, 1994). Some of the computed
results generated during this consortium activity through the 1990s are used here to
discuss the simulation steps. In addition, analysis of the shuttle flowliner in the fuel
feed line, performed in early 2000s, is presented to illustrate complex internal flow
issues involving pump flows.

7.3 Mathematical Formulation for a Steady Rotating Frame of Reference

Geometrically, one of the simplistic pump configurations is shown in Fig. 7.2. Basically, the flow passes through the main impeller passage, usually at a constant speed. In this configuration, there is no interaction between stationary and rotational components, such as the interaction between the impeller and diffuser or between the inlet guide vane and impeller. Stationary shrouds can be treated as a moving boundary. Compared to a full pump assembly, this type of simplified configuration is easier to handle for obtaining experimental measurements, and thus convenient to use as the first step for validating pump computational procedures. In this section, we present a mathematical formulation in a steady rotating frame of reference. This formulation will be used for validation in the next section.

The computational procedure used in this chapter is based on the artificial compressibility method discussed in Chapter 4. In a steadily rotating frame of reference,

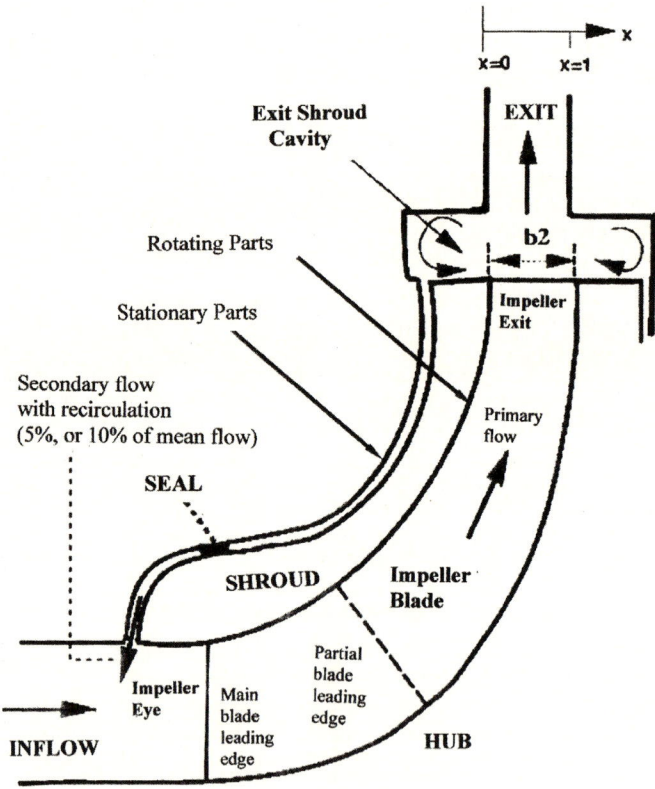

Fig. 7.2 Schematic view of a generic pump impeller cross-section

the governing Equation (4.20) is modified to include the centrifugal force and the Coriolis force as source terms.

$$\frac{\partial \hat{D}}{\partial \tau} + \frac{\partial}{\partial \xi}\left(\hat{E} - \hat{E}_v\right) + \frac{\partial}{\partial \eta}\left(\hat{F} - \hat{F}_v\right) + \frac{\partial}{\partial \xi}\left(\hat{G} - \hat{G}_v\right) = \hat{S} \qquad (7.1)$$

If the relative reference frame is rotating around the x-axis, this term is given by:

$$\hat{S} = \begin{bmatrix} 0 \\ 0 \\ \Omega\left(\Omega y + 2w\right) \\ \Omega\left(\Omega z - 2v\right) \end{bmatrix}$$

where Ω is the rotational speed. The source term can be set to zero in most cases other than for obtaining rotational steady solutions. Relative velocity components are written in terms of absolute velocity components u_a, v_a, and w_a as:

$$u = u_a$$

$$v = v_a + \Omega z$$

$$w = w_a - \Omega y$$

An unfactored implicit scheme can be obtained from the governing equations by linearizing the flux vectors about the previous time step. Then the following delta form of governing equation is obtained:

$$\left[I + \alpha \Delta \tau J\left(\delta_{\xi_i}\left(\hat{A}_i - \Gamma_i\right) - \hat{H}\right)\right]\left(D^{n+1} - D^n\right) = -\Delta \tau J\left[\delta_{\xi_i}\left(\hat{E}_i - \hat{E}_{vi}\right)^n - \hat{S}\right]$$
$$(7.2)$$

where

$\hat{A}_i =$ Jacobian matrices of $\hat{E}_i$
$\hat{H} =$ Jacobian matrix of the source term $\hat{S}$

$$\Gamma_i D^{n+1} = \left(\frac{\nu}{J}\right)\nabla \xi_i \cdot \left(\nabla \xi_j \frac{\partial}{\partial \xi_j}\right) I_m \frac{\partial D}{\partial \xi_I}$$

$\delta_\xi =$ finite difference form of $\dfrac{\partial}{\partial \xi}$
$\alpha = 1/2$ for trapezoidal, or 1 for Euler implicit
$I_m = diag\,[0, 1, 1, 1]$

This equation is iterated in pseudo-time until the solution converges to steady state, at which time the original incompressible Navier-Stokes equations are satisfied. A direct inversion of Equation (7.1) would become a Newton iteration for a steady-state solution. In three dimensions, however, direct inversion of a large block-banded matrix of the unfactored scheme would be impractical. Implicit methods discussed in Chapter 4 can be applied to this formulation. The computed results presented in this chapter have been obtained using upwinding and line relaxation schemes with the one-equation turbulence model by Baldwin and Barth (1991).

7.4 Validation of Simulation Procedures Using a Steadily Rotating Inducer

One of the difficulties facing the validation of a pump flow simulation procedure comes from the scarce data available for validating detailed flow features. For the pump CFD consortium activity mentioned above, a special pump inducer geometry with a high flow coefficient was developed and experimentally studied by Rocketdyne (see Garcia et al., 1992). An inducer that provides a sufficient pressure rise to the pump inlet is crucial to prevent or minimize cavitation on the impeller blades, and is therefore a critically important element in a rocket engine pump design. The Rocketdyne inducer is designed to deliver 2,236 gal/min with a design speed of 3,600 rpm. The tip diameter of the inducer is 6 inches. As shown in Fig. 7.3, the upstream section of the inducer is composed of a 10-inch-long straight channel. The tip clearance is 0.008 inches and the tip-leakage effect had to be included in the computational validation. The Reynolds number was based on 1 inch and the average inflow velocity of 339.6 inch/s was 191,800. The computational validation steps are presented here to illustrate the procedure.

As the first step of the validation, rotational steady or ensemble-averaged solutions are computed. Therefore, it is possible to represent the flow with one blade passage that is one-sixth of the cross-section of the tube. An H-H grid topology with grid dimensions of 187 × 27 × 35 was used, as shown in Fig. 7.4. The H-type surface grid was generated for each surface of the passage by using an elliptic grid generator. The interior region was filled by using an algebraic solver coupled with an elliptic smoother. It is assumed that similar grids can be generated using other

Fig. 7.3 Rocketdyne inducer geometry used for validating computational procedure

Fig. 7.4 Surface grid for the
Rocketdyne inducer

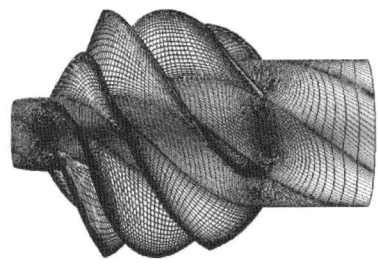

existing grid generation software. The bull nose of the inducer (front of the inducer
shaft in Figs. 7.3 and 7.4) was treated as a rotating wall and the cavity section at
the center of the bull nose was neglected. Geometric detail of this region can be
included by adding one more zone. However, the cavity region is not expected to
change the flow details in the inducer passage. Figure 7.4 shows the surface grid of
the inducer configuration.

Since one passage is computed, periodic boundary conditions are imposed on
the circumferential direction. For the rotational steady solution, the iterative process
is considered converged when the maximum residual drops at least five orders of
magnitude. The convergence history is shown in Fig. 7.5.

Validation in a qualitative sense can be done using three-dimensional flow
visualization.

For more quantitative validation, experimental measurements were made on a
number of cross sections, as shown in Figs. 7.6 and 7.7. Relative total velocity and
relative total angles obtained from the current computation are compared with those
obtained from experiments in Figs. 7.8, 7.9, 7.10 and 7.11. The radial velocity was
not measured experimentally. Therefore, the total velocity is defined as the combination of the axial and tangential velocity components. The flow angle is computed

Fig. 7.5 Convergence history
for Rocketdyne inducer
validation computation

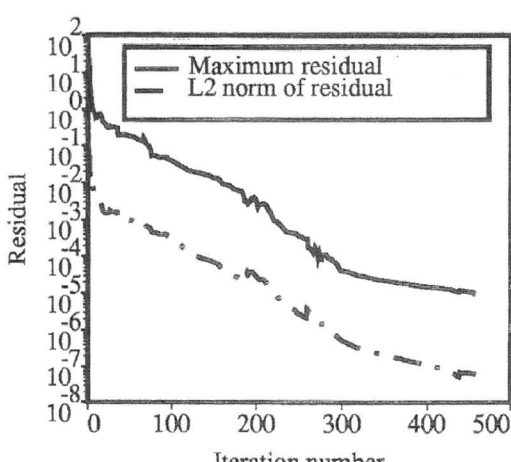

Fig. 7.6 Schematic
representation of
measurement locations along
the streamwise direction

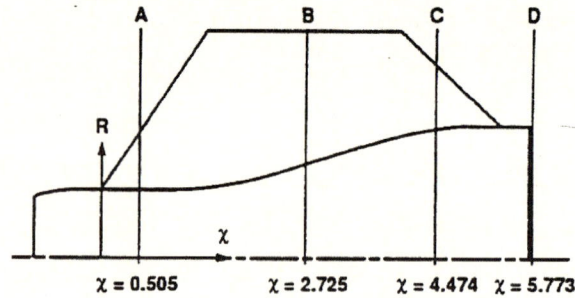

as the angle between the axial velocity and the total velocity. The circumferential
angle in all four figures is measured from the suction side to the pressure side of
the blade. The symbols represent the experimental measurements and solid lines
represent the computed results.

The computations and experiments compare well near the leading edge (Plane A)
from the hub to the tip region. Inside the blade passage (Plane B), computed and
experimental results compare fairly well. The biggest discrepancy occurs near the
hub region. The computation under-predicts the axial velocity near the suction side

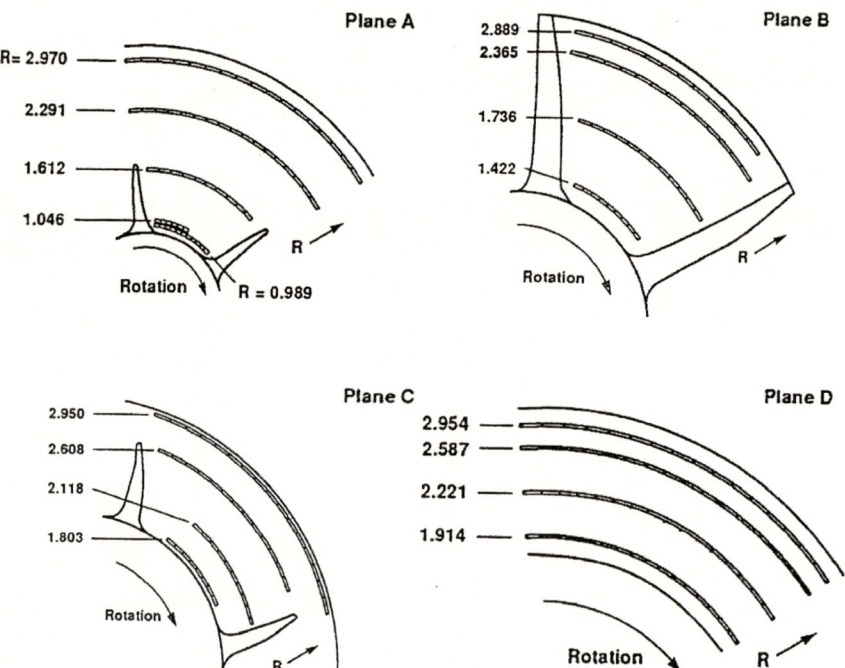

Fig. 7.7 Schematic representation of measurement locations at radial positions designated as
Planes A, B, C, and D along the streamwise direction

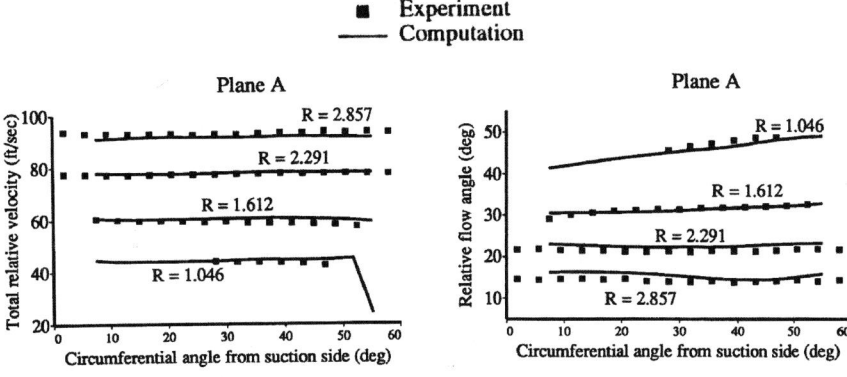

Fig. 7.8 Comparison of experimental measurements and computed results: relative total velocity and relative flow angle in Plane A

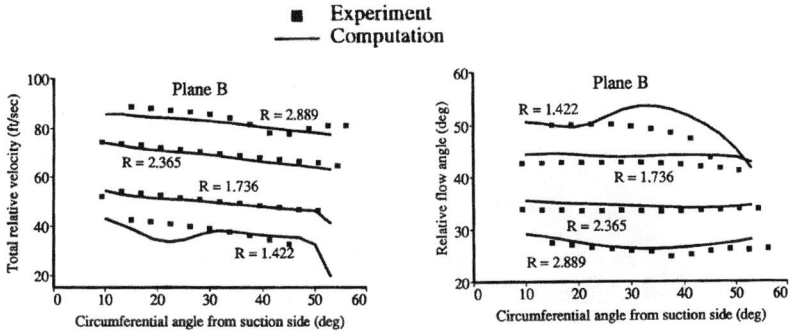

Fig. 7.9 Comparison of experimental measurements and computed results: relative total velocity and relative flow angle in Plane B

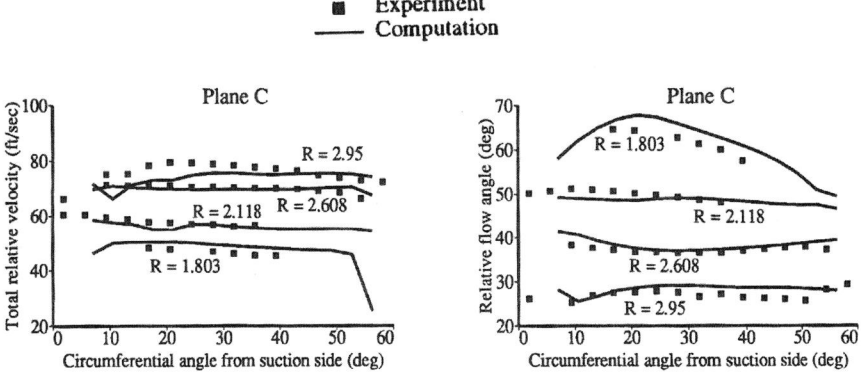

Fig. 7.10 Comparison of experimental measurements and computed results: relative total velocity and relative flow angle in Plane C

■ **Experiment**
───── **Computation**

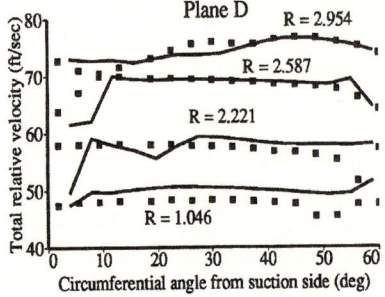
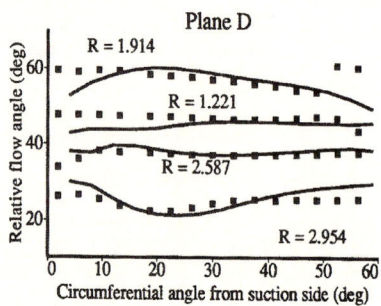

Fig. 7.11 Comparison of experimental measurements and computed results: relative total velocity and relative flow angle in Plane D

of the hub where secondary flow is significant. The accuracy of the computed results may be improved by increasing the grid resolution for the boundary layer region. In the core and near the tip region, the agreement is much better.

As we move downstream (Plane C), the error at the hub section decreases significantly, as seen in Fig. 7.10 (R = 1.803). In the core section, computed results follow the trend measured in experiments. Near the casing wall (R = 2.95), 5–8% differences are observed where the computed results over-predict the wake strength.

In Fig. 7.11, results are compared in the mixing region (Plane D). The location of the wake in the core region indicates a 5- to 10-degree difference between the numerical results and the measurements (R = 2.221 and R = 2.587). One possible reason for this discrepancy may come from the clocking error between the computational grid and the location of the measurements during post-processing. In Planes A, B, and C, the suction-side and pressure-side boundaries are clearly defined. However, in Plane D, it is assumed that the suction side starts at the same angle as the blade trailing edge. The computational grid does not follow the same circumferential angle as the blade trailing edge in the wake region. Therefore, it is possible to have slight differences between the computational and the experimental suction-side locations, causing the clocking uncertainties.

The flow field in this validation problem is very complicated, and predicting the turbulent flow structure in and around the blade passages is highly challenging, at best. In demonstrating the current computational procedure, turbulence is modeled by a one-equation model using an eddy viscosity hypothesis. However, the flow field includes high-speed rotational effect, juncture flow region and tips vortex flow, and is very likely to have significant non-equilibrium turbulence. Thus it is not expected to resolve the flow field accurately with the type of model we used in the current computations. Rather, it has been shown how one can utilize a Navier-Stokes solver to capture important viscous flow features in an ensemble-averaged sense. Even though the overall comparison of the integral quantities looks reasonable, more validation is necessary to assess whether the computational procedure is capable of predicting flow physics in detail, especially phenomena involving unsteady flow. In engineering, it is of critical importance to obtain computed results in a timely fashion for analysis and design. Therefore, validation computations were performed using design conditions. In off-design conditions, one may face additional complexity, such as massive separation. Interpretation of the computational results of this flow may require a deep understanding of the flow features and an assessment of the validity and limitations of the results.

7.5 Application to Impeller Simulation

The computational procedure discussed in the previous section is applied to an impeller simulation in this section. Baseline cases are selected from the class of pumps used in the SSME High Pressure Fuel Turbopump (HPFTP). The SSME fuel pump is shown on the left-hand side of the engine powerhead in Fig. 6.1.

7.5.1 SSME Impeller

The water test data of the SSME HPFTP impeller are used for validating the simulation procedure. Water test conditions are shown in Fig. 7.12. From the water test, flow data were collected at the exit of the impeller. Thus, the computed results are compared to the data at the impeller exit between two long blades. Rotational

Fig. 7.12 SSME High Pressure Fuel Turbopump (HPFTP) impeller geometry and water rig test conditions

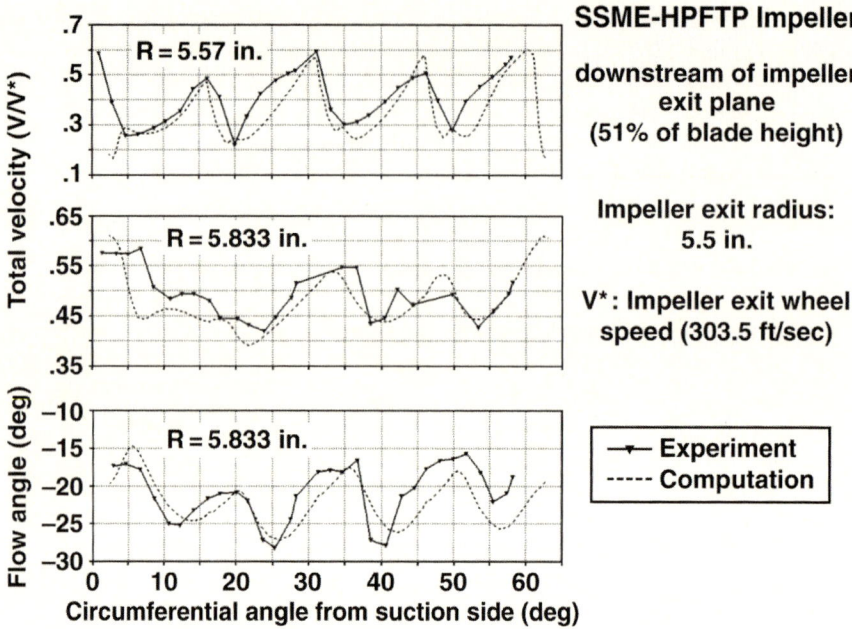

Fig. 7.13 Comparison of computed results and water test data downstream of the impeller exit for the SSME HPFTP

steady results are compared. However, the flow field in several regions such as the blade-hub junction, tip region, and entrance region to the diffuser may involve some unsteadiness. Full validation of all the features is difficult, partially because the experimental data are limited to ensemble-averaged quantities. In Fig. 7.12, the hub surface is colored by computed static pressure to aid in visualizing the flow field.

As shown in Fig. 7.13, the flow field was measured at two locations just downstream of the impeller exit at the half-height of the blade at the exit. The circumferential angle was measured from the suction side of a long blade to the pressure side of the next long blade. The fluctuating velocity and flow angle represent a partial blade and two short blades within this one flow passage.

The results in an ensemble-averaged sense compares reasonably well, considering that the flow field is quite complicated. Based on this comparison, flow visualization of the computed results can be utilized to shed light on the dynamics of the secondary flow and tip vortex flow. Three-dimensional computations require large grids and, consequently, a large amount of computer time is needed. To solve big problems, parallel implementation of the solver is necessary and various strategies for parallel implementation can be applied. Some of these will be explained in conjunction with the computation for the SSME flowliner, which will be discussed in a later section.

7.5.2 Advanced Impeller

During the CFD consortium activity (Garcia et al., 1992, 1994), a series of flow analyses inside an advanced impeller geometry were performed to verify the design. The computational procedure described above was applied to this configuration. Comparison between the computed results and experimental data is presented here (Kiris and Kwak, 1994b).

The computational grid and pump geometry are shown in Fig. 7.14. The impeller design flow rate is 1,205 gal/min with a design speed of 6,322 rpm. The Reynolds number for the computation was 181,283 per inch. The computation was performed using the formulation in a steady rotating frame of reference.

In Fig. 7.15, the circumferentially averaged meridional velocity is plotted at the impeller exit. The relative x-distance is measured from shroud to hub, where x = 1.0 corresponds to the hub location. As sketched in Fig. 7.2, the exit shroud and hub cavity start at the impeller exit, and in the present case, x-distance starts from −0.5 at the shroud cavity, extending to 1.5 at the hub cavity. The average meridional velocity, Cm, is non-dimensionalized by the wheel speed of 249.5 ft/s. The meridional velocity distribution for 5 and 10% recirculation from the exit shroud cavity are also plotted. The recirculation is due to leakage from the exit shroud cavity back to the impeller eye. This recirculation causes the peak velocity at the impeller exit move toward the hub. However, the overall impact of the leakage on the solution is minor at the impeller exit. The test data show that the peak velocity is closer to the center of the b2 width compared to the computed data. At the time the computation was performed, the recirculation in the hub cavity was not included. Since the leakage at the hub cavity leads to a stronger recirculation, this causes the shift in the velocity peak toward the center of the b2 width.

In Fig. 7.16, blade-to-blade velocity distributions at the impeller exit are plotted. This plot shows how the velocity is distorted at the impeller exit. The wake shows jet-like flow patterns in both locations between the full and partial blades. In a pump

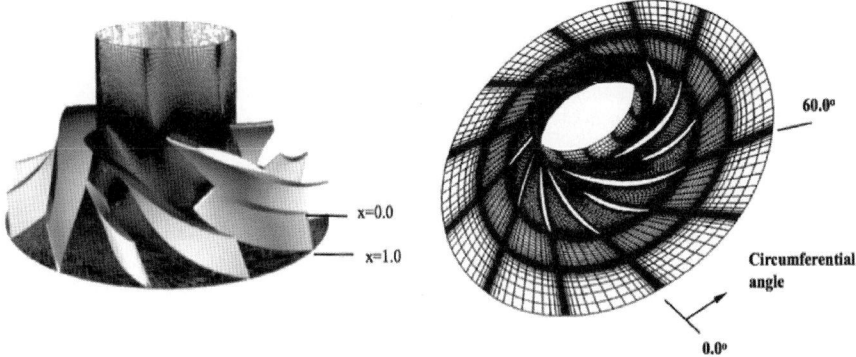

Fig. 7.14 Advanced pump impeller geometry and computational grid on the hub surface

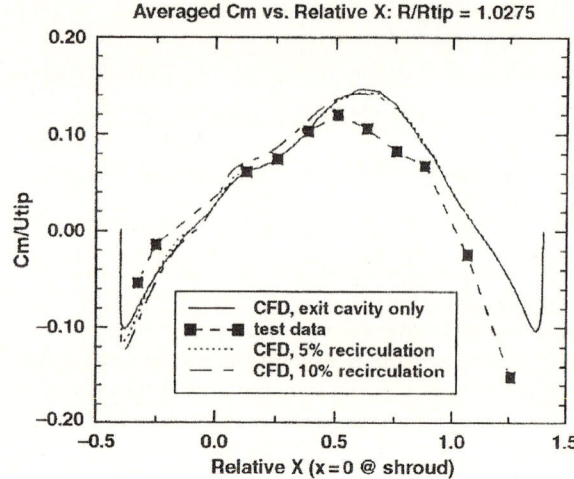

Fig. 7.15 Comparison of circumferentially averaged meridional velocity at the impeller exit

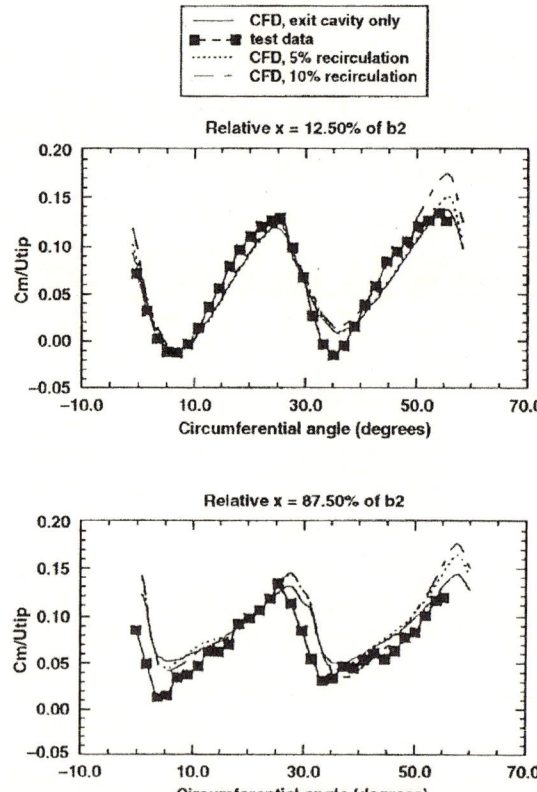

Fig. 7.16 Comparison of blade-to-blade meridional velocity at the impeller exit

assembly, this jet-like flow imposes unsteady load in the diffuser vane usually placed right after the impeller exit. Overall, the numerical results compare fairly well with experimental data for this impeller-only configuration.

7.6   Simulation of a Complete Pump Geometry

The geometry of an oxidizer or a fuel pump in a liquid-propellant rocket engine has both rotating and stationary components. As illustrated in Fig. 7.17, a typical pump has various components such as a flow straighter or inlet guide vane, inducer, impeller, and diffuser. The geometry handling is complicated, especially when relative motion is involved between rotational and stationary components. The flow is complex and unsteady, requiring time-dependant simulations. In this section, computational aspects of this multi-component simulation are discussed. In the previous section, impeller-only geometry from the SSME fuel pump was used to validate a rotational steady flow computation. In this section, the baseline geometry is selected again from the class of pump similar to the SSME High Pressure Fuel turbopump (HPFTP). The arrangement of the inlet guide vane (IGV), impeller, and diffuser is illustrated using the SSME-rig1 shuttle upgrade geometry in Fig. 7.17. We will refer to this baseline test geometry as the SSME HPFTP for discussing computational procedures.

7.6.1   Geometry and Computational Grid

In the baseline configuration, the impeller consists of 6 long blades, 6 medium blades, and 12 short blades, as shown in Fig. 7.18. The inlet guide vane (IGV) has

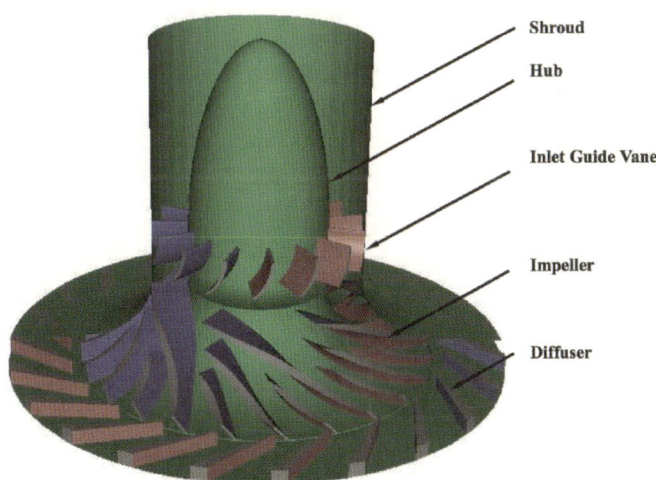

Fig. 7.17   Schematic of SSME-rig 1 shuttle upgrade pump geometry

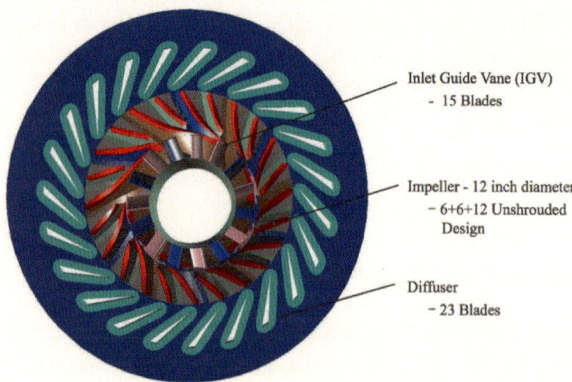

Fig. 7.18 Baseline water test geometry of the SSME HPFTP class pump

15 blades and the diffuser has 23 blades, as shown in the figure. This is a simplified model extracted from the actual multi-stage fuel turbopump and is used for water tests to collect validation data. This model computation ignores cavitation and does not include upstream and downstream manifolds and ducting. Depending on the flow solver, the computational grid for this configuration can be either a structured or unstructured grid. Since we are using a structured-grid-based solver, an overset grid system is employed for ease of generating grids at the expense of interpolation in the overlapping regions. Other grid arrangements can be created to capture essentially the same features of the flow.

For the geometry including the IGV and diffuser blades, flow data needs to be transferred correctly through the interface between the rotating impeller and the stationary IGV, or between the impeller and diffuser blades. Since this pump geometry requires 360-degree computations, an overset grid approach can be utilized to generate component-level grids independently from the neighboring geometry, offering maximum flexibility. Here, for connecting regions between the rotating and the stationary blades, a ring grid idea can be applied in the interface region. As sketched in Fig. 7.19, ring grids are included to cover the gap between the stationary and moving regions. For example, a ring grid fills the gap between the impeller region and the diffuser blades with some overlap on both sides. With this arrangement, impeller grids and diffuser grids can be generated independently without constraints, and the two regions connected by the ring grid.

Using this approach, grids were generated for each region. In the example shown here, a total of 34.3 million grid points are used. The IGV grid consists of 23 overset zones with 6.5 million grid points, and the diffuser has 31 zones with 8.6 million grid points (Fig. 7.20).

The grid for the impeller is the most involved. As shown in Fig. 7.21, the overset grid is set up around three different blades of the impeller to resolve the flow features. Near the tip region the grid resolution must be increased to capture the leakage and tip vortex flow. Generating this grid system is very time consuming,

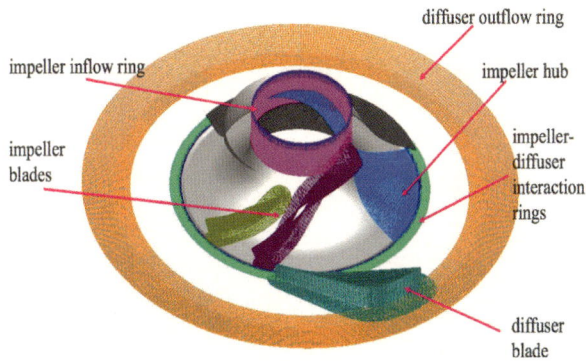

Fig. 7.19 Ring grid arrangement for a generic impeller-diffuser geometry

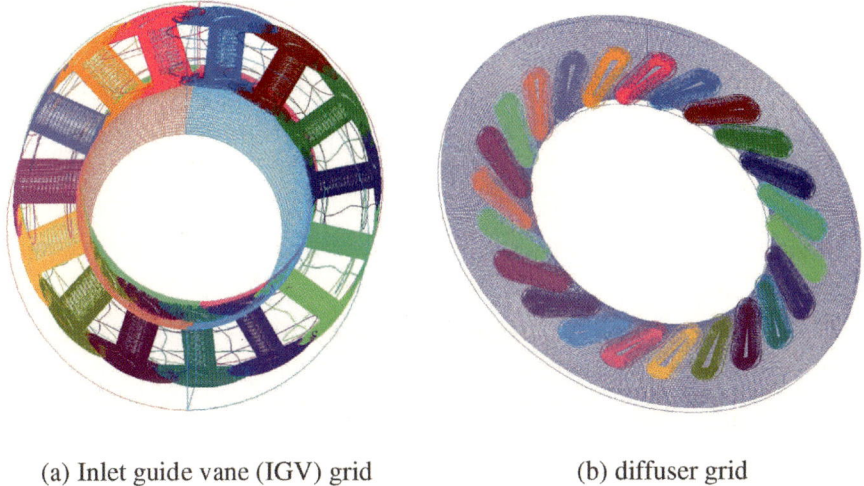

(a) Inlet guide vane (IGV) grid   (b) diffuser grid

Fig. 7.20 Overset grid arrangement for (**a**) IGV and (**b**) diffuser for the SSME HPFTP water test geometry

so an accelerated procedure is preferred for simulating pump flow, in order to evaluate the effects of geometric variations such as different blade shapes and blade angles. One approach, developed in conjunction with the generic turbopump geometry shown in Fig. 7.17, is to automate the grid generation process through scripting. This technique accelerates changing the number of blades and shapes from weeks to several minutes.

7.6.2 Issues Related to Large-Scale Computations

To achieve reasonable turnaround for computing the pump flow, especially for computing unsteady flow, using grids of this magnitude requires parallel processing.

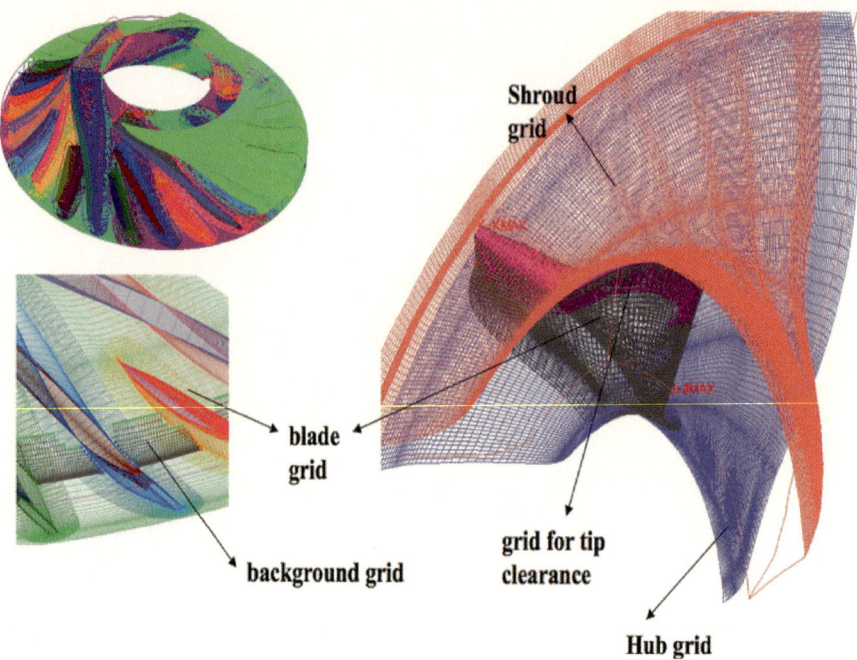

Fig. 7.21 Geometry and overset grid arrangement for the SSME fuel pump impeller geometry

The specifics for parallel implementation of any given code depends on the computer architecture and the flow solver being used. As the grid size becomes larger and requires a large number of processors, scalability becomes one of the primary issues. Here, we illustrate how we performed parallel computations using the SGI Origin 3000, the production computer available at the time the SSME HPFTP class pumps were simulated.

Two different approaches using the INS3D code are explained here. The first is a hybrid MPI/OpenMP approach and the second is Multi-Level Parallelism (MLP) developed at NASA Ames (Taft, 2000). The approach is to use message-passing interface (MPI) for inter-zone parallelism, and to use OpenMP directives for intra-zone parallelism. The INS3D-MPI code is based on the explicit message-passing interface across MPI groups and is designed for coarse-grain parallelism. The primary strategy is to distribute the zones across a set of processors. During the iteration, all the processors are to exchange boundary condition data between processors whose zones shared interfaces with zones of other processors. A simple master-worker architecture was selected because it is relatively easy to implement and it is a common architecture for parallel CFD applications. All I/O was performed by the master MPI process and data were distributed to the workers. After the initialization phase was complete, the program began its main iteration loop.

The MLP approach differs from the MPI/OpenMP approach in a fundamental way, in that it does not use messaging at all. All data communications at the coarsest and finest level are accomplished via direct memory referencing instructions.

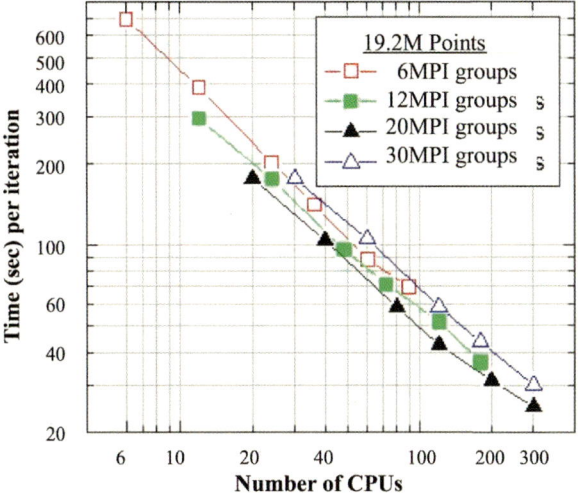

Fig. 7.22   Computing time (in seconds) per iteration for the SSME impeller using MPI/OpenMP

However, note that this can only be executed on shared-memory computers. The coarsest-level parallelism is implemented by spawning independent processes via a standard UNIX fork. The advantage of this approach over the MPI procedure is that the user does not have to change the initialization section of a large code. Libraries of routines are used to initiate forks, to establish shared memory arenas, and to provide synchronization primitives. The boundary data for the overset grid system is updated in the shared memory arena by each process. Other processes access the data from the arena as needed.

The performance of both the MPI/OpenMP and MLP approaches, are compared using the SSME impeller with 19.2 million grid points. In Fig. 7.22, the scaling performance of the MPI/OpenMP approach is shown. In Fig. 7.23, scalability using the MLP approach is shown for the same grid arrangement.

Using the MLP approach, time-accurate computations for the SSME-rig 1 configuration were performed on the SGI Origin 2000 and 3000 platforms. Computation of this case was started with the flow at rest, and the impeller began to rotate impulsively. It took three full rotations before the flow was established. A total of 128 CPUs of the Origin 3000 system were used with 34.3 million grid points. It took 3.5 days to complete the computation.

The computing platform and speed are continuously advancing, and these numbers are provided only to give the order of magnitude of the problem. As shown here, the computational approach for the design and analysis of turbopump systems require large computing resources and turnaround time. Therefore, relatively low-fidelity methods have often been used. With increased computer capability, the resolution for both space and time can enhance the ability to predict unsteady flow and capture more accurate flow physics.

With the large number of grid points and unsteady data, post-processing becomes another challenge in simulating a complete pump configuration. One avenue was

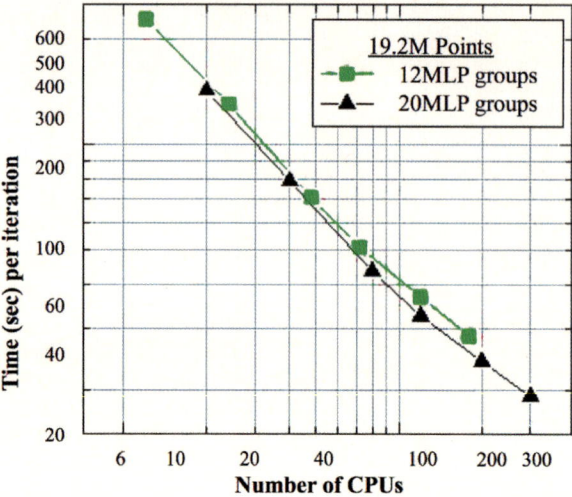

Fig. 7.23 Computing time (in seconds) per iteration for the SSME impeller using MLP

to visualize the flow in unsteady mode, which will provide analysts with valuable insights into the dynamic flow phenomena. Figure 7.24 shows snapshot from the unsteady computed results to illustrate the flow details one can obtain from the unsteady flow simulation. More examples can be found in Kiris et al. (1993, 2008) and Kiris and Kwak (2000, 2002). Further details on parallel implementation of pump flow computational procedures will be discussed in Section 7.8.

7.6.3 Issues Related to Flange-to-Flange Simulation

We have seen that the impeller-only flow field has been compared fairly reasonably with experimental data. However, this comparison is only a partial validation. In realistic cases, full flange-to-flange simulation is desired with the full range of pump speed. Other flow features such as tip vortex interaction with flow through impeller passages, vibration due to a fluctuating flow field, and fluid dynamic loads on full and partial blades require more extensive experimental data. To use CFD as a design tool, more complete validation is needed to verify the prediction capability for the full operational range. For example, the procedure validated for design flow conditions should produce equally valid results for off-design conditions. Unless the simulation procedure correctly represents the flow physics, it is difficult to assess geometric variations and operating conditions. In particular, the turbulence modeling should include non-equilibrium turbulence as well as rotational effects. Since most high-speed pumps encounter cavitation, a multi-phase capability will become important at some point in the design process.

Even though the flow simulation procedure for pumps has not matured enough to design a pump completely based on CFD, real impacts can be made in the areas of pump development and retrofitting. To date, CFD has been utilized in analysis

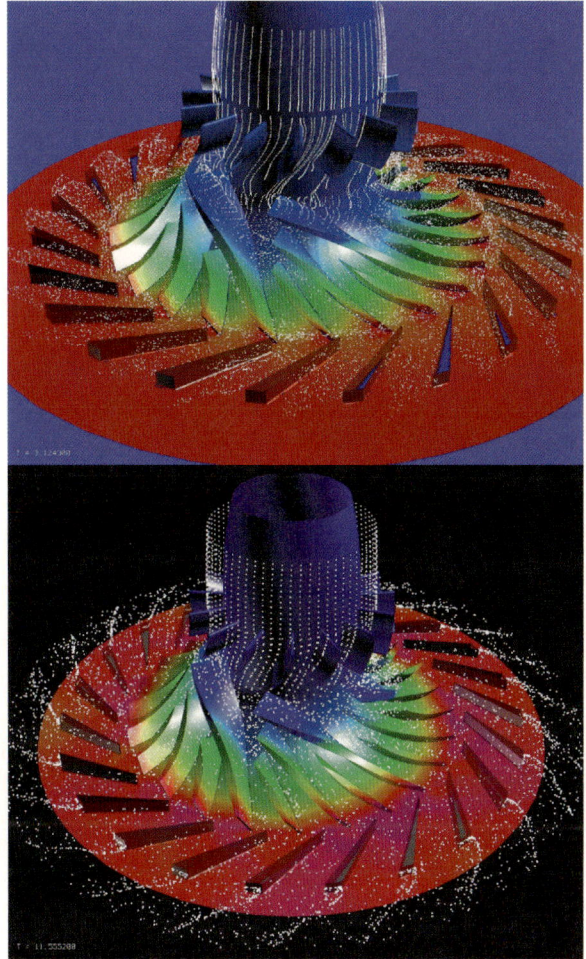

Fig. 7.24 Snapshot of particle traces and surface pressure from unsteady computations

and parametric studies of various pump-related flow features. In the next section, we discuss one such example that played a crucial role in determining the shuttle's flight rationale in conjunction with NASA's Return to Flight Program in the early 2000s.

7.7 High-Fidelity Unsteady Flow Application to SSME Flowliners

In the shuttle Main Propulsion System (MPS), there are two 17-inch diameter feedline manifolds, one for liquid oxygen (LOX) and another for liquid hydrogen (LH2). These feedlines contain three outlets each, which are connected to each SSME by

12-inch diameter feedlines providing LOX and LH2 to the engine (see Fig. 7.23). Fuel enters the orbiter through gaseous and liquid fuel branches. In each liquid hydrogen branch, LH2 enters the low-pressure fuel turbopump (LPFTP) when the prevalve is open. The entire feedline system in the powerhead can be found in the literature, and we explain the LH2 feedline leading to the flowliner region to define the current CFD simulation task.

The shuttle orbiter LPFTP is an axial flow pump similar to the inducer used for validation of the simulation procedure in Section 7.4. The LPFTP is driven by a two-stage turbine powered by gaseous hydrogen. The main function of LPFTP is to boost the pressure of the liquid hydrogen supplied to the high-pressure fuel turbopump (HPFTP), from approximately 30–276 psi. This increase in inlet pressure to HPFTP will permit the HPFTP to operate at high speeds with minimum impact from cavitation. The LPFTP operates at the level of 16,000 rpm and is approximately 18 by 24 inches in size. The flowliner is located in the connecting region between the LH2 feedline and the LPFTP, just upstream of the pump, to accommodate gimbaling of the main engine.

7.7.1 Description of the Flow Simulation Task

In May of 2002, three cracks were found in the downstream flowliner at the gimbal joint in the LH2 feedline of the SSME #1 of orbiter OV-104. Subsequently, inspections of the feedline flowliners revealed that all orbiters were found to have LH2 feedline flowliner cracks. To produce a correct flight rationale, it became necessary to investigate this issue. To identify primary contributions to the cracking, characterizations of various elements were performed, such as structural dynamics and material and fluid dynamics involving LPFTP. The flow simulation presented in this section is a part of this combined effort.

The LPFTP creates transient flow features such as reverse flows, tip clearance effects, secondary flows, vortex shedding, junction flows, and cavitation effects. Flow unsteadiness originating from the orbiter LPFTP inducer is one of the major contributors to the high-frequency cyclic loading that results in high cycle fatigue damage to the gimbal flowliners. The reverse flow generated at the tips of the inducer blades travels upstream and interacts with the bellows cavity. The dynamic environment with limited means of validation makes it very difficult to simulate the flow. In order to characterize various aspects of the flow field near the flowliner and determine vibration frequencies generated from unsteady flow, full-scale tests were carried out using the configuration sketched in Fig. 7.25, as well as a single flowliner model. The computational model of the test article is shown in Fig. 7.26 with an enlarged view of the flowliner region. The flight hardware feeds LH2 into three main engines. Therefore, the data obtained using the test articles with one LPFTP are to be used for validating the simulation procedure. Then the resulting procedure can be extended to analyze various flight conditions. In this section, we discuss various computational issues associated with generating analysis data.

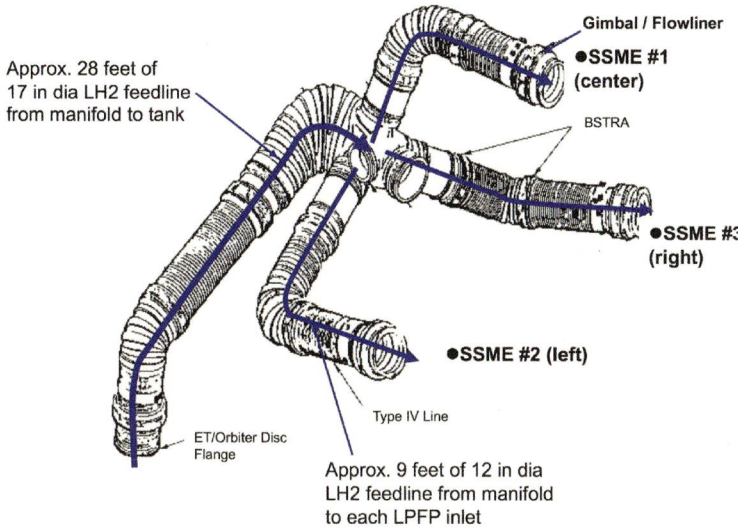

Fig. 7.25 Sketch of the fuel feedline arrangement in the shuttle orbiter Main Propulsion System

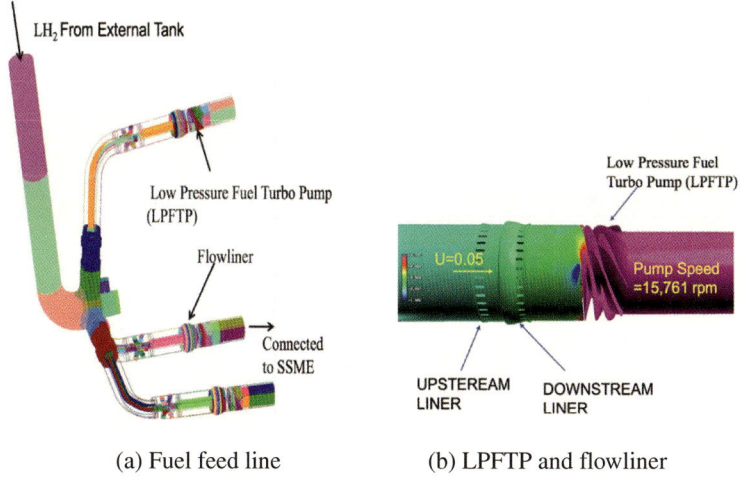

(a) Fuel feed line   (b) LPFTP and flowliner

Fig. 7.26 Computational model and flow conditions of the shuttle flowliner

7.7.2 Computational Model and Grid System

In order to characterize various aspects of the flow field near the flowliner, several computational models have been developed and high-fidelity computations carried out. The computations include a straight pipe model with the LPFTP inducer, and the LPFTP inducer with the addition of upstream and downstream flowliners

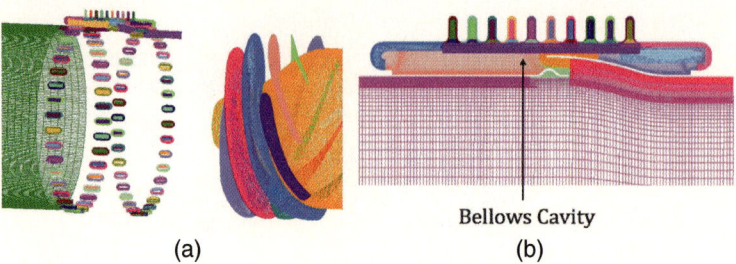

(a)   (b)

Fig. 7.27 Overset grid arrangement for the shuttle flowliner computation: (**a**) surface grids for the LPFTP inducer and the liquid LH2 flowliner; (**b**) details of the flowliner overset grid system

with 38 slots (Fig. 7.26b), an overhang area between the liners, and the bellows cavity (Fig. 7.27).

The first computational model (Model I) includes the LPFTP inducer with four long and four short blades, and a straight duct, which extends four duct diameters upstream of the inducer. The bull nose of the inducer and split-blades are included in the model. The objective of studying the inducer model alone is to compare unsteady pressure values against existing data. To resolve the complex geometry in relative motion, an overset grid approach is employed. The geometrically complex body is decomposed into a number of simple grid components. Connectivity between neighboring grids is established by interpolation at each grid outer boundary. The addition of new components to the system and simulation of arbitrary relative motion between multiple bodies are achieved by establishing new connectivity without disturbing the existing grids. This computational grid has 57 overset zones with 26.1 million grid points.

The second computational grid system (Model II) is based on the first, with the addition of the flowliner geometry. The grid system includes 38 upstream slots, 38 downstream slots (see Fig. 7.26), the overhang area between liners, and the bellows cavity (see Fig. 7.27). This model is very similar to the ground test article. It consists of 264 overlapped grids with 65.9 million grid points. Details of the grid system are shown in Fig. 7.27a, b. The flowliner component consists of an axisymmetric chamber around the external wall of the pipe, and two rows of slots in the streamwise direction. Each slot is a rectangular-shaped hole with rounded corners. On the outside wall of the chamber are the bellows, which are shaped like 10 periods of a sine wave. The bellows cavity is connected to the duct via the overhang area and the slots. Two-dimensional overset grids are first created for the bellows, sidewalls, and the overhang area of the bellows cavity. These are then revolved 360° to form the volume grids. Each slot consists of a body-conforming grid and a warped Cartesian core grid in the middle of the hole. The flowliner component alone contains 212 grids and 41 million points.

The orbiter fuel feedline manifold grid system consists of an inflow pipe, the manifold, three exit pipes with elbows to the main engines, and two short exit

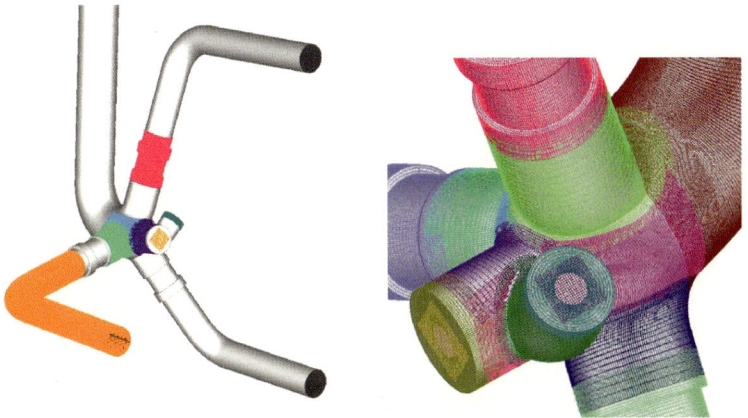

Fig. 7.28   Detail of the computational model for the orbiter fuel feedline manifold

pipes; one for the recirculating pump and the other for the fill and drain line (see Fig. 7.28). The recirculating pump and the fill and drain line are not included in the computational model, so we have closed these two exit pipes.

All pipes are connected to the manifold via internal collar grids. The upstream side of the inflow pipe and the downstream side of the exit pipes are modeled by body-conforming O-grids with a singular polar axis running down the core of the pipe. In the regions of the inflow and exit pipes near the manifold, the singular axis is avoided by adding a warped Cartesian core grid. Body-conforming grids are used for the walls of the manifold. A series of uniform Cartesian grids are used to occupy the core of the manifold. The entire grid system consists of 38 grids and 12 million points (see Fig. 7.28). A separate computational model was generated for the representative manifold test article. The computational grid representing the test article is created using O-H grids consisting of six overlapping zones and a total of 7.1 million points

In order to speed up and automate the grid generation procedure, a script system was developed to automatically and rapidly perform the various steps prior to the use of the flow solver. Just as in the impeller simulation in Section 7.5, special procedures were implemented to automatically create grids for each component type. The component types included in the script are blade, pipe, ring, nose, flowliner, and strut. The blade component is one of the most common parts of a liquid rocket subsystem and may contain multiple sections of one or more sets of different blades, for example, inducer, impeller, and diffuser. The pipe and ring components are used to connect different blade components. Pipes can be straight or curved and are bounded by the shroud. Rings can only be straight and are bounded by both the hub and the shroud. The nose component is a cap that fits at the start or end of the hub. The flowliner is a highly complex part with bellows and slots. The strut component consists of multiple blades connected to brackets at the shroud end and a central hub at the other end. The strut component was not used in the test article; it *was* used in the flight configuration, which is not included in this section.

7.7.3 Computed Results

The incompressible Navier-Stokes flow solver based on the artificial compressibility method was used to compute the flow of liquid hydrogen in each test article. All computations included tip leakage effects with a radial tip clearance of 0.006 inches; a pump operating condition of 104.5% rated power level (RPL); a mass flow rate of 154.7 lbm/s; and a rotational speed of 15,761 rpm. The problem was non-dimensionalized with a reference length of one inch and reference velocity equal to the inducer tip speed. The Reynolds number for these calculations is 3.6×10^7 per inch. Liquid hydrogen is treated as an incompressible single-phase fluid.

The past decade has seen considerable progress in the development of engineering CFD models for the multiphase flows characteristic of cavitation. The most practical approach among these is the homogeneous-mixture model, wherein the liquid-vapor mixture is treated using individual transport equations for each phase, and source terms are employed to describe the phase-change process. In spite of the progress made in multi-phase simulations methods in recent years, cavitation remains an extremely complex physical phenomenon and quantitative prediction is still a major challenge. Our current computations were performed with a single-phase assumption to provide a baseline solution for this complex flow. The validated cavitation model can then be implemented to produce results relative to single-phase results.

Initially, the flow is at rest. Then, the inducer is rotated at full speed. Mass flow is specified at the inflow, and characteristic boundary conditions are used at the outflow. Simulations for 14 inducer rotations were completed for Model I, and 12 inducer rotations were completed for Model II. The time history of non-dimensional pressure difference from INS3D calculations (Model I) at a location where experimental measurements are taken is plotted in Fig. 7.29a. Even though computed results have not fully converged to periodic solution in time and may still show evidence of start-up transients, the dominant 4 N (4 times the rotational speed) unsteadiness at a fixed location is seen in Fig. 7.29a. In Fig. 7.29b, maximum and minimum pressure values are recorded from the experimental data. Comparisons between CFD results and hot fire test data measured at a location near the downstream liner (see Fig. 7.26) also show good correlation in the non-dimensional pressure amplitudes.

Quantification of flow features in this unsteady environment is very difficult. Identification of sources and magnitudes of vibration is extremely important to the safety and reliability of the vehicle components involving the pump. The amplitude and the frequency of the unsteadiness due to backflow could be calculated as shown in Fig. 7.30. However, vibration at higher frequencies, possibly due to acoustics and cavitation, could not be resolved with the current computation and still remains a simulation challenge.

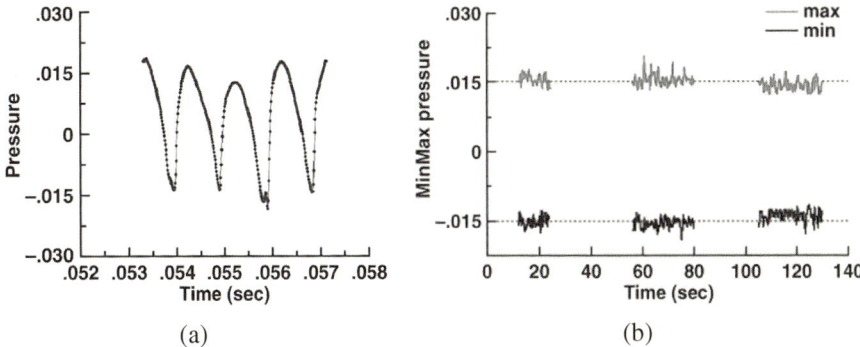

(a)   (b)

Fig. 7.29 Time history of non-dimensional pressure during one inducer rotation (Model I, 14th inducer rotation), and Min/Max values of non-dimensional pressure from hot fire measurements near downstream liner

In addition to the quantification of flow features, visualization techniques provide valuable insights into the flow phenomena. In Fig. 7.30, an instantaneous non-dimensional pressure map from the Model II computation is shown on the inducer surface. Blue indicates the least value and magenta indicates the greatest value. The pressure difference between the pressure side (facing downstream into the pump) and suction side (facing upstream) of the inducer blade is clearly visible in these pictures. Backflow near the inducer blade tip is caused by this pressure difference.

In Fig. 7.31, an instantaneous axial velocity map is shown on a vertical plane. Since the number of slots (38) produces a lack of symmetry on the vertical plane. To create this figure, the data on each of the structured overset grids are cut vertically and projected onto an unstructured two-dimensional surface. Inherent in this process is the creation of small discontinuities in the contours between overset grids that do not line up with one-to-one matching in the selected two-dimensional plane.

The contours show strong reverse flow regions coming from the blade, traveling through the overhang region and creating a jet-like flow on the order of 10%

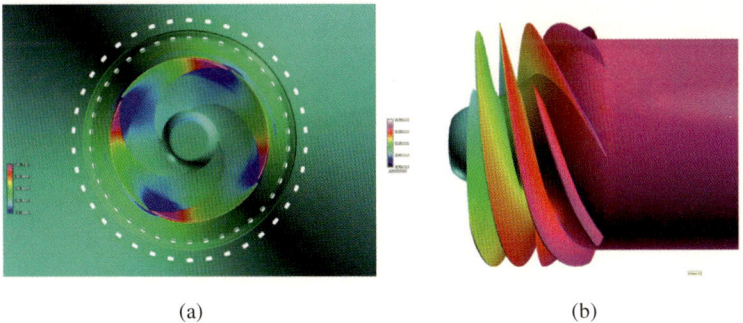

(a)   (b)

Fig. 7.30 Instantaneous surface pressure map on inducer (Model II): (**a**) view from upstream of the LPFTP looking into the flow direction; (**b**) side view of the pump inducer

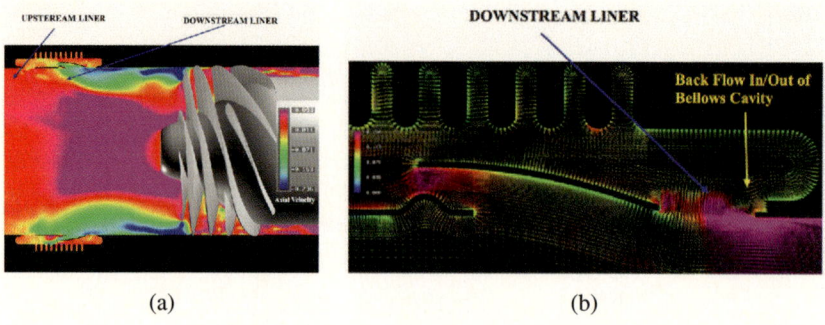

Fig. 7.31 Velocity at an instantaneous time: (**a**) axial velocity map in a vertical cut; (**b**) velocity vectors colored by total velocity magnitude in the bellows cavity region

of the inducer tip speed. The backflow regions travel up from the inducer blades to the upstream flowliner. The region of reverse flow extends far enough upstream to interfere with both downstream and upstream flowliners in the gimbal joint. Positive axial velocity values are colored in red/magenta and negative axial velocity values are colored in blue. Axial velocity values are non-dimensionalized by tip velocity. The strong interference between the backflow in the duct and the flow in the bellows cavity is clearly visible. The backflow velocity magnitude reaches 15–20% of the tip velocity magnitude in the overhang area between the upstream and downstream liners. It should be noted that this interaction is unsteady and backflow travels in the circumferential direction, as well. Due to strong interactions in the overhang area, flow is excited in the bellows area, which results in time-dependent recirculation regions. This observation can be seen in Fig. 7.31b, where strong jet flow with velocities of about 10–15% of the inducer tip speed penetrates directly into the bellows cavity, resulting in strong unsteady recirculation regions in the cavity. The time-dependent interaction between the duct and the bellows cavity can be one of the major contributors for high cycle loading. Figure 7.31 also shows that modeling the gap in the overhang area between flowliners is very important. Jet-like flow in the overhang area pushes the fluid in the bellows cavity toward the duct through the slots. Without proper modeling of this detailed geometry, one cannot obtain fine-scale flow unsteadiness on the liner. This transient phenomena creates an unsteady pressure-loading spectrum on the flowliner surfaces. Backflow also causes pre-swirl to occur in the flow approaching the inducer.

Visualization of unsteady phenomena can best be done using animation, or a series of instantaneous snapshots. In Fig. 7.32a, b, particles are released from the upstream slots and evolve for five inducer rotations. The colors of the particles represent forward flow (blue) and backward flow (red). We see from the figures that the particles are driven toward the center of the duct and travel to the inducer, where some of the particles are trapped into the backflow regions. In Fig. 7.32c, d the particles are released from the downstream slots and evolve for five rotations. In Fig. 7.32c, the particles are colored the same as in the previous two figures. We observe a much more complicated flow structure in which many of the particles

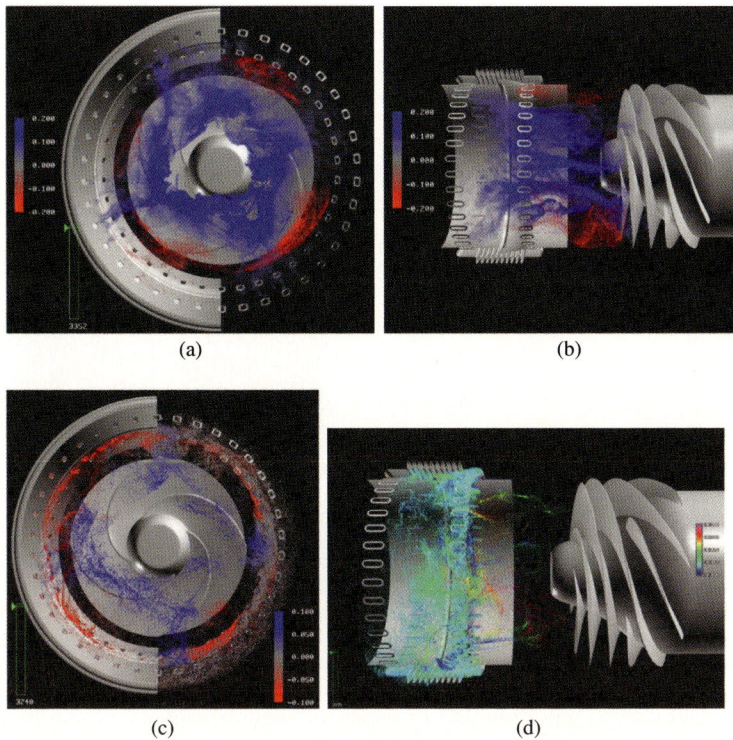

(a)   (b)

(c)   (d)

Fig. 7.32 Instantaneous snapshots of particle traces colored by axial velocity values

travel into and out of the bellows cavity. In Fig. 7.32d, the particles are colored with axial velocity in order to show their presence in the bellows cavity. When the particles are released from the downstream slots they are under the influence of the backflow and swirl, resulting in fewer particles traveling toward the inducer.

In the Space Shuttle, the LH2 feeds into the fuel pump through a three-pronged manifold. This arrangement will cause non-uniform inflow into the flowliner. For system-level analysis, one should take this into account. Additionally, the geometry for the orbiter fuel feedline manifold and the experimental test article are different, as shown in Fig. 7.33. Therefore, the flow field through the manifold and the test article is computed to characterize the similarities and differences between the two configurations.

Initially, steady-state calculations were conducted for both the orbiter manifold and the representative test article. The calculations for the orbiter manifold did not converge to a steady solution because of high grid resolution, which captures the fine-scale, unsteady details that exist in high Reynolds number flows. Instead, time-accurate calculations were performed, and the mean flow results are presented here. For both of these computations, the same non-dimensionalization is used as in the flowliner analysis, including a Reynolds number of 36 million based on the inducer

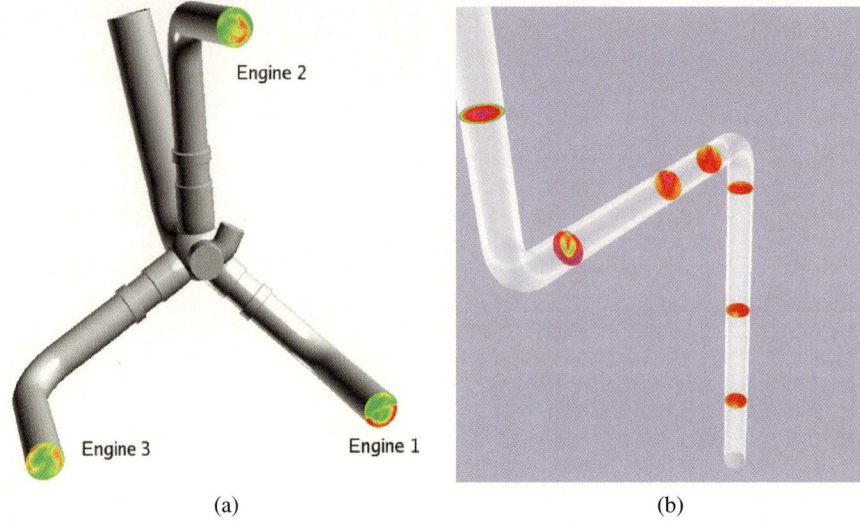

Engine 2

Engine 3   Engine 1

(a)   (b)

Fig. 7.33 Comparison of total velocity for orbiter manifold and the test article: (**a**) total velocity map of the orbiter manifold at the inlet to the LPFTP; (**b**) total velocity at various cross sections of the test article

tip speed and the reference length of an inch. Consistent inflow and exit boundary conditions are used for the test article and the orbiter manifold, such that the test article exit and the three engines downstream of the orbiter manifold receive the same mass-flow rate. At the inflow, the mass-flow rate is specified with a corresponding turbulent velocity profile and the pressure is calculated through the characteristic relation. At the exit, the pressure is extrapolated and the mass-flow rate is enforced.

The orbiter manifold displayed in Fig. 7.33a is qualitatively different for each of the three outlet sections, and none of these sections are represented well by the test article results. Figure 7.33a also shows that at the outer wall of the engine 1 feedline the velocity is large, while the high velocity is more uniformly distributed around the entire wall of the engine 3 feedline, and that near the walls of the engine 2 feedline, high-low velocity regions are not well developed. One of the reasons for the different velocity profiles at the three engine feedlines is the difference in lengths of the these feedlines. The velocity profiles for each of the three engines are different as they leave the manifold. In Fig. 7.33b, the velocity magnitude is displayed at various cross-sections of the test article. The flow direction is from the top of the figure to the bottom. The velocity profile is uniform before the first turn. It then becomes non-uniform after the first turn, but returns to a more uniform distribution toward the end of the pipe. The inflow profiles into the pumps for the three engines are different. This may explain one of the reasons why the flowliner failure rates vary among three orbiters. Since the test article does not reproduce inflow conditions from any of the three engines, the analysis of the test article does not provide a one-to-one picture of the real case in a strict sense. However, data from the test article provide detailed flow features for validation of the simulation procedure, increasing the confidence level of the simulation results for a real case.

7.8 Some Aspects of a Parallel Implementation

The total number of grid points for pumps and flowliner computations is very large, requiring large amounts of memory and computing time. To perform the computation within a reasonable turnaround time, parallel implementation of the flow solver is necessary. For almost all large-scale applications, parallel computing is necessary given the computer architecture we used and processor speeds. Here, we will illustrate a few strategies for parallel computing using the solver used for flowliner computations. This section is intended to illustrate some aspects of parallel computing, and the specifics discussed here may have to be modified depending on the flow solvers and computer architecture being used.

Two distinct parallel processing paradigms have been implemented into the INS3D code. These include the Multi-Level Parallelism (MLP) and the MPI/OpenMP hybrid parallel programming models. Both models contain coarse- and fine-grain parallelism. Coarse-grain parallelism is achieved through a UNIX fork in MLP and through explicit message passing in the MPI/OpenMP hybrid code. Fine-grain parallelism is achieved using OpenMP compiler directives in both the MLP and MPI/OpenMP hybrid codes. The multi-level parallel organization for INS3D is shown in Fig. 7.34.

Both the MLP and MPI/OpenMP approaches use a group-based data structure for global solution arrays. Computations on a single node of the parallel platform at NASA Ames Research Center (the Columbia platform) have been carried out using the MLP and MPI/OpenMP hybrid approaches. At the time (2004), Columbia was a 10240-processor supercluster consisting of 20 SGI Altix nodes with 512 processors each. In the MLP implementation, all data communication at the coarsest and finest parallelization levels is accomplished via direct memory referencing instructions. The coarsest level parallelism is supplied by spawning independent processes via the standard UNIX fork. A library of routines is used to initiate forks, to establish shared memory arenas, and to provide synchronization primitives. The MLP code

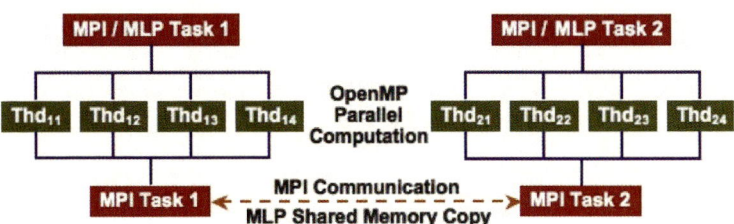

Fig. 7.34 MLP and MPI hybrid parallel organization for INS3D

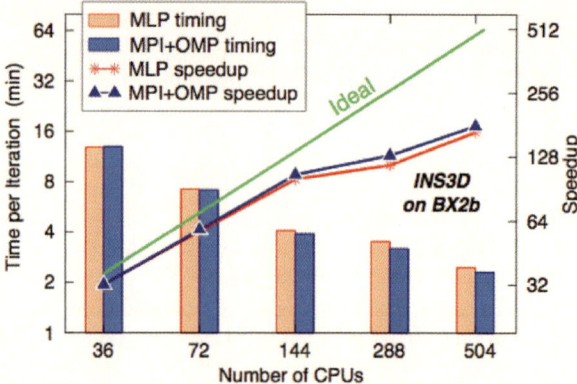

Fig. 7.35 INS3D-MLP and MPI+OMP performance on NASA Ames' Columbia platform

uses a global shared memory data structure for overset connectivity arrays, while the MPI/OpenMP code uses local copies of the connectivity arrays, providing a more local data structure.

Computations were performed to compare the scalability between the MLP and MPI/OpenMP hybrid versions of the solver on the Columbia system using the BX2b processors. Initial computations using one group and one thread were used to establish the baseline runtime for one physical time step, where 720 such time steps are required to complete one inducer rotation.

In Fig. 7.35, both the time per iteration (in minutes) and the speedup factor for the MLP and MPI/OpenMP hybrid implementations are displayed. Here, 36 groups have been chosen to maintain good load balance for both versions. Then, the runtime per physical time step is obtained using various numbers of OpenMP threads (1, 2, 4, 8, and 14). The scalability for a fixed number of both MLP and MPI groups and varying OpenMP threads is good, but begins to decay as the number of OpenMP threads becomes larger. Further scaling can be accomplished by fixing the number of OpenMP threads and increasing the number of MLP/MPI groups until load balancing begins to fail. Unlike varying the OpenMP threads, which does not affect the convergence rate of INS3D, varying the number of groups may deteriorate the convergence rate. This will lead to more iterations even though faster runtime-per iteration is achieved. Figure 7.35 shows that the MLP and MPI/OpenMP codes perform almost equivalently for one OpenMP thread, then as the number of threads are increased the MPI/OpenMP hybrid version of the code begins to perform slightly better than the MLP version. This advantage can be attributed to having local copies of the connectivity arrays in the MPI/OpenMP hybrid code. Achieving consistent scaling between the MPI/OpenMP and MLP versions of INS3D is promising, as the former version is easily portable to other platforms.

Performance results of the INS3D MPI/OpenMP code on multiple BX2b nodes on Columbia are compared against single node results. The results include running the MPI/OpenMP version using two different communication paradigms, master-worker communication, and point-to-point communication. The runtime per

Fig. 7.36 Performance of INS3D across multiple BX2b nodes using NUMAlink4 and InfiniBand interconnects (MPI point-to-point communication)

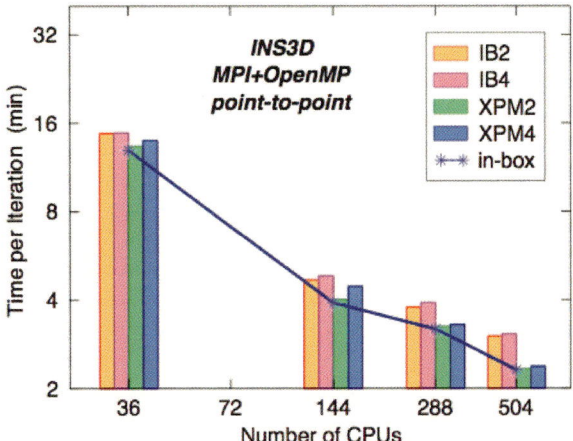

physical time step is recorded using 36 MPI groups and 1, 4, 8, and 14 OpenMP threads on one, two, and four BX2b nodes. The communication between nodes is achieved using the InfiniBand and NUMAlink4 interconnects denoted IB and XPM, respectively. Figure 7.36 contains the results using the point-to-point communication paradigm. When comparing the performance attained using multiple nodes with that of the single node, we observe that the scalability of the two-node and four-node runs with NUMAlink4 interconnects is similar to the single-node runs, which also use NUMAlink4. When using InfiniBand interconnects we observe a 10–29% increase in runtime per iteration on two- and four-node runs. The difference in runtime per iteration between four-node runs and two-node runs decreases as the number of CPUs increases.

Figure 7.37 shows the results using the master-worker communication paradigm. The first observation is that the time per iteration is much higher using this communication protocol compared to the point-to-point communication. We also see a

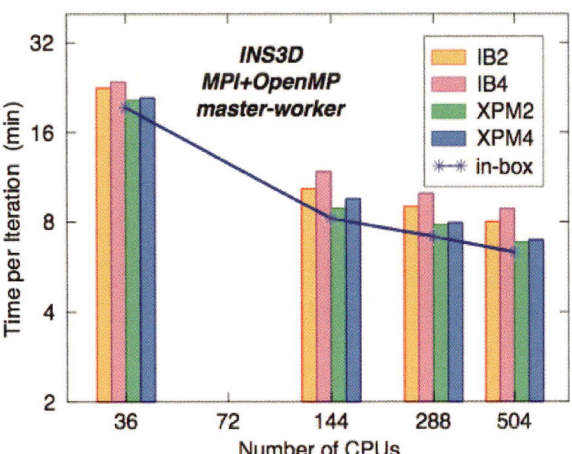

Fig. 7.37 Performance of INS3D across multiple BX2b nodes using NUMAlink4 and InfiniBand interconnection (MPI master-worker communication)

more significant deterioration of the scalability between the single- and multiple-node runs—even for the two- and four-node runs with NUMAlink4 interconnects, where almost no difference is observed when using point-to-point communication. Using NUMAlink4 interconnects, we observe a 5–10% increase in runtime per iteration for one to two nodes and an 8–16% increase using four nodes. This is because the master resides on one node, and all the workers residing on the other nodes must communicate with the master using the interconnect. Alternatively, when using point-to-point communication, many of the messages are bound to the node from which they are sent. In fact, the MPI groups can be manipulated so that a minimum number (as low as one in many cases) of messages must be passed between each node. Note that this optimization has not been utilized here and will be studied further. An additional 14–27% increase in runtime is observed when using InfiniBand interconnects instead of NUMAlink4—a similar increase was observed when using point-to-point communications, shown in Fig. 7.37.

The parallel implementation explained above illustrates strategies for enhancing the performance of large-scale CFD computations, and optimal choice will depend on the flow solver and the particular computer architecture on hand.

Chapter 8
Hemodynamics

Since our primary interest has been in human space flight, biomedical performance of humans during space flight and post-flight recovery, especially for long-duration missions, has been an important aspect of space exploration.

During space flights, astronauts are exposed to hostile environment such as the damaging effects of strong radiation, bone and muscle loss due to altered gravity, and the difficulties of living in a confined space. Astronauts' performance and post-flight recovery data have been recorded since the early days of space flight. Unfortunately, available flight data is limited to the maximum duration of 6-month stays on the International Space Station. For longer space travel, extrapolation of the current data, either from flight or ground-based (e.g., from artificial gravity experiments) is difficult. Modeling these and many other factors have been considered in conjunction with long-duration human space flight (see White and Avener, 2001, for an overview). To date, however, biomedical performance modeling for astronauts is accomplished primarily via empirical correlations.

One of the important aspects of the human performance model is related to the fluid dynamics of blood circulation. Under altered gravity conditions, blood circulation can be significantly modified compared to its circulation on Earth, which can in turn affect biomedical performance of astronauts in space. This motivated us to apply the incompressible flow methods discussed previously to human circulatory system simulations.

In this chapter, we present computational procedures for extending the incompressible flow computation capability to blood flow simulation, focusing on humans. For comprehensive human circulatory systems simulations, human physiological aspects must be included in some depth. Since this chapter is intended for use by CFD practitioners, possibly collaborating with bio-medical researchers, our discussion is limited to obtaining biomedical engineering solutions within a reasonable computing time and amount of effort.

Computing the human circulatory systems is very involved and requires multi-disciplinary modeling. For example, computational hemodynamics for studying arterial or vascular disease should include blood rheology, arterial wall structure, and general cellular biology. However, the intention of this chapter is narrowly focused on blood flow simulation with regard to human space flight. Extending the procedure discussed here to general biomedical research will naturally require

D. Kwak, C.C. Kiris, *Computation of Viscous Incompressible Flows*, Scientific Computation, DOI 10.1007/978-94-007-0193-9_8, © US Government 2011

further development of multi-disciplinary computational hemodynamics, including extensive physiological aspects. For the rest of this chapter, the term "hemodynamics" is used to represent hemodynamics with our limited scope.

We discuss the hemodynamics of altered gravity first, followed by applications for the development of prosthetic devices. The material presented here is based on studies performed by the authors and their co-workers. Further details can be found in individual publications cited in this chapter, for example, Kim et al. (2004, 2006), Kiris et al. (1990, 1997, 1998), and Kwak et al. (1988).

8.1 Issues in Computational Hemodynamics for Humans

Computational hemodynamics in humans has been of considerable interest for many years, since it offers the potential for big payoffs in such areas as analysis and treatment of atherosclerotic diseases, surgical planning, long-term biomedical studies, development of mechanical heart assist devices, and biomedical study of the physiological functions of the brain, kidneys, and other organs. Hemodynamics is a closely coupled problem of multiple disciplines such as complex geometrical systems, biochemistry, control, the cardiovascular system, and more. Blood flow involves a wide range of geometric and time scales from large vessels to micron-scale capillaries, eventually leading to molecular diffusion of various scalar quantities. In biomedical areas, "in vivo" measurements are difficult, making computed results—especially for unsteady, three-dimensional problems—very valuable.

Computational modeling, simulation, and analyses for treatment of the human circulatory system are all of major interest to medical researchers. Handling of any of those subjects in a comprehensive manner requires major effort, and is beyond the scope this chapter. The genesis and motivation of this chapter stemmed from an attempt to extend viscous incompressible flow simulation capabilities to benefit human space flight. As such, the methods and applications presented here are intended to help understand the alteration of physiological phenomena due to changes in gravitational force, especially through altered blood circulation, and its impacts on human biomedical performance in space. Eventually, it is hoped that the analysis can be used to develop countermeasures for risks encountered in space, and to extend that knowledge to human lives on Earth.

Since human subjects vary in geometry and physical conditions, without all the parametric detail it is very difficult to develop analytical models for circulation systems.

Here, we attempt to build some analytical basis for developing a predictive model for human performance under a different gravitational environment. We propose to use the circulatory system as a medium to connect various biomedical performance models. Important features that need to be included in circulatory system modeling are arterial wall motion due to fluid-wall interaction, shear thinning effects of blood flow, and a boundary condition procedure.

We next review some of the basic features that need to be considered.

8.1.1 Geometry of the Human Vascular System

A complete model of the human circulatory system is beyond the current modeling and simulation capability, although it may not be necessary or practical to include all details of a vascular bed. Depending on the particular goal at hand, it is more realistic to design a high-fidelity local model, where an accurate boundary condition procedure becomes necessary at truncated locations. The size of blood vessels in human varies from a few centimeters in diameter for the aorta to a few microns in capillaries. For example, an ascending aorta starts with a blood vessel approximately 1.25 cm in radius and gradually tapers down about 7 cm downstream to approximately 1.14 cm in radius. Actual sizes vary depending on the subjects, as well as where and how measurements are made.

One possible modeling approach is illustrated in Fig. 8.1, where the human circulatory system is modeled by a vascular bed. In the figure, the impact of gravitation is illustrated for the typical human in a standing position. At the truncated location or in a more conventional sense at outflow boundaries, the boundary conditions can be derived from an electrical circuit model of this vascular bed. An alternative is to design a one-dimensional model combined with an auto-regulation model derived from physiological knowledge. Some of the specific modeling issues are reviewed next.

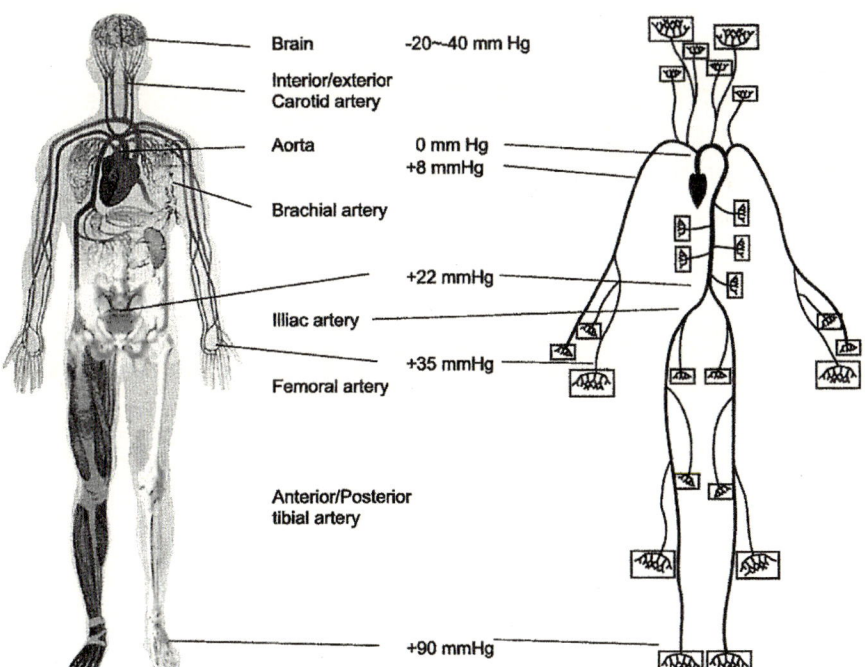

Fig. 8.1 Schematic of a human vascular system model

8.1.2 *Modeling Non-Newtonian or Stress-Supporting Flow*

Newtonian flow assumption is usually valid in large arteries where shear stress is high. However, in low shear regions (such as in capillaries) non-Newtonian characteristics become increasingly significant. In general, it is safe to include a non-Newtonian model in the governing equations.

8.1.3 *Turbulence Model*

In large arteries, the Reynolds number of the flow is in the laminar regime. However, the flow may show some turbulent behavior locally; for example, near heart valves. When prosthetic devices are used the flow may become locally turbulent.

8.1.4 Geometry *and Morphology*

Since the circulatory system varies among populations, one of the first issues in human blood flow simulation is the definition of geometry. Patient-specific geometry data can be used to model a representative geometry. In order to capture flow dynamics that closely resemble real phenomena, variations in blood vessel sizes and shapes have to be included in modeling. Detailed analysis of local flow is just as informative as the entire circulation simulation. Regardless of how inclusive the model is, it is realistic to truncate the vascular tree at manageable locations. Special boundary conditions are required at the truncated locations to correctly represent physiological conditions of the truncated region. For example, one can develop a detailed model for a combined cardiovascular and brain geometry, and then impose boundary conditions at truncated location derived from a vascular bed model.

8.1.5 *Arterial Wall Model*

Arterial walls move as blood flows. The wall motion depends on the wall structure and the local load from the flow in such a way that flow is distributed as dictated by human physiological needs. This motion has to be incorporated in the computational procedure. For example, relative blood vessel wall diameter change due to the heart pulse is found to be up to about 10% in common carotid arteries of young people (Reneman et al., 1986). Also, Giller et al. (1993) reported that the mean diameter change in the large cerebral arteries (internal carotid, middle cerebral, and vertebral artery) is less than 4%, but the smaller arteries such as the anterior cerebral artery showed 21% diameter changes to the mean change in a blood pressure of 30 mmHg with 16 mmHg deviations. In a standing posture on Earth, gravity pulls blood down to the feet. The blood pressure in the feet can be about 100 mmHg higher than that of the heart, whereas it is 20–40 mmHg lower in the brain (see Fig. 8.1, above).

In space flight condition, however, blood pressure equalizes and becomes uniform throughout the body. Consequently, under microgravity, arteries will contract in the lower body and dilate in the upper body.

Detailed high-fidelity computations have been performed to characterize local phenomena. However, for a complete circulatory systems simulation, it is realistic to design a simplified model assuming that the blood vessel is a thin-walled and linearly elastic channel (Caro et al., 1978; Steinman and Ethier, 1994). In practice, when the wall is sufficiently thin, this wall motion algorithm produces very similar results to those obtained from finite element methods for computing arterial wall displacements (Perktold and Rappitsch, 1995; Zhao et al., 2002). It is expected that for general biomedical studies, this approach will produce first-order estimates on arterial wall displacements. Even though this model is a simplification of wall physics, it shows sufficient evidence for arterial contraction and dilatation due to changing conditions such as gravitational forces, which should be sufficient to assess the overall impact of circulation on other biomedical functions of the body.

8.1.6 Boundary Conditions

The complete circulatory system includes large numbers of branching arteries down to the capillary level and then to veins. Explicit modeling of this complete system, starting from and returning back to the heart will be nearly impossible and may not be necessary for practical purposes. Since the entire human vascular network is enormous in size and complexity, minor arteries such as arterioles, venules, and capillaries need to be truncated to perform the numerical simulation at a computationally manageable level. At the truncated position where flow information is not available, outflow boundary conditions are necessary.

Several models have been reported in the literature for designing truncated boundary conditions. Lumped models utilizing the analogy of arterial networks to electric circuits have been used to provide boundary conditions for three-dimensional computations (for example, Quarteroni et al., 2000; Formaggia et al., 2002; Ferrandez et al., 2002) have defined a novel boundary condition using a control theory to simulate the peripheral resistance of the cerebrovascular tree and its auto-regulation function. Olufsen et al. (2000) developed a vascular tree model. In yet another approach, a vascular bed model was adopted to impose pressure boundary conditions at truncated boundaries (Cebral et al., 2000).

All of these models were designed to account for the resistance at the truncated locations such that blood flow distribution to various parts of the body is reasonable. The main question then is how to impose boundary conditions to predict blood supply to various parts of the body when conditions are not normal, such as under microgravity, during lengthy air travel in a confined space, or under abnormal physiological situations. The downstream boundary conditions are challenging for incompressible flow computations in general. Yet blood flow at truncated locations are not "downstream" in a strict sense. Flow resistance in arteriolar beds varies dynamically such that the resulting blood flow rate is maintained at near constant. One option for incorporating this feedback mechanism is to model this by an arteriolar auto-regulation algorithm (AAR), as will be discussed later in this chapter.

8.1.7 Cardiovascular Model

Many researchers have studied cardiovascular modeling over many years. For system-level simulations, simple models have been used, while higher fidelity models are generally intended for studying the heart itself. We assume that heart models are available to readers, so in this chapter, cardiac output is assumed as given boundary conditions for simulating other parts of the circulatory system. Readers are referred to the literature for a more comprehensive heart model (for example, Hunter et al., 1997; Smith et al., 2002).

8.1.8 Brain Model

There have also been several numerical studies on brain circulation. Cebral et al. (2000) simulated the blood flow in patient-specific cases taken from magnetic resonance angiograms (MRA) as a planning tool for neuro-surgical and interventional procedures. Ferrandez and David (2000) and Ferrandez et al. (2002) have shown that the variants of arterial geometry and modeling of the auto-regulatory mechanisms are crucial in determining the correct amount of blood supply to the brain. Later in this chapter, an example is presented for simulating local blood flow in the major arteries supplying blood to the brain.

Specific shapes and connections of the arteries in the brain vary among the human population (Alpers et al., 1959; Zhao et al., 2002). Three-dimensional reconstruction techniques have been used to obtain the subject-specific vasculature from magnetic resonance imaging (MRI), magnetic resonance angiogram (MRA), and computed tomography (CT); for example see, Taylor et al. (1999), Cebral et al. (2000), Quarteroni et al. (2000) and Steinman et al. (2002).

Computational simulations coupled with these medical imaging techniques can provide physicians with patient-specific information to predict the outcome of surgical procedures. In addition, flow variables that are difficult to measure in vivo can be calculated using real geometries. Considering the geometric variations, a computational approach will be of enormous value in mapping a wide spectrum of flow conditions complementing experiments using in vitro models. In a similar manner, this kind of information can be used for planning astronaut-specific space travel and for designing countermeasures.

In this chapter, a basic formulation will be presented first, followed by discussion of specific issues related to human models. In addition, there are special circumstances where blood flow simulation can be of significant value in developing prosthetic devices. For many decades, there have been attempts to use mechanical devices as a replacement or temporary measures in various medical treatments. The flow involving machines such as ventricular assist devices and mechanical heart valves can substantially alter blood circulation. The blood flow simulation aspects of these applications are different from natural circulation modeling, and will be discussed at the end of the chapter.

8.2 Model Equations for Blood Flow Simulation

Until it reaches the capillaries, blood flow can be assumed to be continuum and incompressible. Thus, for simulations involving most arterial networks, it is suitable to use the three-dimensional, unsteady, incompressible Navier-Stokes equations given by Equations (2.1) and (2.2). The gravitational source terms, g_j, are added to Equation (2.2) to account for cases where different postures on Earth and in altered gravity during space flights are important, resulting in the following form:

$$\frac{\partial u_i}{\partial t} + \frac{\partial (u_i u_j)}{\partial x_j} = -\frac{\partial p}{\partial x_i} + \frac{\partial \tau_{ij}}{\partial x_j} + g_j \tag{2.2a}$$

With a fixed reference pressure such as the one at heart level, the blood pressure gradient due to height differences must be added to the boundary conditions. The blood flow is largely laminar except at the localized region of turbulence near the heart valves and possibly around the stenosed bifurcation region due to deposits. Even in those situations, the flow is not likely to be fully turbulent, and is very chaotic (see Berger and Jou, 2000). Therefore, the shear stress tensor, τ_{ij}, can be approximated as below, only accounting for blood rheology without including turbulence stresses:

$$\tau_{ij} = 2\nu S_{ij} \quad \text{for Newtonian flow} \tag{2.3a}$$

$$\tau_{ij} = 2\eta S_{ij} \quad \text{for non-Newtonian flow} \tag{2.3b}$$

where

$$S_{ij} = \frac{1}{2}\left(\frac{\partial u_i}{\partial x_j} + \frac{\partial u_j}{\partial x_i}\right) \tag{2.4}$$

$$\eta = fcn(\dot{\gamma})$$

$$\dot{\gamma} = \text{Shear rate} = \sqrt{2\left(S_{ij}S_{ij} - S_{kk}^2\right)}$$

This is a simplified expression where non-Newtonian behavior of the blood is accounted for only through shear stress-dependent viscous stress terms. Some simplified models via apparent viscosity models are reviewed below.

8.2.1 Blood Flow Model

Blood consists of suspended particles such as red cells, white cells, and platelets in plasma, an aqueous solution. Red cells are the largest in concentration, are deformable and, at low strain rate, form aggregates, called *roleaux*. Formation of these aggregates results in increased viscosity at low strain. When the strain increases, these aggregates break down and the blood flow exhibits shear thinning effects, as illustrated in Fig. 8.2. In this figure, eddy viscosities from experimental data available in the literature (Chien, 1975; Merrill, 1969; Thurston, 1979), from two different non-Newtonian models and from a Newtonian flow model, are plotted to illustrate shear stress-dependent behavior of blood flow.

In a Newtonian flow, the stress is a linear function of the rate of strain. For a stress-supporting medium, the stress and strain have non-linear dependence. Various types of visco-elastic fluids exist where constitutive expressions have been developed in the past. In blood flow, for simplicity of the applications, empirical models are often used to account for the shear-thinning effects only. In most circulatory system simulations, these models provide adequate means to describe the pseudo-plastic behavior in computing dynamic aspects of the blood flow. Two models are described below that can be used to capture the change in viscosity due to the shear thinning effect of blood.

A model by Carreau-Yasuda (Bird et al., 1987) accounts for shear-thinning effects and was not designed for capturing visco-elastic behavior. The blood viscosity, η, is expressed in terms of the deformation tensor, $\dot{\gamma}$, as:

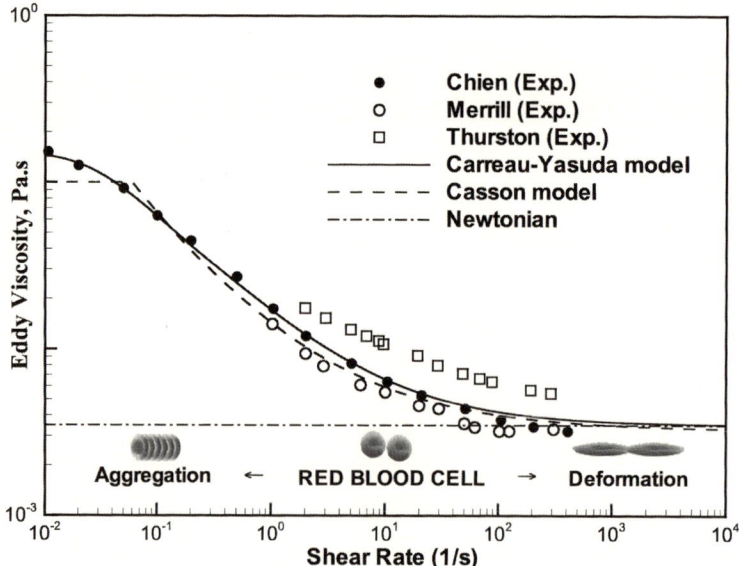

Fig. 8.2 Shear-thinning effect of non-Newtonian blood model compared to experiments

$$\eta\left(\dot{\gamma}\right) = \eta_{\infty} + (\eta_0 - \eta_{\infty})\left[1 + (\lambda\dot{\gamma})^a\right]^{\frac{n-1}{a}} \tag{8.1}$$

Where λ and a define the width of the transition region from the Newtonian to power-law regions, and $n-1$ is a power-law slope, as shown in Fig. 8.2. The constitutive parameters of human blood, based on the experimental measurements (see Chien, 1975; Merrill, 1969) are given by:

$$\begin{aligned} \mu_{\infty} &= 0.00348\,Pa \cdot s, \quad \mu_0 = 0.1518\,Pa \cdot s, \\ \lambda &= 40.0s, \quad a = 2.0, \quad n = 0.356 \end{aligned} \tag{8.2}$$

As shown in Equation (8.1), the model quantifies the shear thinning behavior of blood flow with asymptotic apparent viscosities, η_0 and η_{∞}, at zero and infinite shear rates, respectively.

Another model is given by Casson (see, for example, Perktold et al., 1991). An extended version is given below:

$$\eta\left(\dot{\gamma}\right) = \left(\frac{1}{\dot{\gamma}}\left[C_1\left(Ht\right) + C_2\left(Ht\right)\sqrt{\dot{\gamma}}\right]^2, \eta_{\max}\right) \tag{8.3}$$

where Ht is the hematocrit, C_1 and C_2 are coefficients determined for $Ht = 40\%$ as $C_1 = 0.2(\text{dyn/cm}^2)^{1/2}$ and $C_2 = 0.18(\text{dyn}\cdot\text{s/cm}^2)^{1/2}$ based on the experimental data (Merrill, 1969). To avoid extreme values at lower shear rates, the apparent viscosity is confined within $\eta_{\max} = 0.1518\,\text{Pa} \cdot \text{s}$.

As compared with several different experimental datasets in Fig. 8.2, the Carreau-Yasuda model is smoothly varying at the low shear region compared to the Casson model. Numerous other models exist, most of which are designed to reproduce empirical results for particular applications being studied. Since this chapter is written as an extended application for a general incompressible flow simulation, we use the above two models for computing examples listed here.

For circulatory system simulations, there are many modeling aspects coupled to hemodynamics. The simplified expressions presented here to account for non-Newtonian effects are adequate to illustrate the simulation procedures. As is the case for any physical model or equation of state, new models should be used whenever a more advanced prediction capability is needed.

8.2.2 Deformable Wall Model

Arterial walls have a finite thickness with layers of material. The outermost layer is known as the tunicia externa and the innermost layer in contact with the blood flow is the tunicia intima. The internal cavity through which the blood flows is known as the lumen. Descriptions of the anatomical construction of blood vessel walls and associated properties such as elastic behavior and distensibility are found in many medical references. As a first approximation for blood flow simulation, the arterial wall can be assumed to be a thin-walled, linearly elastic channel.

This is a practical approach for circulation modeling, however, the effect of wall motion to flow needs to be accounted for. For a general circulatory system simulation with a truncated network, wave reflection phenomena have to be included in the model. Compared to structural computation of the wall using finite element methods (Perktold and Rappitsch, 1995; Zhao et al., 2002), this assumption provides a first-order, yet realistic approximation to the complex behavior of the arterial wall (Caro et al., 1978; Steinman and Ethier, 1994).

Due to tethering to the surrounding structure, the arterial motion in the longitudinal direction is thought to be minimal (Milnor, 1989). Thus, assuming that the blood vessels are circular and have thin elastic walls with negligible longitudinal displacements, the increment of the blood vessel radius, r, can be estimated by:

$$\frac{\Delta r_i}{r} = \frac{1 - v_P^2}{E} \frac{r}{h} (p_w - \bar{p}_w)_i = D_w (p_w - \bar{p}_w)_i \tag{8.4}$$

where p_w and $\bar{p}_w$ are transmural and reference pressures at the wall, respectively, and the arterial wall distensibility D_w is defined as:

$$D_w = \frac{1 - v_P^2}{E} \frac{r}{h} \tag{8.5}$$

where h is the wall thickness, E and v_p are the elastic modulus and a Poisson's ratio of the arterial wall, respectively. Although the arterial wall material shows non-linear elastic characteristics, it can be modeled as a linear or constant elastic material over the normal physiological range (Zhao et al., 2002). In a simplified study, a constant material property can be assumed for both normal and microgravity conditions. For a typical blood vessel wall, the elastic modulus of 3.0×10^5 Pa and a Poisson's ratio of 0.49 can be assumed, which represents a nearly incompressible isotropic wall property (Steinman and Ethier, 1994; Perktold and Rappitsch, 1995). For the deformable wall, the no-slip boundary condition has to be replaced by the moving wall boundary condition using the Equation (8.5). This model can be used to account for a fluid-vessel wall interaction in a circulation simulation.

8.2.3 Vascular Bed Model

The primary interest of the circulatory system simulation in this chapter is to study the impact of altered gravity on human physiological function—both the overall impact and local phenomena. The human arterial network consists of a series of bifurcating trees, a cross-sectional area of which increases from approximately 5 cm^2 at the aortic root by two orders of magnitude smaller at the arterioles (Caro et al., 1978).

Starting from the pulsatile inflow at the aortic root, the blood flow simulation of the entire circulatory system is quite involved and includes complex geometry, a moving wall, and reflected pressure waves moving backward toward the upstream direction. To make the circulatory system simulation computationally manageable,

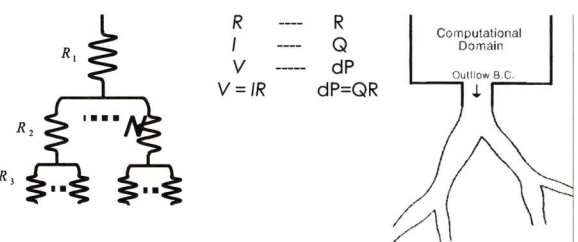

Fig. 8.3 Model of arterial network

minor arteries such as arterioles and capillaries need to be truncated. This requires an outflow boundary condition procedure at the truncated position. High-fidelity local solutions can be obtained with this approach.

There are several different approaches to modeling the arterial network. Olufsen et al. (2000) used a one-dimensional tree for large arteries, with structured trees for small arteries to provide outflow boundary conditions for the large ones (see Fig. 8.1 for a schematic of vascular trees). Another approach is to use an analogy between the arterial network and the electrical circuit, similar to the one illustrated in Fig. 8.3 (see Quarteroni, 2001). In this approach, flow resistance corresponds to electric resistance, flow rate corresponds to electric current, and pressure drop corresponds to electric voltage. The truncated artery is assumed to divide into N branches of the same size; for instance, N equals two for bifurcation and three for trifurcation. Under this assumption, the outflow boundary condition, especially for pressure, can be approximated by utilizing the electrical circuit analogy and the Poiseuille's theorem (Nichols and O'Rourke, 1998; Cebral et al., 2000).

The pressure drop in each branch can be expressed with the Poiseuille's formula as follows:

$$\dot{Q} = \frac{\pi \, r^4}{8\mu L}\Delta p \Rightarrow \dot{Q} = \frac{\Delta p}{R}, \quad \dot{Q}_{k+1} = \frac{1}{N}\dot{Q}_k \qquad (8.6)$$

where $\dot{Q}$ is mass flow rate, p is pressure, and R is flow resistance. At the kth bifurcation (or trifurcation), its mass flow rate is N times the flow rate through the $(k-1)$th branches. Assuming that the flow resistance ratio f is constant, the pressure drop at each level can be expressed in a geometrical series form:

$$\Delta p_{k+1} = R_{k+1}\dot{Q}_{k+1}, \quad \Delta p_k = R_k \dot{Q}_k$$

$$\Delta p_{k+1} = \frac{1}{N}\left(\frac{R_{k+1}}{R_k}\right)\Delta p_k = f \, \Delta p_k, \quad f = \frac{1}{N}\left(\frac{R_{k+1}}{R_k}\right)$$

$$\Delta p_{Total} = \sum_{k=1}^{\infty} \Delta p_k = \frac{1}{1-f}\Delta p_1 = \frac{8\pi \mu L_1}{(1-f)A_1^2}\dot{Q}_1 \qquad (8.7)$$

$$\therefore \quad \Delta p = \frac{8\pi \mu L}{(1-f)A^2}\dot{Q}$$

where A and L are the sectional area and length of the vessel, respectively. The flow resistance ratio f is generally unknown and should be properly determined to avoid an unrealistic pressure drop at the capillary bed. The optimal value of f can be determined from the arteriolar auto-regulation model explained in the next section.

This is just one approach that can be chosen. A similar idea can be applied where the vascular trees are modeled by a series of one-dimensional pipe flows with different sizes and lengths In these models, one needs geometry information on the representative vascular dimensions as well as a model of the physiological response at the truncated location.

8.2.4 Arteriolar Auto-Regulation Model

In the above, we discussed the need for constructing downstream boundary conditions to simulate blood flow using a truncated vascular bed model. This accounts for only the geometrical impact of truncating the network. In addition to this peripheral resistance, the human body shows a form of homeostasis called "auto regulation." Flow resistance in arteriolar beds varies dynamically by dilating or constricting mechanisms of the blood vessels, so the blood flow rate is maintained nearly at constant for a certain range of the perfusion pressure. This "auto" mechanism is pronounced in the kidney, heart, and brain such that the body can maintain a stable blood flow to support these vital organs.

In order to incorporate this feedback mechanism in the arteriolar bed, the arteriolar auto-regulation (AAR) model can be developed. In their series of computational efforts, David and his co-workers (for example, Ferrandez et al., 2000, 2002) developed auto-regulation models for numerical simulation of cerebral circulation. The model was used to study abnormalities of the Circle of Willis (CoW) in the human brain.

Following this idea, an AAR model is illustrated next, which is to be incorporated into the vascular bed modeling in Equation (8.7). Using the flow rate at the nth time step, the outflow pressure at the next time step is updated with the flow resistance ratio f until the flow rate satisfies the reference flow rate, $\dot{Q}_{ref}$. Equation (8.7) is rewritten as shown below.

$$p_e^{n+1} = p_{ref} + \frac{8\pi\mu L}{(1-f^{n+1})A^2}\dot{Q}_e^{n+1} = p_{ref} + R_e^{n+1}\dot{Q}_e^{n+1}$$

$$R_e^{n+1} = \frac{8\pi\mu L}{(1-f^{n+1})A^2}, \quad f^{n+1} = 1 - 0.5/a^{n+1}$$

$$a^{n+1} = a^n + k_t\frac{(\dot{Q}_e^{n+1} - \dot{Q}_{ref})}{\dot{Q}_{ref}} k_t = \alpha\Delta t/T$$

(8.8)

In this model α is selected in the range of 3.6–8.0, which produces reasonable auto-regulatory response, and Δt and T are the physical timesteps and period of the heartbeat, respectively. The velocity components at the outflow boundary are extrapolated from the interior domain. The above model is presented as an example, and we expect to see further development in modeling the auto-regulation mechanism as numerical simulation of the circulatory system advances. Even with this simplified modeling approach, numerical simulation can provide much information not readily available from empirical means.

8.3 Validation of the Simulation Procedure

To illustrate how the above modeling approaches work, as well as to validate the algorithm and simulation procedures, two test problems, namely, a carotid bifurcation and a circular tube with 90° bend, are computed next. Those two problems are chosen because they are commonly encountered building-block configurations in circulatory system modeling.

8.3.1 Carotid Bifurcation

Anatomically, a pair of common carotid arteries (CCA) arises from the ascending aorta of the heart and connects to the brain through the neck. The CCA are divided into internal (ICA) and external carotid arteries (ECA), respectively. A model geometry used for the experiment by Gijsen et al. (1999a, b) is sketched in Fig. 8.4. In their experiment, a blood analogy fluid (KSCN-X; KSCN-Xanthan gum solution) was used to mimic the shear thinning property of blood.

For the validation computations, the same flow condition is used as in this experiment. For the non-Newtonian model, both the Carreau-Yasuda and Casson models are implemented.

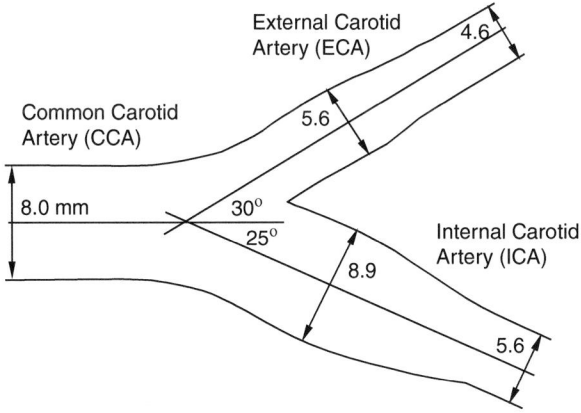

Fig. 8.4 Schematic definition of a carotid arterial bifurcation

The constitutive parameters of the Carreau-Yasuda model in Equation (8.1) are modified based on the experimental data, as in the following.

$$\eta_\infty = 0.0022\,Pa \cdot s, \quad \eta_0 = 0.022\,Pa \cdot s,$$
$$\lambda = 0.11s, \quad a = 0.644, \quad n = 0.392 \tag{8.9}$$

Likewise, a parameter of the extended Casson model is modified as $C_1 = 0.4(dyn/cm^2)^{1/2}$ for this blood analogy fluid.

The Reynolds number based on the CCA diameter is 270, and the flow division ratio of ECA over CCA is 0.45. The inflow in the CCA is assumed to be fully developed, and multiple outflow boundary conditions for the ICA and ECA are determined using the AAR algorithm in Equation (8.8) based on the flow division ratio. Since both models produced very similar results for this case (Kim et al., 2004), the Carreau-Yasuda model is used hereafter in all the validation computations presented for non-Newtonian flows.

There are many options for selecting a flow solver to simulate this problem. During the early development of CFD codes in the late 1970s and 1980s, many flow solvers were developed and, later, fully developed commercial software became available to users. It can be safely assumed that almost all organizations have either in-house developed or commercial codes available to compute the current problem, incorporating some of the models discussed above.

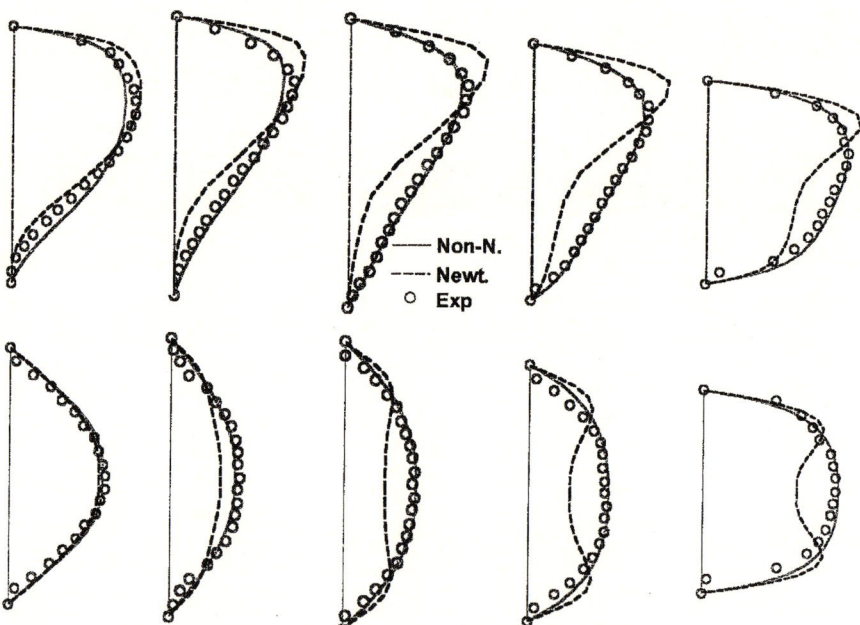

Fig. 8.5 Axial velocity profile on the symmetry (*upper*) and its perpendicular plane (*lower*) in the ICA shown in Fig. 8.4 at Re=270, compared with experimental data by Gijsen et al. (1999a)

For the computations presented here, a parallel version of the INS3D code (see, for example, Kwak et al., 1986; Rogers et al., 1991a; Kiris et al., 2002) has been extended with the above numerical modeling approaches. Computed results are compared with experimental data for both steady and unsteady non-Newtonian cases. The steady non-Newtonian flow in a carotid bifurcation, illustrated in Fig. 8.4, was calculated as the first validation problem.

In Fig. 8.5, computed axial velocity profiles are compared to experiments at five different positions along the centerline of the ICA, equally separated from the apex by a distance equal to the CCA diameter. Both non-Newtonian and Newtonian results are plotted. Newtonian results show significantly different velocity profiles from the non-Newtonian case. Axial velocity profiles are skewed toward the flow divider, and the increase of the sectional area results in an adverse pressure gradient in the sinus of the ICA. The flow reaccelerates after passing through the maximum diameter region.

8.3.2 Circular Tube with 90° Bend

For unsteady code validation, the pulsatile non-Newtonian flow in a 90° circular tube was simulated with the same flow condition as in the experiment by Gijsen et al. (1999). The same blood-analogy fluid used in the steady bifurcating flow discussed above was also used in this case. The Reynolds number based on the tube diameter and diastole velocity is 300. The Womersley number based on the tube radius is 14. The Womersley number (a dimensionless number showing the relation between pulsatile frequency and the viscous effects) is defined as $\alpha = r\sqrt{\omega/\nu}$ where ω is the angular frequency of the oscillation. A typical Womersley number for the human aorta is approximately 15. The geometry for the experiment and the computation is shown in Fig. 8.6. The tube radius and its centerline radius of curvature are 4 and 24 mm, respectively.

For pulsatile inflow boundary conditions, the experimental waveform is regenerated using twelve harmonics based on the Fourier theorem (Nichols and O'Rourke, 1998) as shown in Fig. 8.7. Computed results are compared with the experimental data at three different phases: end diastole, peak systole, and begin diastole. In Fig. 8.8, axial velocity profiles are compared between computations and experiments on the plane of symmetry and the plane perpendicular to the symmetry plane. In all three phases, the computed results compare reasonably well with the experimental data.

8.3.3 Effect of Arterial Wall Distensibility

To include the wall motion accurately in the flow simulation, the complete fluid-structure interaction needs to be modeled with anatomically correct representation of the wall construction. In realistic circulatory systems simulations, however, an approximate model with thin-wall assumption can be used. Considering the

Fig. 8.6 Circular tube with 90° bend: radius, r = 8 mm; curvature, R/r = 6.0; diastole velocity, U = 7.8 m/s; KSCN-X density = 1, 410 kg/m³; KSCN-X viscosity = 2.9 cPoise = 0.0029 Pa s; Re based on tube diameter and diastole velocity = 300

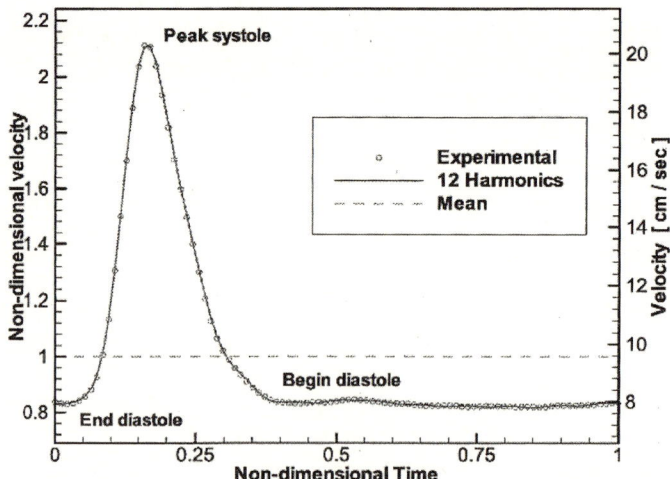

Fig. 8.7 Inflow waveform regenerated using 12 harmonics used for unsteady flow through a circular tube with 90° bend

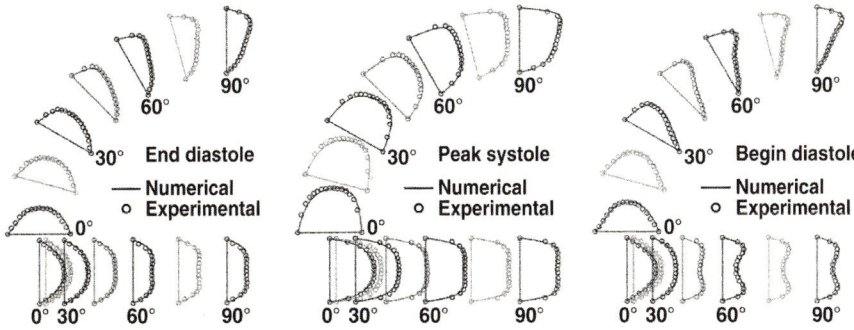

Fig. 8.8 Axial velocity profiles on the plane of symmetry and the plane perpendicular to the symmetry plane at three different phases of a pulse; comparison between computations and experimental data by Gijsen et al. (1999b)

complexity of the arterial network, in addition to individual variations, having a simulation model for a reasonable turnaround of analysis tasks will be of considerable value. Here, we illustrate how the simplified process can be used to study the effect of wall motion using the benchmark problem of pulsating flow through a carotid bifurcation.

The pulsatile flow of blood-analogy fluid through the carotid bifurcation model used above was computed with the mean flow rate of 8 ml/s in the common carotid artery (CCA), with a Reynolds number of 388. Figure 8.9 shows a moving wall grid at the maximum displacement location. Also shown is the baseline rigid-wall grid for comparison.

For this carotid bifurcation model, the arterial wall thickness is 0.3 mm for the CCA, and 0.24 and 0.21 mm for the internal (ICA) and external carotid artery

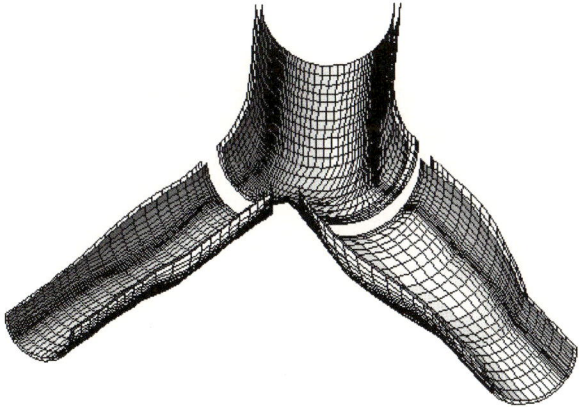

Fig. 8.9 Distensible wall due to fluid-wall interaction: inner grid–baseline rigid wall grid; outer grid–distensible wall at its maximum displacement position

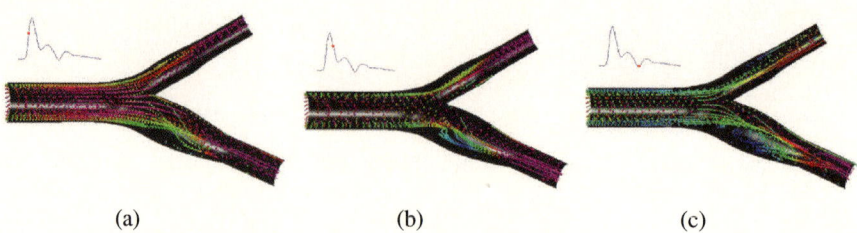

| (a) | (b) | (c) |

Fig. 8.10 Streaklines through a carotid arterial bifurcation at three different phases in pulsatile flow rate—the particular time where the particle traces are shown is indicated as a dot in the pulsatile inflow profile: (**a**) systolic acceleration phase (t/T = 0.28); (**b**) systolic deceleration phase (t/T = 0.36); (**c**) minimum flow rate phase (t/T = 0.58)

(ECA), respectively. The elastic modulus E is given by $3.0 \times 10^6 \mathrm{dyn/cm^2}$ and the Poisson's ratio ν_p is 0.49, which represents a nearly incompressible isotropic wall property (Steinman and Ethier, 1994; Perktold and Rappitsch, 1995). Wall motion was computed using Equation (8.4).

The magnitude of wall motion is illustrated in Fig. 8.9, where the maximum displacement position is shown relative to the baseline wall grid. A smooth connectivity in wall thickness between the CCA and its branches is enforced. The bifurcating apex is geometrically constrained to prevent unrealistic rigid body motion.

Figure 8.10 shows particle traces at three different points in time to represent different phases of the pulsatile flow: (a) systolic acceleration (b) systolic deceleration and (c) minimum flow rate. During a pulse cycle, the distensible wall case shows a maximum displacement of about 8% of the vessel diameter at the sinus of the ICA. At the systolic deceleration phase, the increased vessel diameter reduces the axial velocity profiles compared with the rigid wall results. Much like the steady-state results, a strong skewing occurs toward the flow divider walls. After the peak systole, a secondary reversed flow occurs and extends along the sinus until the

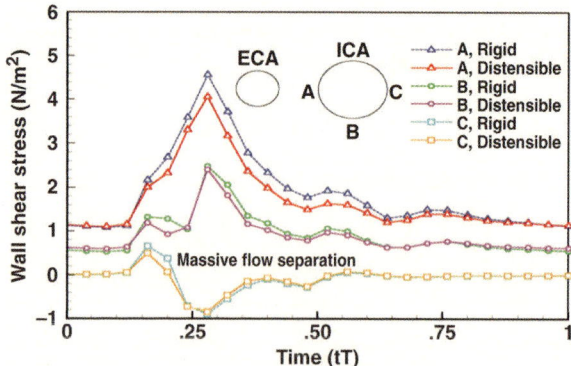

Temporal wall shear stress during a heart pulse (Re=388)

Fig. 8.11 Temporal wall shear stress during the pulse cycle at three different points around the ICA sinus (ICA: internal carotid artery, ECA: external carotid artery)

diastole begins. As shown in Fig. 8.11, the wall distensibility due to the pulse allevi-ates the amplitude of wall shear stress locally up to 16% compared with the results under the rigid wall assumption. The temporal wall shear stress at point C in the ICA sinus indicates the progress of massive flow separation and reattachment on the outer wall of the sinus during $t/T = 0.22 - 0.54$ in both the distensible and rigid wall cases.

8.3.4 Effects of Altered Gravity on Blood Circulation

In general, gravity has non-negligible effects on blood circulation. Even on Earth, blood circulation to the brain, for example, is affected by the posture of a person. In space flight, altered gravity has even greater impacts on humans. Astronauts experience a stressful environment of microgravity in which body fluid shifts to the upper body. Blood circulation as well as body fluid distribution undergoes sig-nificant adaptation during in-flight and post-flight recovery periods. Much study of physiological changes under weightlessness has been performed since the early days of the US space program. In particular, cardiovascular research in conjunction with the Space Shuttle program has included diverse physiological functions affected by the nervous system, including heart rate, blood pressure, hormone release, and respiration. The altered cardiac output, due to deconditioning in space and readjust-ment on Earth, does impact the blood circulation in the human body. In particular, the altered blood supply to the brain and consequent oxygen delivery to certain parts of the brain makes non-negligible impact on the health and safety of astro-nauts. Thus, it is essential to understand what happens to the arterial wall mechanics and resulting blood flow patterns under various gravitational forces. Human per-formance under altered gravity is far more involved a process than altered blood circulation. However, in this section, altered blood flow under different gravitational conditions is discussed using the same carotid bifurcation model as reviewed above.

The effect of wall distensibility under different gravity conditions is illustrated next (Figs. 8.12 and 8.13). Three different cases are computed, namely, under microgravity (or approximately supine posture under normal gravity), standing, and hand-standing conditions.

For the standing case, the normal one-G gravity force (1G) is applied downward. The consequent arterial contraction leads to an increase in the magnitude of flow velocity to maintain the constant flow rate. For the hand-standing case, the gravity force is applied toward the head, that is, negative G (or –1G.). The arterial dilatation results in less flow separation and flow recirculation than observed for the 1G cases. Figure 8.11 shows temporal wall shear stress distribution at a systolic decelera-tion phase under three different gravity conditions. Compared with the microgravity case, the 1G case shows about a 6.2% decrease in the CCA diameter, whereas the –1G case shows about a 7.2% increase in the CCA diameter. The reverse flow zone in the ICA sinus becomes narrower as the diameter increases. Throughout the pulse cycle, it was observed that gravitational variation has a significant influence on the arterial deformation and on the resulting changes in velocity profile and wall shear stress distribution.

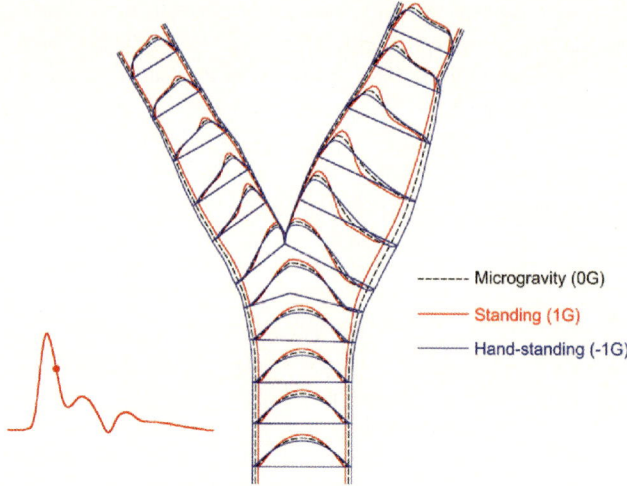

Fig. 8.12 Effects of gravitation on wall motion and axial velocity profiles at the systolic deceleration phase, as indicated with a *dot* on the pulse cycle

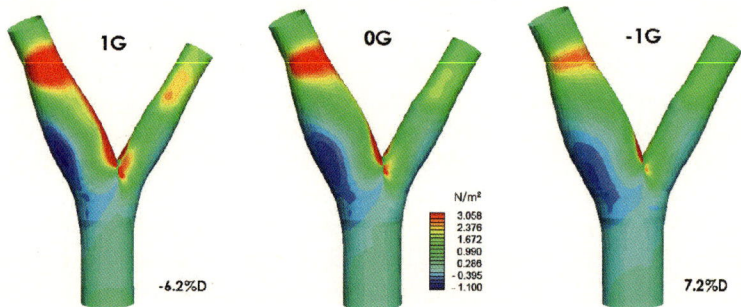

Fig. 8.13 Gravitational effect on wall shear stress distribution during the systolic deceleration phase. Percent change of diameter with respect to microgravity case is indicated

8.4 Blood Circulation in the Human Brain

Blood circulation in the human brain directly affects the physical ability of an individual, and, therefore, circulation simulations of the brain are of significant interest to performance prediction under altered gravity. From a biomedical point of view, blood flow in the brain can be directly relevant to explaining the impact of stroke, which is one of the major causes of death and long-term disability. We discuss a computational procedure to illustrate a potential application of an incompressible flow simulation approach as applied to brain circulation. This can be regarded as a building-block for simulating the entire circulatory system for the purpose of predicting human performance related to space flight.

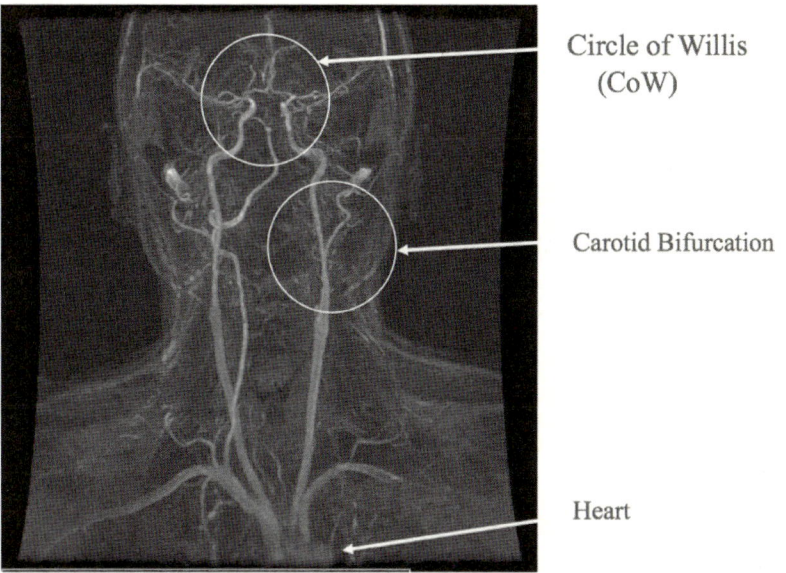

Circle of Willis
(CoW)

Carotid Bifurcation

Heart

Fig. 8.14 Circle of Willis (CoW) in the brain; the arterial network is obtained from a clinical MRI, which shows the blood vessel from the aorta to the CoW. Source: Tim David, U. Canterbury, NZ

In Fig. 8.14, a clinical MRI shows large arteries extending from heart to brain. Anatomically, two arterial pairs supply blood from the heart to the brain. One pair is the internal carotid arteries (ICA) and the other is the vertebral arteries. Some of the arteries are shown in Fig. 8.15 schematically in an idealized configuration of the human brain. The vertebral arteries are distally combined into the basilar artery that ends by dividing into the two posterior cerebral arteries (PCAs). The left and right ICA and the basilar artery are connected to an important part of the brain, the so-called Circle of Wills (CoW). The CoW sits at the base of the brain, and its main function is to distribute blood evenly throughout the brain.

8.4.1 Collateral Circulation Under Auto-Regulation

To provide a fundamental understanding of the mechanism of collateral circulation under auto-regulation, an idealized CoW configuration was designed based on anatomical measurements (Alpers et al., 1959; Gray, 2000) with minor arteries truncated. The inset in Fig. 8.15 shows a Chimera overset grid with ten domains for this idealized configuration, which results in a total of 0.3 million-grid points. An overset grid approach was used in this simulation, mainly to use the existing flow solver. However, other grid topologies like an unstructured grid would be very effective for this type of geometry. Both the ICA and the basilar artery have the same inflow rate of 3.5 ml/s for this configuration. The Reynolds number based on the ICA diameter is 240.

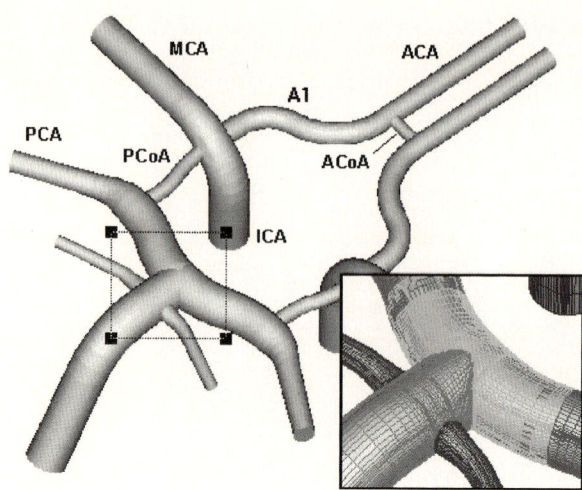

Fig. 8.15 Chimera overset grid with 10 domains for an idealized Circle of Willis configuration. ACA: Anterior Cerebral Artery, ACoA: Anterior Communicating Artery, ICA: Internal Carotid Artery, MCA: Middle Cerebral Artery, PCA: Posterior Cerebral Artery, PCoA: Posterior Communicating Artery

Even though the circle is completely connected in the ideal CoW, approximately 50% of the population has an incomplete circle. Anatomical variations in the communicating arteries are common, such as a missing or enlarged PCoA or missing A1 segment. When one or more of the main arteries in the brain is stenosed or even missing, the distal smaller arteries can receive blood from the other arteries through the CoW. To simulate this interesting mechanism of "collateral circulation" under auto-regulation, the left ICA is presumed 20% stenosed. This means that only 80% of the normal supply of blood is delivered to the CoW through the left ICA, as shown in Fig. 8.16.

Unlike the balanced configuration case, the mass flux through the posterior communicating arteries (PCoA) and anterior communicating artery (ACoA) is considerably increased to compensate for the deficiency in the left middle cerebral artery (MCA). On the other hand, the mass flux through the proximal part (A1 segment) of the left anterior cerebral artery (ACA) is decreased by 26% in order to distribute the blood as evenly as possible.

In Fig. 8.17, the time-dependent auto-regulatory process is shown using the AAR algorithm given in Equation (8.8). The ratio of the reference flow rates among the MCA, PCA, and ACA were set to be 6:4:3 to maintain a negligible mass flux through the PCoA.

From the computational experiment, it is observed that this AAR algorithm is robust and consistent for a wide range of physical time steps ($dt = 0.05T$ to $0.25T$). The optimal value for α was chosen to be 8.0 so that the left MCA and ACA have regained their initial (or reference) flow rates within about 10 s after a sudden stenosis in the left ICA. The present simulation illustrates how collateral circulation in the brain occurs under auto-regulation.

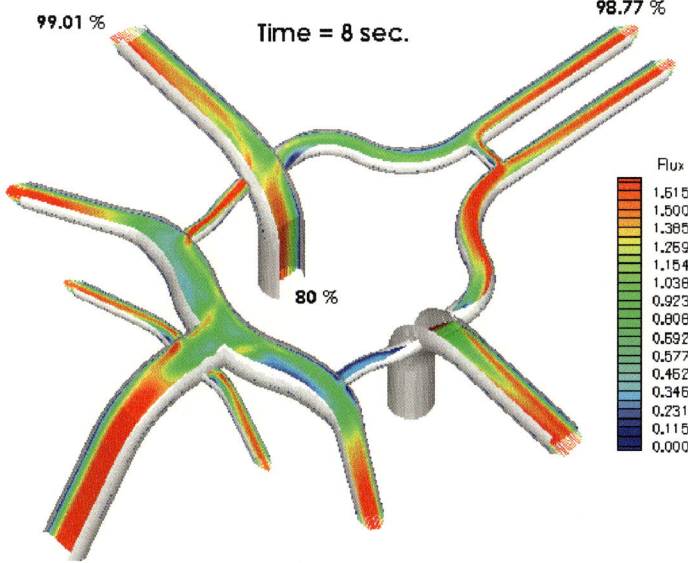

Fig. 8.16 Collateral circulation with the left internal carotid artery 20% stenosed

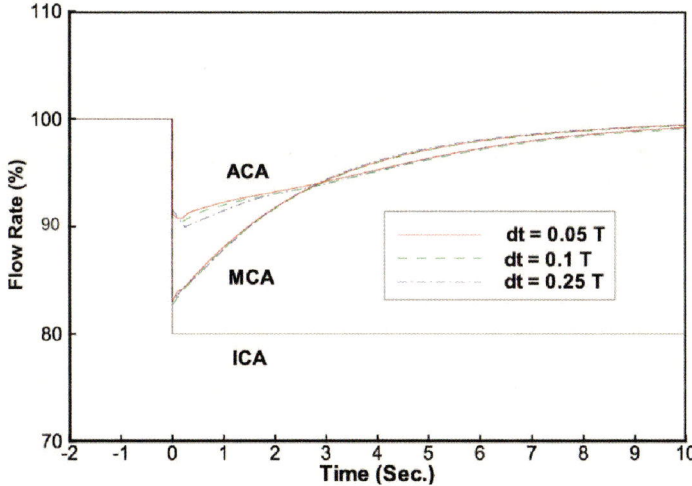

Fig. 8.17 Percent changes of flow rate in left, middle, and anterior cerebral arteries under auto-regulation

8.4.2 *Extraction of Geometry Data from Anatomical Picture*

An anatomically realistic CoW geometry was reconstructed three-dimensionally from human-specific magnetic resonance angiography (MRA) using image segmentation techniques, as illustrated in Fig. 8.18. First, the raw MRA images were

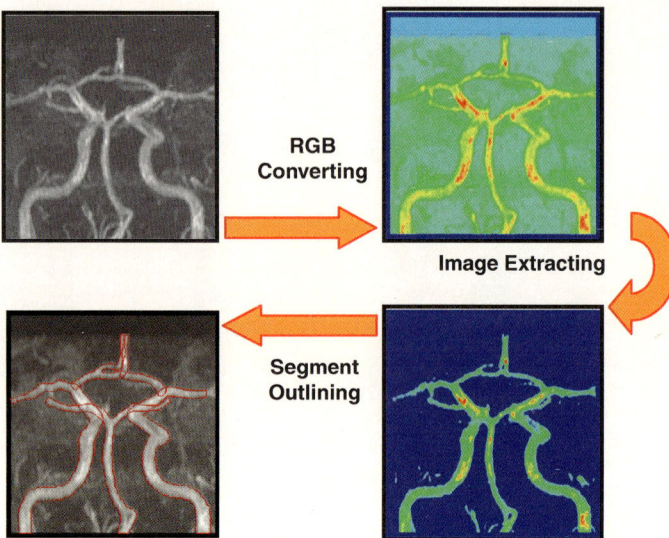

Fig. 8.18 Image segmentation from a magnetic resonance image for a human-specific Circle of Willis (MRA provided by John Fink and Mike Hurrell at Christchurch Hospital, New Zealand)

converted to the RGB graphic file format for efficient numerical treatment. After extracting the segments of interest by filtering the voxels with intensities below a certain threshold, a segment-outlining algorithm was used to display the extracted objects on each sectional layer, using very little computer memory. Then a three-dimensional CoW was reconstructed, as shown in Fig. 8.19. This process illustrates just one way of digitizing an image such as those obtained from MRA. Since imaging technology advances rapidly, more advanced methods will become available in the future.

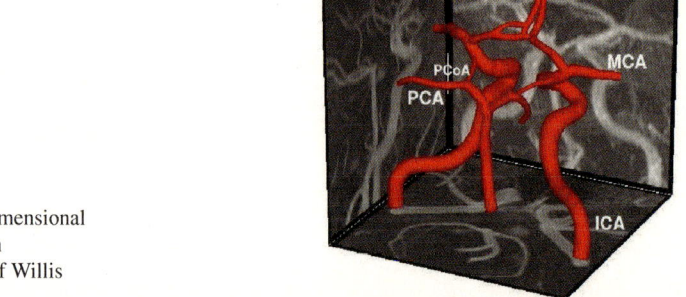

Fig. 8.19 Three-dimensional reconstruction of an anatomical Circle of Willis configuration

8.4.3 Effects of Gravitational Variations

To study the impacts of gravitational variations on blood circulation in the brain, a computational model of a CoW is first constructed from a clinical image. Various off-the shelf grid generators can be used to generate a grid compatible with a particular flow solver. In this example, a Chimera overset grid system with 31 domains was generated. Even though both left and right PCoAs are invisible in the original MRA because of very low blood flux, they were added to the grid system by artificially connecting the MCA and PCA in order to obtain a complete Circle of Willis configuration. Spatial dimensions are normalized by an ICA diameter of 5.6 mm. The total grid size was approximately 1.2 million node points. The mean flow rates in the ICA and the basilar artery are 3.5 and 2.1 ml/s, respectively. The Reynolds number based on the ICA diameter is 240. At the outflow boundary, the auto-regulation algorithm given by Equation (8.8) was used.

Computations for microgravity (0G), standing (1G), and hand-standing ($-$1G) postures were performed to demonstrate the effects of gravitational variation on the brain circulation. Figure 8.20 shows the time-averaged velocity magnitude and flow direction through this subject-specific CoW model under microgravity, standing (1G), and hand-standing ($-$1G) postures, respectively. Despite changes in perfusion pressure to a certain extent, the auto-regulation mechanism maintains nearly a constant blood supply to the brain. Therefore, inflow rates through the left and right ICA and the basilar artery were assumed the same under three different gravity conditions. For distensible wall motion, local changes of the wall distensibility factor, D_w in Equation (8.5) were given by introducing different ratios of vessel radius and wall thickness to each arterial component. Based on the experimental measurement of Giller et al. (1993), large cerebral arteries are assumed to have about two to four times smaller D_w compared with smaller arteries such as the ACA. Compared with the ACA, the wall distensibility factors of the ICA, basilar artery, and MCA are given by 3.5, 2.5, and 2 times smaller values, respectively. Wall distensibility at the arterial conjunctions was neglected to avoid unrealistic rigid body motion.

In the standing posture (1G), due to height difference, the static pressure at the exit boundary in the ACA is lower than that of the heart level by about 30 mmHg. Compared with the microgravity case, standing posture leads to about a 16.3% increase in ACA diameter, 6.9% increase in the MCA diameter, and 10% increase in the PCA diameter (Fig. 8.20). Due to the arterial contraction, the magnitude of flow velocity in each artery should be increased to supply constant blood flow rates toward the major parts of the brain. The AAR algorithm was used to simulate the auto-regulatory blood flow through this realistic CoW model.

On the other hand, compared with the microgravity case, the increased pressure in the hand-standing posture ($-$1G) causes about a 15% increase in the ACA diameter, 6.9% increase in the MCA diameter, and 9.6% increase in the PCA diameter (Fig. 8.20). With the arterial dilatation, the magnitude of flow velocity in the hand-standing posture was decreased to maintain the same flow rates with the microgravity and standing-posture cases. Similarly, wall shear stress distribution under micro- and normal gravity conditions can be compared. Overall, the

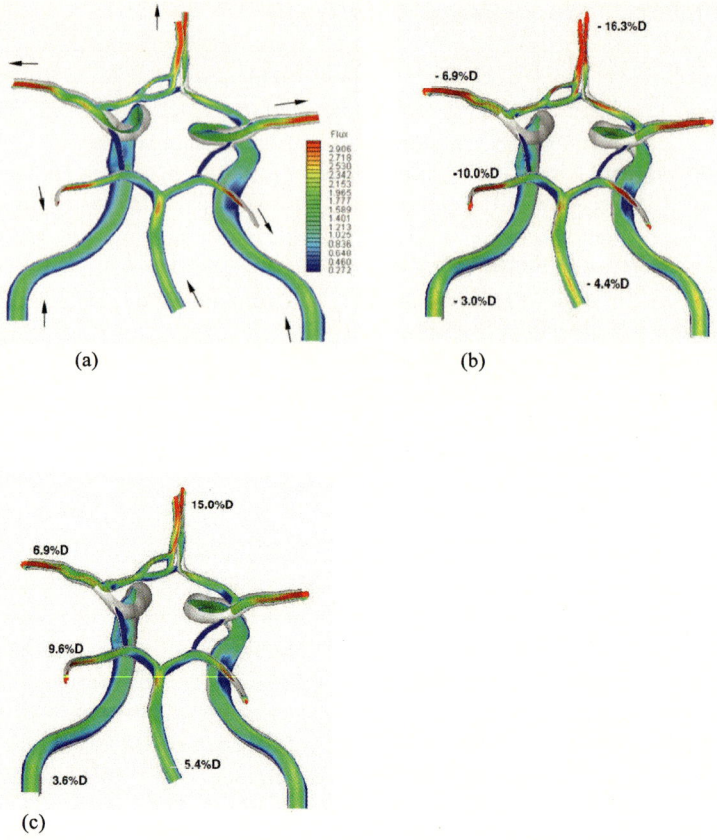

Fig. 8.20 Time-averaged blood flow within compliant walls; changes in vessel diameters are indicated in the figure: (**a**) microgravity (0G); (**b**) standing posture (1G); (**c**) hand-standing posture (−1G)

standing posture under normal gravity (1G) leads to high wall shear stress values in the smaller arteries (that is, ACA and PCA) with large wall distensibility. However, both the left and right PCoAs have very low wall shear stress distribution because of the negligible flow rate. As already observed in the carotid artery model, it is reconfirmed that the altered gravity has considerable effects on the arterial deformation and consequent flow patterns.

This study illustrates the possibility of using numerical simulation to analyze local blood flow in detail. Combined with an arterial network model, one can probe local flow phenomena to unravel various performance issues. Even though some basic building-block steps are explained here, a tremendous amount of research still lies ahead to develop human performance models during long-duration space flight, possibly including astronaut-specific anatomical data. It is hoped that this type of simulation capability can provide a significant part of the so-called "Digital Astronaut" model in the future.

8.5 Simulations of Blood Flow in Mechanical Devices

Simulation of blood flow in the presence of artificial devices, such as artificial heart valves and ventricular assist devices, poses unique challenges where boundary conditions for resolving flow through these devices have been defined from blood flow in natural hearts and arteries. Many computational technology features developed for aerospace applications can be extended to device modeling. However, characterization of blood flow involving artificial devices poses different challenges. For example, artificial devices create turbulence that in turn affects hemolysis (red blood cell damage), and blood exposed to an artificial surface for a certain period of time can have damaging effects on blood cells.

Multi-scale issues can be present, since devices can be operated at much different scales both in frequency and amplitude from natural organs. Also, the moving geometry needs to be defined: it can be either prescribed or determined depending on the force balance between mechanical energy input and physiological input generated from natural blood flow conditions. In general, however, from a device development point of view, local blood flow simulation in and around mechanical devices is better defined in geometry and flow conditions. Overall impacts on the body circulation with and without a device will require the entire circulatory system simulation coupled to the specific device under consideration.

Several different types of mechanical devices have been developed to date. Some of the most frequently used devices are discussed next from the point of view of how flow simulation can contribute to the development of those devices.

8.5.1 Artificial Heart Valves

Artificial heart valves have been widely used since the early 1960s to replace or to assist natural ones. However, prosthetic devices are generally less efficient than natural organs and are accompanied by various problems. Serious problems are often due to modified a flow field created by these artificial devices. Major problems related to fluid dynamics include: (1) thrombus formation due to secondary and recirculation regions; (2) hemolysis or red cell damage due to high shear regions created by these mechanical devices; and (3) a large pressure drop across these devices causing additional load on the heart.

Several early experimental studies on commonly used valves, for example by Yoganathan et al. (1979a, b), pointed out that stagnation and recirculation regions associated with these devices create adverse effects on blood flow. Although the experimental studies played an important role in designing mechanical devices, they can provide flow characteristics for only limited regions of the flow field. In addition, accurate experimental measurements are difficult because of the moving boundaries involved in using these devices. To complement this, quantifying the flow field through numerical simulations will provide design engineers with significant insight into the characteristics of the flow field involving these devices. At the same time, a numerical approach offers valuable opportunity to optimize design configurations, thereby reducing the number of clinical tests.

Computational studies of blood flow through hearts and heart valves have been performed for many decades. Notable work has been done by Peskin and his co-workers (for example, Peskin, 1982, Peskin and McQueen, 1980, 1989, and McCracken and Peskin, 1980). Their main focus was on the natural heart and heart valves, combining Eulerian flow equations and the Lagrangian description of walls and valves. Readers interested in natural hearts and valves are referred to their work.

Many different types of artificial valves have been developed over many years. In this section, blood flow problems associated with mechanical valves are discussed, where body motion is prescribed. To describe the computational procedure, a tilting disk valve is selected here. There are numerous previous computations on disk-type valves. For example, Idelsohn et al. (1985) modeled the flow through the Kay-Shiley caged-disk, Starr-Edwards caged-ball, and Bjork-Shiley tilting disk valves, and compared their performances. Turbulent flow through tri-leaflet aortic heart valves was simulated by Stevensen et al. (1985). Many early numerical studies prior to the mid-1990s neglected valve opening and closing. This is probably because the simulation technology involving unsteady flow and moving boundaries was not mature enough at that time, and the computer processor speed was slow for the unsteady simulation. In addition, the computer processor speed was too low to simulate the unsteady flow with moving bodies in relative motion.

To illustrate the computational procedure, the Bjork-Shiley tilting-disk heart valve is computed next. The tilting disk is placed in front of the sinus region (for example) of the human aorta, as shown in Fig. 8.21. The aortic root has three sinuses about 120° apart. The tilting-disk valve model used in this computation is simplified by assuming that the sinus region of the aorta has a circular cross section. The cage and strut holding the free-floating disk inside the sewing ring are not included in the geometry. It is also assumed that the walls do not have an elastic deformation. The channel length is taken to be five aorta diameters long.

In Fig. 8.21a, three different positions of a tilting disk valve are shown from a computational model of a valve similar to the one used in the Pennsylvania State University artificial heart. The computational procedure of a piston type device is illustrated using the Penn State heart later in this section. In Fig. 8.21b an overlapping grid arrangement is shown to deal with bodies of relative motion. The main grid covers the entire aorta from entrance to exit and contains 17,199 points, which are distributed as $63 \times 21 \times 13$ in the stream-wise (ξ), circumferential (η), and radial (ξ) directions, respectively. On this stationary main grid, a secondary

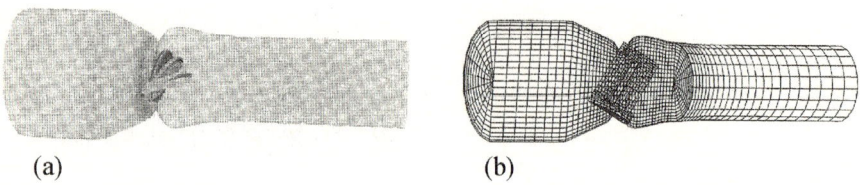

(a)   (b)

Fig. 8.21 Tilting disk valve: (**a**) geometry showing three positions of valve motion; (**b**) illustration of overset grid arrangement

grid consisting of 4,725 points distributed as $25 \times 21 \times 9$ points in three directions
is overlaid, which wraps around the tilting disk and moves with the disk. In the
Chimera grid-embedding technique, grid points that lie within the disk geometry
and outside the channel grid are excluded from the solution process. These excluded
points are called "hole points," and the immediate neighbors of the hole points
are called "fringe points." The information is passed from one grid to another via
fringe and grid boundary points by interpolating the dependent variables. A tri-linear
interpolation scheme is used in the current example.

Computationally, the hole and fringe points are differentiated from regular points
using an IBLANK array in the flow solver. For hole, grid boundary, and fringe points
the IBLANK is set to zero, otherwise it is set to one. In order to exclude the hole and
grid boundary points from the solution procedure, the coefficients of the system of
algebraic equations and the right-hand terms are multiplied by the IBLANK value.
If the grid point is a hole, an outer boundary, or a fringe point, then the value of
(1-IBLANK) is added to the main diagonal of the matrix equation.

Results of the steady flow computations using the overset grid topology described
above are presented next. The computational model is non-dimensionalized using
the entrance diameter as the reference length and the average inflow velocity as
the unit velocity. The inflow and outflow boundaries are truncated shorter than the
experimental studies in order to minimize computational and memory requirements.
In addition, the exact shape of the sinus region of the aorta geometry was approximated since the exact geometry data used in the experiments was not available.
These differences between the experiment and the computational model can add
some uncertainties in comparing the computed results with the measurements.

Steady-state computations are performed for the 30° disk orientation for
Reynolds numbers in the range of 2,000–6,000, for which experimental data are
available. A mixing-length model designed for internal flow by Chang and Kwak
(1988a) is used. In Fig. 8.22, the pressure drop across the disk valve is plotted

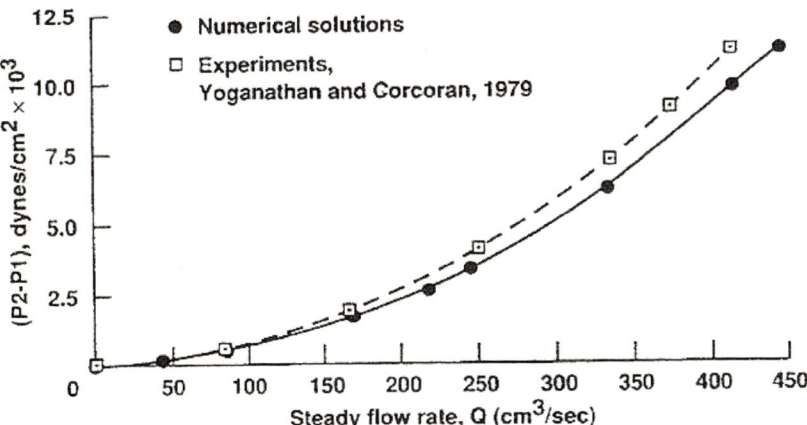

Fig. 8.22 Pressure drop across the tilting disk valve at 30° disk orientation versus steady-state
flow rate

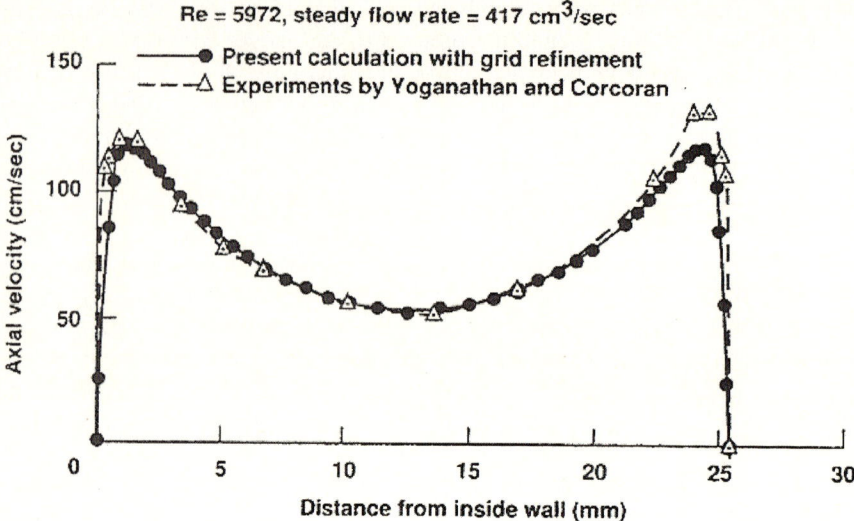

Fig. 8.23 Comparison of computed and measured axial velocity profiles at 42 mm downstream from the disk in horizontal plane

against flow rates of various physiological interests. Then the computed and measured axial velocity profiles at 42 mm downstream from the disk are compared, as shown in Fig. 8.23. Computations overestimated boundary layer development compared to experiments.

In Fig. 8.24, general characteristics of the flow field are visualized, where it is shown that the flow is directed toward the upper part of the aorta generates vortices in the sinus region of the aorta, and thus generates a large separated region along the lower wall of the aorta. Since separated and low-flow regions have potential for thrombus formation, clotting may occur on the upper sinus region and the lower wall of the aorta. Figriola and Mueller (1981) also presented mean velocity profiles, which show flow characteristics similar to those indicated in the present computations, at several locations. Particle traces in Fig. 8.24b indicate that the flow does not separate adjacent to the tilting disk. The tilting disk divides the flow into a major

(a)   **(b)**

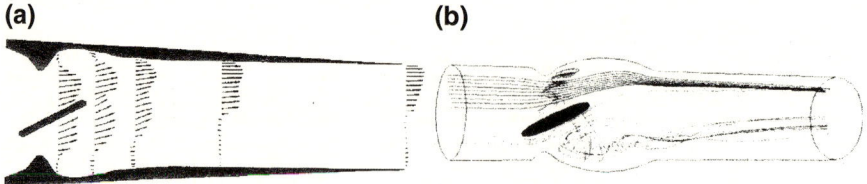

Fig. 8.24 Visualization of computed results for the tilting disk valve at 30° orientation: (**a**) velocity vector (**b**) particle traces showing major and minor flow regions

flow region along the upper wall of the tube and into a minor flow region along the lower wall. As a result, separation, reverse flow, and swirling motion mostly occur in the minor flow region. This validation computation gives the basis for applying the same procedure for simulating the piston-type artificial heart, which includes similar valves.

8.5.2 Ventricular Assist Devices

Approximately 20 million people worldwide suffer annually from congestive heart failure (CHF), a quarter of them in America alone. In the United States, an alarmingly low 2,000–2,500 donor hearts are available each year (c. year 2000). One potential approach to improve this situation is to use a mechanical device to boost or create blood flow in patients suffering from hemodynamic deterioration; that is, loss of blood pressure and lowered cardiac output. The goal of this device can be to replace the natural heart with a total artificial heart, or to assist an ailing heart. In either approach, the device can be used to bridge the gap while waiting for a matching donor heart for transplantation. However, to ease the shortage of donor hearts, making these devices suitable for long-term or permanent use would be an ultimate goal.

Another benefit of assist devices is the potential for providing time for the natural heart to recover. In some patients, it has been observed that the natural heart can recover by unloading the pumping requirement through the use of a VAD. In what conditions this might happen is not very well quantified at this time and should involve the physiological particulars of patients, among other factors. The challenge is to design a device that can deliver the required blood circulation while not adversely impacting human physiological conditions. Since the computational aspects for developing either a total artificial heart or an assist device are largely the same, mechanical blood pumping devices are represented generically by a VAD in this chapter.

The requirements of a VAD related to fluid dynamics are demanding, for example: simplicity and reliability; small size for ease of implantation; pumping capacity to supply 5 l/min of blood against 100 mmHg pressure; high pumping efficiency to minimize power requirements; and minimum hemolysis and thrombus formation. In addition to fluid dynamics issues, many other important aspects must be taken care of such as material compatibility to humans, controls, and implantation procedures. Due to the complexity of flow physics and delicate operating conditions, an empirical approach to quantify the flow phenomena in a VAD is not straightforward, very time consuming, and expensive—especially to study many design variations. Herein lies the potential utility of computational simulation tools for the development of these devices. In this sub-section, the discussion is focused on how fluid dynamics issues of the VAD can be resolved via a computational approach.

Computational flow analysis of blood flow through mechanical devices such as VADs and artificial valves is very challenging (Kiris et al., 1991). Flow is unsteady and involves moving parts. For a complete analysis of a VAD, human circulatory

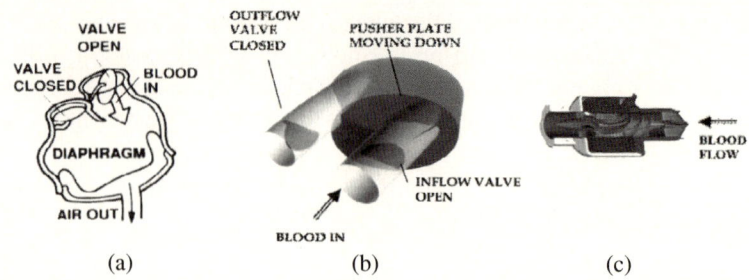

Fig. 8.25 Schematic of mechanical heart assist devices: (**a**) sketch of a pulsatile device with diaphragm driven by compressed air; (**b**) piston type; (**c**) schematic of an axial flow pump

system simulations have to be coupled to the device in use. However, for the purpose of developing mechanical components, a truncated circulation system can be modeled. For example, empirical inflow condition can be specified at the inlet of a VAD. Even with this type of simplification, the computational approach can produce flow field data in great detail, to help obtain a better understanding of the dominant flow physics produced by an artificial device. In particular, computational analysis can be utilized to optimize the design of mechanical devices at a significantly reduced cost and timeframe than that required by an empirical approach. In addition to the geometric and operational complexities, these devices introduce a variety of flow phenomena that do not normally exist in a natural heart. These include transition, turbulence, boundary layer separation, rotational effects, tip vortex, and reverse flow phenomena. While a moving wall is one of the great challenges in natural circulatory system simulation, unnatural flow phenomena are added challenges for simulation created by insertion of artificial devices in the natural system.

There are largely two types of VADs commonly used to date. As sketched in Fig. 8.25, these are (1) pulsatile devices, and (2) axial flow pumps. One example from each category will be discussed in some detail to illustrate the computational procedure in general. However, readers can apply a similar approach to meet their specific design and/or analysis task objectives. Since the computational tools are derived from those developed under aerospace vehicle and propulsion systems development, tools used for VAD development are extensions of procedures discussed earlier in this monograph. In the next two sections, we discuss simulation procedures representing a pulsatile and axial flow pump for the Penn State heart and De-Bakey VAD.

8.5.2.1 Pulsatile Devices

Some of the earlier models of artificial hearts and assist devices have used a pulsatile mechanism supposedly mimicking the pumping action by a natural heart. However, blood pumped by a heart was later found to be more of a wringing action than piston-type pumping. The success rate of these types of devices has been low, so understanding the flow phenomena will be of significant value toward finding improvements. Here, we attempt to explore how numerical simulation can be used as one way to expedite development and improvement of these devices.

Background

Since the inception of these types of devices, several problems have been found related to fluid dynamics of the blood flow created by the pulsatile pumping mechanism. Research and development efforts were made to improve the flow quality. In addition, development of biocompatible material and control systems is essential to manage blood damage at a low enough level for human applications.

Experimental investigations on these devices are difficult and limited and many fluid dynamic aspects are yet to be studied. Blood flow simulation through these devices is very complicated in many respects, especially when a full operational geometry is to be modeled. The fluid may exhibit significant non-Newtonian characteristics locally. In these devices, the flow can be highly chaotic or become turbulent, thus red blood cells may be damaged as they go through high-shear or turbulent flow regions; the downstream region of an artificial valve is an example of this. The flow is unsteady, possibly periodic, and very viscous and incompressible. This problem is very much interdisciplinary and an attempt for a complete simulation would be a very formidable task. However, an analysis based on a simplified model may provide much-needed physical insight into the mechanics of the blood flow through these devices.

The formulation of the flow solvers described earlier can be applied to analyze pulsatile as well as other types of ventricular assist devices. A full simulation of viscoelastic flow is very difficult because of the nonlinearities of the fluid. However, as a first step toward full simulations, non-Newtonian effects of the blood flow can be simplified by a constitutive model for the viscous stresses, or by the non-Newtonian model discussed earlier in this chapter.

There are also variations in pumping mechanism. In one approach, the blood in a diaphragm is pumped by pressurized air supplied externally. In Fig. 8.25, a VAD arrangement is sketched where different types of devices can perform the blood-pumping function. Simulation of a pneumatically driven diaphragm requires a moving boundary procedure where the boundary itself should be determined by the motion of two fluids. In another approach, a piston-type pusher plate pumps the blood. One developed at Penn State utilizes an electric motor-drive pusher plate. This design can be used as an assist device or, when two chambers are combined, can form a total replacement heart. The present demonstration calculation is performed on one chamber of the Penn State artificial heart. The geometry of a computer model and grid topology chosen is described next.

Geometry and Grid of the Penn State Artificial Heart

The model of the Penn State artificial heart in its entirety, including the blood, sack poses very difficult problems from a computational standpoint. For the present computation, a simplified model is constructed that excludes the internal blood sack. In

Fig. 8.26 Overset computational grid arrangement for the Penn State artificial heart model, showing zonal and overlapped grid regions

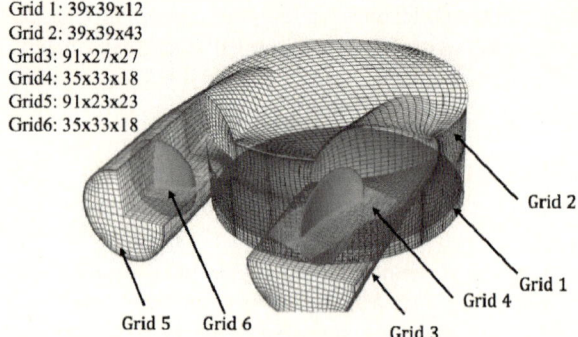

Grid 1: 39x39x12
Grid 2: 39x39x43
Grid3: 91x27x27
Grid4: 35x33x18
Grid5: 91x23x23
Grid6: 35x33x18

Fig. 8.26, surface grids of the chamber, valve region and inflow and outflow tubes are illustrated. The heart is composed of a cylindrical chamber with two openings on the side for valves. The pumping action is provided by a cylindrical piston that moves up and down inside the chamber. The actual heart device has a cylindrical tube extending out of each of the valve openings. The inflow (mitral) and outflow (aortic) tubes contain concave tilting disks that open and close to act as valves. As the piston reaches its bottom-most position, the outflow valve will open and the inflow valve will close. In the computational model, tilting-disk mitral valve rotational motion in time was obtained from the experimental data provided by Penn State. The aortic disk valve rotation in time was obtained from mitral valve rotation with a phase difference.

A pusher plate whose velocity is sinusoidal in time provides the pumping action in the model. The pusher plate diameter is 7.26 cm, and the stroke length is 2.28 cm. The problem is non-dimensionalized with the inflow tube diameter of 2.54 cm and a unit velocity of 20 cm/s. The Reynolds number is based on the unit length and velocity was 900. The computation was started from rest and requires about four cycles to reach a periodic flow. For each cycle, the pusher plate motion requires 240 time steps. At each time step, the artificial compressibility method requires 10–20 sub-iterations to drive the maximum divergence of the velocity down to at least two orders of magnitude (see, for example, Rogers et al., 1989).

In the computations, as the piston reaches its topmost position, the outflow valve closes and the inflow valve opens instantaneously. In the actual heart device, the piston moves through the entire chamber volume, which includes most of the valve openings. One can handle this motion in several ways. In the present computational example this problem is tackled through use of a Chimera grid scheme (see Benek et al., 1985). This approach offers the possibility of using an existing flow solver based on a structured-grid approach at the expense of introducing mesh-related "bookkeeping" complexity.

Some gridding details are given next. Readers interested in using different approaches, such as compressing grids, may wish to skip to the discussion on computed results.

In order to handle geometric complexity and the moving boundary problem, a zonal method and an overlapped grid-embedding scheme are employed in the present example. In the zonal method, the computational domain is divided into several simple sub-domains. The overlapped grid-embedding scheme allows sub-domains to move relative to each other, and provides great flexibility when boundary movement creates large displacements. In the Penn State heart model shown in Fig. 8.26, Grid 1 is generated for the pusher plate and moves with it, Grid 2 fills the chamber and remains stationary, Grid 3 and Grid 5 are for the inflow and outflow tube extensions, respectively, and Grid 4 and Grid 6 wrap around tilting disks and move with the disks. Grid points for the tubes and grid points for the chamber are overlapped on three common planes where patched zonal boundary conditions are used. At these interface boundaries, the grid points for the tubes start three stencils inside the chamber outer boundaries. In general, this overlapped-grid embedding scheme can be employed in a similar arrangement. To illustrate the level of grid densities, the breakdown of the grid points for each region are given here: Grid 1, $39 \times 39 \times 12 = 18,252$; Grid 2, $39 \times 39 \times 43 = 65,624$; Grid 3, $91 \times 27 \times 27 = 66,339$; Grid 4, $35 \times 33 \times 18 = 20,790$; Grid 5, $91 \times 23 \times 23 = 48,139$; Grid 6, $35 \times 33 \times 18 = 20,790$—which results in a total of 239,713 grid points. The grid sizes represent grid numbers in the ξ, η and ζ directions in generalized coordinates, respectively.

Computed Results

The computed results presented here are given just to illustrate one possibility of how CFD can be utilized for analyzing this type of device. Experiments are difficult and measured data are not readily available. Considering a wide spectrum of individual variations, for example, inflow and outflow conditions, even qualitative comparisons can shed some light on understanding flow characteristics.

Tarbell and his coworkers at Penn State (see, for example, Tarbell et al., 1986; Baldwin et al., 1989) performed extensive experimental investigations on their artificial heart. The present computations were performed using one of their models, and qualitatively compared in Figs. 8.27 and 8.28. At $t/T = 0.375$, the pusher plate is moving in a downward direction and the mitral valve is in the open position. Here "T" represents the time for one period. In both experiment and computation, a strong swirl motion is observed. However, since the velocity vectors are plotted in two-dimensional planes, three-dimensional structure of the flow in the chamber could not be reconstructed. In addition, the computational study does not include a blood sack inside the chamber, so the comparison between experimental and computational results does not have an exact one-to-one correspondence. As can be seen in the figures, the biggest discrepancy between the two results is the location of the vortex core. Another difference can be seen in the wake region of the mitral valve where the computed wake is not as strong as in the experiment. During the second half of the cycle, the pusher plate moves upward, and the outflow valve is opened. Here, since the inflow valve is closed, residual eddies are quite large near the disk. However, they are quickly weakened as the pressure builds up inside the chamber.

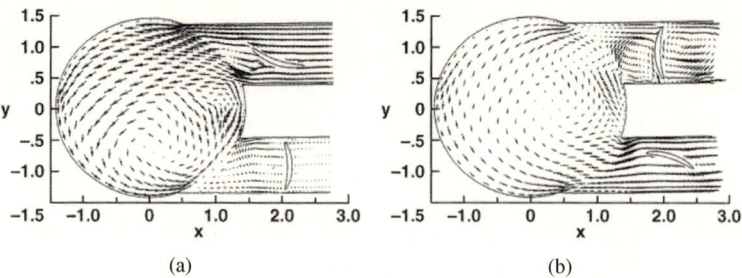

Fig. 8.27 Computed velocity vectors for the Penn State artificial heart chamber: top view in horizontal planes through the center of inflow and outflow tubes and at 3 mm below the top surface of the chamber: (**a**) at t/T = 0.375 (**b**) at t/T = 0.625

Visualizing the computed results is of significant interest for identifying the residence time of a portion of the flow due to recirculation. When red cells are in contact with artificial material over a certain threshhold of time, blood cells can be damaged. Locally turbulent regions can also develop due to the articial environment. When red cells are exposed to a high shear region for a certain period of time, they can also be damaged. Visualizing and quantifing the potential adverse effects will be valuable to developers of mechanical devices. Even though exact prediction of all flow variables is still a challenge, this information can be utilized during the conceptual and preliminary design phases. Once the design configuration is finalized, more inclusive geometry and physics can be modeled in conjunction with high-fideity simulations for fine-tuning the device. Figure 8.29 presents two visualization attempts made in conjunction with the current computational example. The threshold for the high shear region in the visualization is artificially defined to illustrate how computed results can be utilized to identify potential regions of blood damage.

In the present computations, the blood sack is excluded. When the pusher plate moves up, the excess material of the sack is squeezed along the peripheries of the chamber. Blood is not completely washed out from the sack and some small amount still remains in this wrinkled region, causing a potential problem of excess red cell

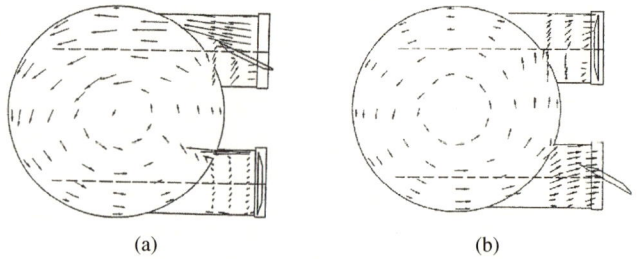

Fig. 8.28 Velocity vectors from experiments by Baldwin et al. (1989) corresponding to the computations in Fig. 8.27

Particle Trace Colored by Vorticity Magnitude

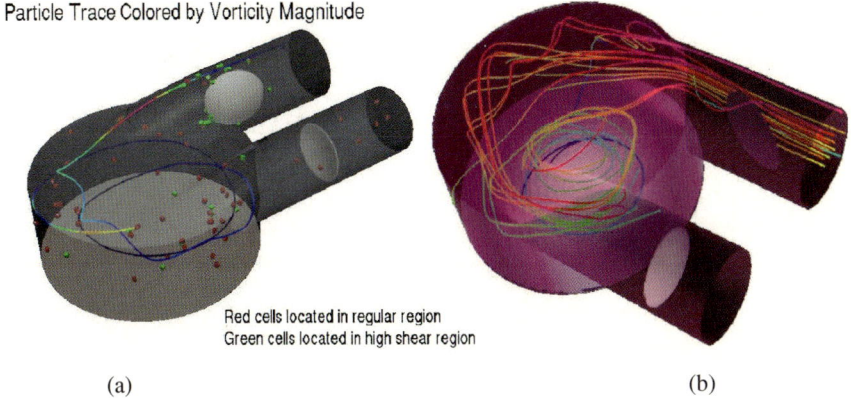

Red cells located in regular region
Green cells located in high shear region

(a)   (b)

Fig. 8.29 Visualization of the computed result at t/T = 0.45 as the pusher plate nears the bottom position: (**a**) particles are *colored* by local shear stress (**b**) 3-D flow visualized by particle traces *colored* by the magnitude of velocity

damage. For a more accurate assessment of this device, high-fidelity simulations that include the detailed geometry of the sack deformation during the pumping cycle are necessary.

8.5.2.2 Axial Flow Pump

One of the difficulties inherent in pulsatile devices comes from the large stroke volume required to pump 5 l/min. The maximum speed of a piston-type pusher plate is mechanically limited. This requires a chamber size big enough to pump the required blood volume. For an electrical, motor-driven VAD such as the Penn State artificial heart, the total volume of the VAD and electrical motor becomes fairly large, requiring a large space in the patient for implantation. For a diaphragm-type pulsatile device, the pumping is done using compressed air, requiring an external compressor that limits patient mobility.

One alternative is to pump the blood by a continuously operating axial flow pump. The conventional idea of mimicking the natural heart's intermittent pumping mechanism is drastically modified in this concept. One major advantage is that the size of the axial flow pump can be small, facilitating human implantation and giving patients mobility. Since the size can be made small, the speed of the pump must then be made high to achieve the required pumping volume. This, in turn, can create a high shear region that can damage red blood cells. Since red cells are resilient against shear stress, an assist device can function as long as the residence time blood cells exposed to the high shear region can be made short. These fluid dynamics features will next be discussed next, in conjunction with axial flow pump development.

Solution Procedure for Axial Flow Pump

As we have said, a small axial pump can be designed—however, to pump the blood volume required, the rotational speed has to be high. Since the shear stress imposed

Fig. 8.30 Schematic of a VAD based on an axial-flow pump

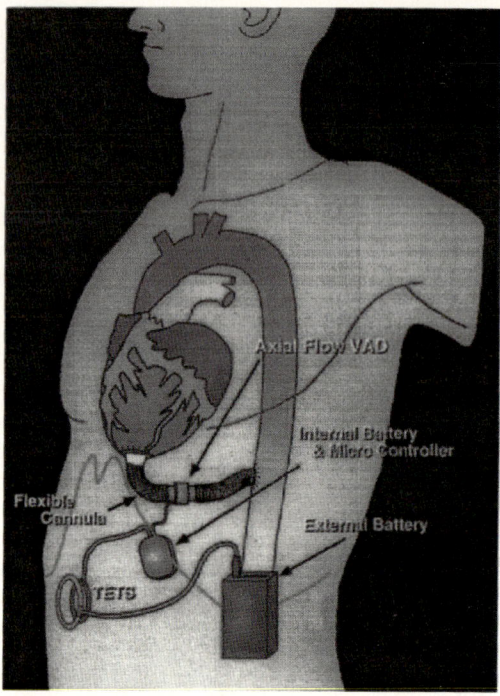

on blood through this device can be high, the residence time of red cells in the device has to be made very short to minimize the damage to blood cells. This is the reason underlying the usage of the axial flow pump as an assist device. An assist device of this type is placed either external to the heart as illustrated in Fig. 8.30 or within the heart, to draw blood from an ailing heart and then to pump blood into aorta.

Blood flow in large vessels, as in this arrangement, is believed to behave as nearly Newtonian fluid. The flow can thus be solved by the incompressible Navier-Stokes equations with a source term, S, added to the momentum equation, as below:

$$\frac{\partial u_i}{\partial t} + \frac{\partial u_i u_j}{\partial x_j} = -\frac{\partial p}{\partial x_i} + \frac{\partial \tau_{ij}}{\partial x_j} + S \qquad (2.2b)$$

When the equations are solved in a steady rotating frame of reference, the centrifugal and Coriolis forces are added as source terms to the governing equations. If the relative reference frame is moving around the x-axis, the source term, S, is given by:

$$S = \begin{bmatrix} 0 \\ 0 \\ \Omega(\Omega y + 2w) \\ \Omega(\Omega z - 2v) \end{bmatrix}$$

where Ω is the rotational speed. Relative velocity components are written in terms of the absolute velocity components ua, va, and wa as:

$$u = u_a$$
$$v = v_a + \Omega z$$
$$w = w_a - \Omega y$$

For component analysis, the rotational steady formulation may provide valuable approximate solutions relatively quickly, compared to full unsteady computations.

DeBakey VAD

To illustrate a step-by-step procedure for developing an axial-flow VAD utilizing simulation, we next discuss the DeBakey VAD development process. In 1989, the DeBakey Heart Center of Baylor College of Medicine (BCM) began developing a new implantable VAD system jointly with NASA Johnson Space Center (JSC) (Aber et al., 1993). This VAD is based on a fast rotating axial pump. In order to deliver the required blood flow rate, the rotational speed is in the range of a rocket pump's operating condition. For that reason, the CFD procedures presented earlier, in conjunction with the rocket pump simulation, were applied to this VAD development.

By probing through computed results, regions of critical design interest were identified, such as regions of high turbulent shear stress, which can damage the red blood cells, and regions of recirculation where blood clots may form. The levels of hemolysis and thrombus formation need to be maintained at a low level in developing mechanical devices. Therefore, the ability to predict local flow quantities could expedite the development of the device.

The baseline VAD impeller was initially designed to achieve the required pumping capacity. This baseline design was first analyzed by solving the incompressible Navier-Stokes equations in a steady rotating frame of reference. This was intended for quick analysis of axial impeller blade performance. For a complete unsteady analysis, rotational and stationary components had to be explicitly included. This required a large amount of computing time and quick turnaround time. For preliminary design studies, quick turnaround is preferred to identify major issues first. In the present example, a solver based on structured grids was applied, as discussed in Chapter 7. Zonal multiblock grids were generated in this component analysis.

As shown in Fig. 8.31, the computational domain is divided into five zones with grid dimensions of $127 \times 39 \times 33$, $127 \times 39 \times 33$, $59 \times 21 \times 7$, $47 \times 21 \times 5$, and $59 \times 21 \times 7$, for Zones 1 through 5, respectively. Zone 1 is the region between the suction side of the partial blade and the pressure side of the full blade; the region between the pressure side of the partial blade and the suction side of the full blade is filled by Zone 2; and Zones 3 through 5 allow tip-leakage effects to be included in the computational study and occupy the regions between the impeller blade tip and the casing. At the zonal interfaces, grid points were matched one-to-one. For all zones, an H-H type grid topology was used. An H-type surface

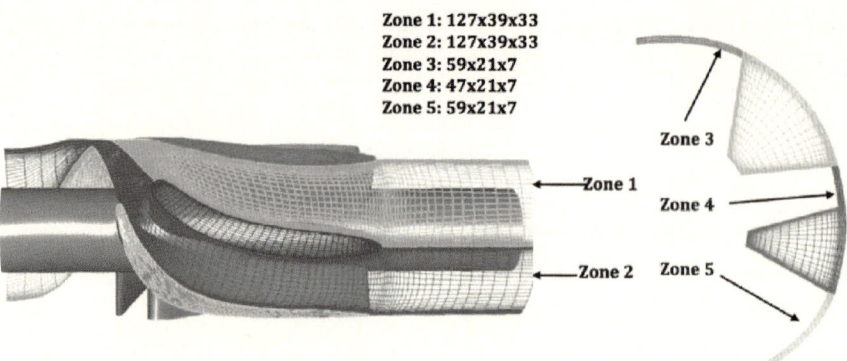

Zone 1: 127x39x33
Zone 2: 127x39x33
Zone 3: 59x21x7
Zone 4: 47x21x7
Zone 5: 59x21x7

Fig. 8.31 Computational grid for the baseline model of the DeBakey VAD: rotational speed = 12,600 rpm; flow rate = 5 l/min

grid was generated for each surface using an elliptic grid generator. The interior region of the three-dimensional grid was filled using an algebraic grid generator coupled with an elliptic smoother. Periodic boundary conditions were used at the end points in the rotational direction. The design flow of this impeller is 5 l/min and the design speed is 12,600 rpm. The problem was non-dimensionalized by the tube diameter (0.472 inches) and the impeller tip velocity. The solution was considered converged when the maximum residual had dropped at least five orders of magnitude.

For assessing the preliminary design concept, the impeller blade design was first optimized to obtain a reasonably performing pump design. For a quick validation of pump performance, a parametric study was done to optimize the impeller blade shape and tip clearance. In real-world application in humans, the inflow condition is the same as pulsatile cardiac output. Therefore, optimization using a mean design speed does not necessarily produce an optimum geometry for the entire cycle. However, since the blade geometry will be fixed while operating conditions might vary, this process is expected to produce the sought-after performance, on the average.

The baseline geometry shown in Fig. 8.31 was analyzed first with different tip clearances. However, a series of clinical tests using the baseline design showed that the performance of this configuration (both in blood cell damage level and the thrombus formation) with varying operating conditions could not be brought up to the level required for human implantation. This prompted design modifications to lower the hemolysis level and blood clotting.

A modified designed was developed that added an inducer similar to the ones used in high-speed rocket pumps. A new design consisting of the baseline impeller plus an inducer was then investigated (Fig. 8.32). Detailed flow features and pump performances are compared for the two designs. The pressure gradient across the blades, and the pressure rise from inflow to outflow were compared. The geometry was then optimized in conjunction with detailed flow analysis.

Fig. 8.32 Pressure surfaces of the baseline (*top*) and the new design (*bottom*)

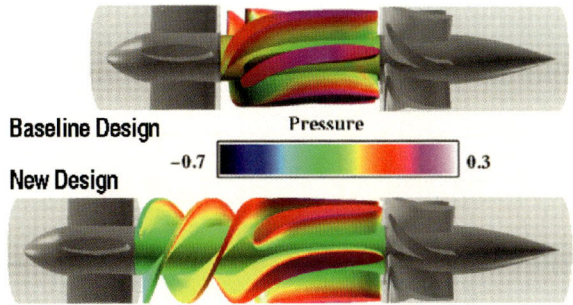

The new configuration was then optimized in several steps. The blade shape was been optimized first, using the following process. The original blade design is referred to as Design I. Design II has less blade curvature than Design I in the trailing edge region, and Design III has more blade curvature than Design I. In Design IV, the blade shape for Design I is maintained and the tip clearance is reduced. In Design V, the hub region has the blade shape for Design I and the tip region has the blade shape for Design II. In this design, the impeller blades have a backward lean near the trailing edge region. In Design VI, the blades have a forward lean, which includes Design III in the hub region and Design I in the tip region. Design VII has a small tip clearance gap with the Design I blade shape and an inducer added to the resulting configuration upstream of the main impeller blades. Not all of these variations are shown in graphical form here, since they are only intended to demonstrate what aspects are considered in changing the blade shapes. Automated techniques can be implemented in the shape optimization of the blades. However, the addition of an inducer—which has the most impact on performance—is not a part of shape optimization. Therefore, it is to be noted that the conceptual design modifications are not entirely an automated process.

In Fig. 8.33, hydrodynamic efficiency, defined as the ratio of real head over the Euler head, is shown for these design variations. The inducer addition (Design VII) clearly shows substantial improvement in hydrodynamic efficiency, and at the same time provides a sufficient pressure rise to suppress cavitation. Using a smaller tip clearance also improves hydrodynamic efficiency. On the other hand, higher blade curvature decreases efficiency due to the separation that occurred near the suction side of the trailing edge in the hub region.

Figure 8.34 shows the circumferentially averaged meridional velocity distribution along the blade height for various designs. All designs, except Design VII, showed backflow near the hub region. The backflow was reduced with a forward blade lean. In addition, the effects of a tapered hub and diffuser angle are combined to minimize the backflow.

As shown in Fig. 8.32, the non-dimensional pressure distributions are compared between the baseline design (Design I) and the new design with the inducer addition (Design VII). The pressure is non-dimensionalized by ρv^2 where v is the impeller tip speed.

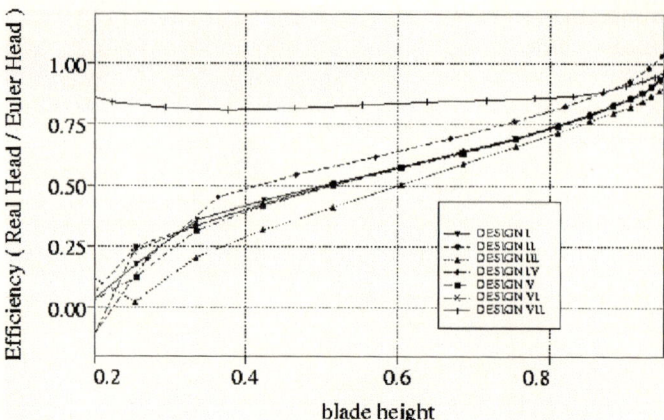

Fig. 8.33 Hydrodynamic efficiency distribution along impeller blade height of various designs

One of the major issues in developing mechanical devices is how to prevent thrombus formation or keep it at an acceptably low level. Therefore, besides improving the pumping efficiency, the design of the VAD requires good wall washing near the solid wall by reducing the stagnation regions. One of the critical regions for potential blood clotting is near the bearing area between the rotating and nonrotating components, as shown by a narrow gap in Fig. 8.35. Clotting can occur in the hub area due to either high shear rate or stagnation, depending on the gap and configuration of the area.

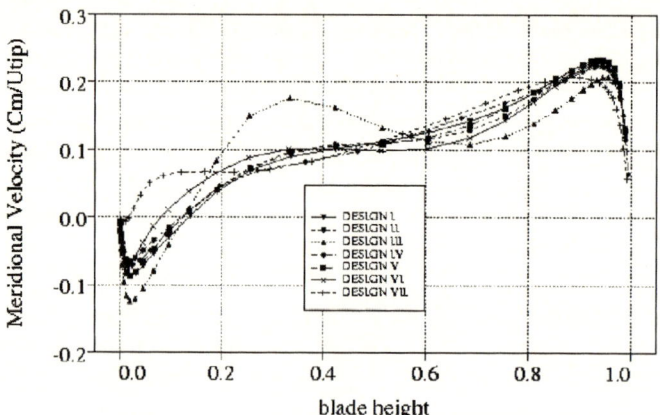

Fig. 8.34 Meridional velocity distribution along impeller blade height of various designs

Fig. 8.35 Velocity vectors inside the initial (*left*) and the final bearing geometry (*right*) of the DeBakey VAD

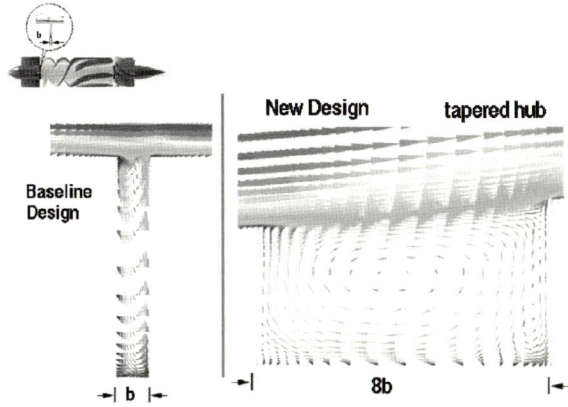

In Fig. 8.35, velocity vectors are colored by velocity magnitude for different bearing designs. Design 1 is the original baseline design with the cavity width of b. This design showed very high shear stresses near the rotating hub face and very stagnant fluid region in the lower portion of the cavity. Gradually increasing the cavity width up to 8×b increased circulation in the cavity substantially. In order to eliminate stagnant areas in the lower portion of the cavity, the hub surface was then tapered, which reduced the cavity height and accelerated the flow near the hub region. This resulted in stronger wall washing in the cavity. An optimized version of this configuration was adopted in the DeBakey VAD design, which has enabled human implantation.

Figure 8.36 shows a schematic of the final VAD design. Areas where CFD analysis contributed to improving the design are summarized in rectangular boxes. The original design goal was to develop a pump that can operate for 2 weeks so that a patient can survive while searching for a matching donor heart. The performance of the new design is substantially better than the baseline design in all aspects of hydrodynamic performance and anti-thrombogenic characteristics. Both the original and the final design configurations were tested in the laboratory at Baylor College of Medicine.

As shown in Table 8.1, red cell damage was reduced by an order of magnitude to an acceptable level for human implantation, and pump operation was not impacted by any thrombus formation. The hemolysis index reported here shows the amount of hemoglobin generated by the pump in grams per 100 l. Destruction of the red blood cells results in the release of hemoglobin. The new design shows a remarkable improvement in performance over the baseline design—a 22% increase in overall pumping efficiency. The original test run was successfully completed for 1-month operation. Overall, the performance of the new design was sufficiently improved, resulting in a device implantable in humans. The first human

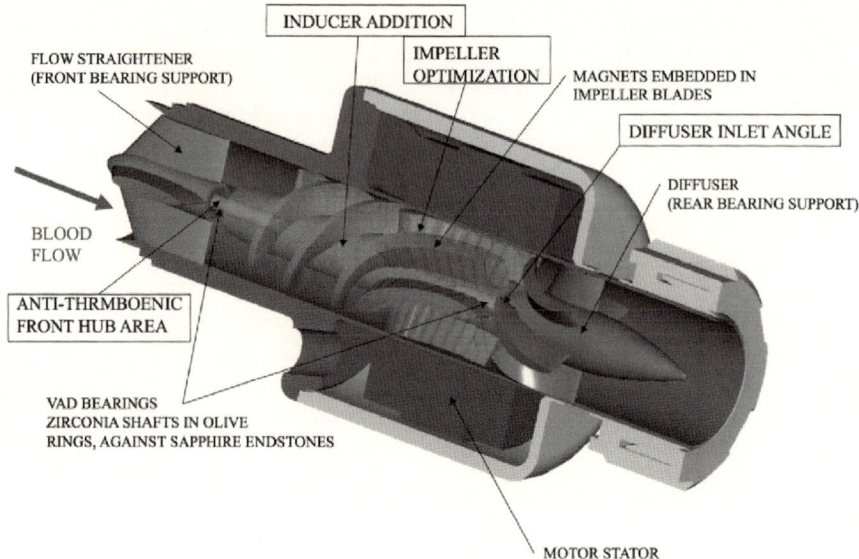

Fig. 8.36 Contribution of CFD analysis to the VAD design

implantation was performed successfully in 1998, followed by many successful implants to other patients since then.

Axial Flow Pump Vs. Pulsatile Device

After a number of human implantations, the longest period the DeBakey VAD has been used reached more than 1 year. This offers a possibility that the VAD can be used as a recovery device, as well as a bridge to transplant. One unresolved question is whether the modified blood pumping, that is, pulsatile vs. continuous pumping, has any significant long-term impacts on other organs such as the kidneys or brain. To date, mechanical assist devices are designed to help end-stage patients temporarily. However, to use these devices for long-term recovery without organ transplant, the impact due to modified pumping needs to be carefully analyzed. Earlier in this chapter, simulation methods for the human circulatory system were discussed as

Table 8.1 Performance comparison of the baseline and improved design of the DeBakey VAD (laboratory and clinical test data provided by R. Benkowski of MicroMed Technologies and Baylor College of Medicine)

Requirements	Baseline design	New design
Pumping efficiency	0.25	0.33
Power required (Watts)	12.60	9.80
Hemolysis index	0.02	0.00
Rotation (RPM)	12,600	10,800
Thrombus formation	Yes	No
Test run time	2 days	30+ days
Human implantation		12+ months

a step toward developing a Digital Astronaut model. A similar approach can be adopted here, this time modified by blood pumping variations instead of altered gravity. The impacts of altered circulation due to mechanical devices on human biomedical performance can then be assessed fully. The challenge for a complete hemodynamic simulation of the human body is enormous, and we have touched just the very beginnings of it.

Closing Remarks

Combining theoretical studies with incompressible flow solutions to solve real-world problems for NASA missions has been our greatest passion for over 20 years. Even more exciting are the fluid dynamics problems that are being solved right now, and those to come. For example, atmospheric physics within weather and climate simulation, oceanography, biological flow, astrophysics, fluid engineering, and basic fluid sciences can also be numerically studied using incompressible flow equations.

We began our discussion with the assumption that unique and valid solutions exist for viscous incompressible equations, i.e., incompressible Navier-Stokes equations. Our main interest was in fluid engineering, and, starting with reasonable initial and boundary conditions we assumed that valid solutions could be obtained. So, we focused on the ways in which we could quickly obtain solutions to fluid engineering problems.

Fluid dynamics of viscous incompressible fluid is mathematically rich. Numerical methods convert mathematical rigor into numerical solutions. We were fortunate to start our task in the early 1980s, at a time when some of the numerical methods were already well developed. Of course, the computer power and simulation technology were still primitive compared to what we see today. Even with those limited capabilities, the computational approach became a viable alternative to empirical approaches for engineering and science.

Since we are solving fluid mechanical problem using numerical methods, but not necessarily solving the governing equations in a strict sense, the question of how to utilize solutions in light of numerical approximations and physical modeling involved is of crucial importance for engineering/mission decisions. This is still true even with increased computational capability and more mature solvers. Accordingly, we discussed the entire process of how we could make use of CFD in solving mission tasks. Our discussions were more from a physics and engineering point of view than from following mathematical rigor—other than touching on salient features of the methods we used.

We have tried to make the material in the book self-contained, especially for those new to incompressible flow methods and simulation procedures, and hope we have presented a reasonably complete process of obtaining CFD solutions in engineering tasks through examples.

D. Kwak, C.C. Kiris, *Computation of Viscous Incompressible Flows*, Scientific Computation, DOI 10.1007/978-94-007-0193-9, © US Government 2011

Future Possibilities and Challenges

There are many important non-aerospace problems we have not discussed where incompressible flow methods are needed, possibly with some additional capabilities. For example, modeling ocean and atmospheric flows, such as those found in climate and weather modeling, will require essentially incompressible flow methods with variable density. Water-cooled nuclear reactor problems can also benefit from incompressible flow methods, with multi-phase modeling included.

Some pacing challenges remain that must be overcome for the more expanded role incompressible flow methods can play in the coming years. These include solution procedures and physics prediction, as well as the human resources aspect and CFD validation.

Solution Procedures

Flow simulations for engineering, especially for space exploration applications, generally involve very complex geometries, flow physics, and flight envelopes requiring substantial computing resources. Specifically, resolving unsteady phenomena is becoming increasingly important in order to fully understand the fluid dynamics issues involved. A typical process of flow simulation, especially for high-fidelity unsteady flow, requires large amounts of both computing time and human involvement in problem set-up and data pre- and post-processing.

Further substantial reductions in overall turnaround time for three-dimensional unsteady flow simulations are required to enable unsteady CFD to become relevant for tomorrow's mission-critical decision-making. A portion of this speedup will come from continued enhancements in computer hardware; however, the remainder must be contributed by advances in grid generation procedures including solution-adaptive grid generation, flow solution algorithms including high-accuracy methods, and more efficient parallel implementations. Some aspects of these are discussed in this monograph. However, many other aspects of these advances—such as finite difference and/or finite volume vs. finite element methods; energy conserving schemes; various grid topologies and variable definitions; assorted iterative schemes; and programming aspects, such as framework—await further discussion.

Prediction of Physics

In order to push the limits of the operational boundary and try bold new ideas, more predictive capabilities will be needed in the future, especially for complicated flows involving transient phenomena, flow separation, tip vortex interactions, and cavitation.

Currently, the relative inaccuracy of physical models is one of the major bottlenecks. Advanced models for flow physics, such as turbulence and transition, chemical reaction, and cavitation physics, must be incorporated into modeling and simulation procedures for accurate prediction of fluid dynamic phenomena. In addition, other quantities such as thermal stresses and structural loads must be coupled to the fluid dynamics models to provide more realistic simulation capabilities in an inherently multi-disciplinary environment. These computations will not only require large computing resources, but massive data storage, as well as efficient data management and analysis technologies.

Perhaps one of the most critical issues remaining in physical modeling remains the turbulence model. These models, routinely used in production CFD codes, are based on equilibrium turbulence (for example, an eddy viscosity model). However, as the requirement increases to apply CFD to engineering problems involving complex flow physics, the correct prediction of transition, time-dependent turbulence, and non-equilibrium phenomena becomes necessary. For example, in aerospace vehicle and systems design, the most productive aspect of CFD applications has often been to predict the relative changes among several design variations.

Even for this trend analysis, advanced turbulence modeling is necessary for consistency of solutions. In high-accuracy turbulence modeling, for example, large eddy simulation could be applied combined with RANS modeling near the wall. For such applications, grid resolution must be high to match the scales of physical motion as well as smooth coupling of two different methods. In addition, the numerical dissipation associated with differencing schemes needs to be minimized so as not to numerically distort large eddy motion. For incompressible flow, kinetic energy conserving schemes can be considered instead of up-winding schemes (Arakawa, 1966; and see Grammeltvedt, 1969, for a review of difference schemes).

These are some of the pacing issues we need to resolve to further expand the simulation capabilities discussed in this monograph. To make these applications feasible, high-fidelity computations using supercomputing resources will play an indispensable role in CFD.

Computational Hemodynamics

One of the most exciting areas we have discussed surrounds the basic computational and physical issues related to simulating blood flow in humans and through mechanical devices. We hope to use the human circulatory system simulation to enhance the biomedical performance prediction capability for astronauts. The so-called "Digital Astronaut" model will predict an individual astronaut's performance for the purpose of developing countermeasures to mitigate risks involved in space flights—especially long-duration space missions. Computational challenges are still enormous, ranging from geometry definition to high-fidelity time-dependent blood flow analysis, preferably in real time. We have barely scratched the surface of the computational hemodynamics for space flight, and hope that "CFDers" continue this challenging and very rewarding research.

Human Resources and CFD Validation

The human resources aspect of applying CFD to mission computing must also be considered. While CFD has advanced remarkably, many challenging cases require experts to produce credible solutions and correct analyses. Computer science can automate a substantial portion of the simulation process, saving significant human time and effort required to obtain reliable solutions. However, a blind application of tools without understanding the capabilities and limitations of the methods involved could lead to catastrophic engineering results. As in many other science and engineering disciplines, CFD researchers and practitioners need to understand the physics and engineering systems being simulated. More rigorous processes and procedures must be developed to validate the CFD solutions and to provide engineering error estimation.

In short, for fruitful application of CFD to engineering problems, one needs to understand the simulation process being applied, and have a thorough understanding of the modeling involved, whether it is a ground-based experiment or a flight vehicle. Future experts must be cultivated in the application of CFD to think through the relevant flow physics and apply the appropriate software and tools to the engineering problem in order to succeed in meeting future challenges.

For Further Reading

Many books and papers related to material are presented in this monograph, but have not been explicitly cited in our discussion. Some of these are listed in the references for further reading. While this list is not exhaustive, we have used these materials during the course of our mission computing tasks and found them highly useful.

References

Abdallah, S.: Numerical solutions for the incompressible Navier-Stokes equations in primitive variables using non-staggered grid, Part I. J. Comp. Phys., 70, 182–192 (1987a)

Abdallah, S.: Numerical solutions for the incompressible Navier-Stokes equations in primitive variables using non-staggered grid, Part II. J. Comp. Phys., 70, 193–202 (1987b)

Aber, G. S., Akkerman, J. W., Bozeman, R. J., Saucier, D. R.: Development of the NASA/Baylor VAD. (1993) (available in NASA Technology 2003: The Fourth National Technology Transfer Conference, 1, 151–157)

Alpers, B. J., Berry, R. G., Paddison, R. M.: Anatomical studies of the circle of Willis in normal brain. Arch. Neurol. Psychiatry, 81, 409–418 (1959)

Arakawa, A.: Computational design for long-term numerical integration of the equations of fluid motion: two-dimensional incompressible flow. Part I. J. Comp. Phys., 1, No. 1, 119–143 (1966)

Arndt, R., Maines, B.: Viscous effects in tip vortex cavitation and nucleation. Proceedings of the 20th Symposium on Naval Hydrodynamics, Santa Barbara, CA (1994)

Armaly, B. F., Durst, F., Pereira, J. C. F., Schonung, B.: Experimental and theoretical investigation of backward facing step flow. J. Fluid Mech., 127, 473–496 (1983)

Baker, C. J.: The laminar horseshoe vortex. Part II. J. Fluid Mech., 95, 346–367 (1979)

Baldwin, B. S., Barth, T. J.: A one-equation turbulence transport model for high Reynolds number wall-bounded flows. AIAA Paper 91-0610 (1991)

Baldwin, B. S., Lomax, H.: Thin layer approximation and algebraic model for separated turbulent flows. AIAA Paper 78-257 (1978)

Baldwin, J. T., Tarbell, J. M., Deutsch, S., Geselowitz, D. B.: Mean flow velocity patterns within a ventricular assist device (VAD). Trans. Am. Soc. Artif. Intern. Organs, 35, 425–433 (1989)

Barth, T. J.: Analysis of implicit local linearization techniques for upwind and TVD algorithms. AIAA Paper 87-0595 (1987)

Beam, R. M., Warming, R. F.: An implicit factored scheme for the compressible Navier-Stokes equations. AIAA J., 16, 393–402 (1978)

Belie, G.: Flow visualization in the space shuttle's main engine. Cover Story, Mechanical Engineering, September 1985 issue (1985)

Benek, J. A., Buning, P. G., Steger, J. L.: A 3-D chimera grid embedding technique. AIAA Paper 85-1523 (1985)

Berger, S. A., Jou, L.-D.: Flow in stenotic vessels. Ann. Rev. Fluid Mech., 32, 347–382 (2000)

Bird, R. B., Armstrong, R. C., Hassagar, O.: Dynamics of Polymer Liquids, Vol. I, 2nd edn., Wiley, New York (1987)

Bradshaw, P.: Effect of curvature on turbulent flow. AGARD-AG-169 (1973)

Braza, M., Chassaing, P., Ha Minh, H.: Numerical study and physical analysis of the pressure and velocity fields in the near wake of a circular cylinder. J. Fluid Mech., 165, 79–130 (1986)

Briley, W. R., McDonald, H.: Solution of the multidimensional compressible Navier-Stokes equations by a generalized implicit method. J. Comp. Phys., 24, No. 4, 372–397 (1977)

Brown, D. L.: Accuracy of projection methods for the incompressible Navier-Stokes equations. In Numerical Simulations of Incompressible Flows, ed. by Hafez, M., World Scientific, Singapore (2002)

Burke, R. W.: Computation of turbulent incompressible wing-body junction flow. Proceedings of the 27th Aerospace Sciences Meeting, Reno, Nevada, January 9–12, AIAA Paper 89-0279 (1989)

Caretto, L. S., Gosman, A. D., Patankar, S. V., Spalding, D. B.: Two calculation procedures for steady three-dimensional flows with recirculation. Proceeding of the 3rd International Conference on Numerical Methods in Fluid Dynamics, Paris, France (1972)

Caro, C. G., Pedley, T. J., Schroter, R. C., Seed, W. A.: The Mechanics of the Circulation, Oxford University Press, Oxford, pp. 243–349 (1978)

Cebral, J. R., Lohner, R., Burgess, J.: Computer simulation of cerebral artery clipping: relevance to aneurysm neuro-surgery planning. Proceeding of the ECCOMAS, September 11–14, Barcelona, Spain (2000)

Chakravarthy, S. R., Anderson, D. A., Salas, M. D.: The split-coefficient matrix method for hyperbolic systems of gas dynamics. AIAA Paper 80-0268 (1980)

Chakravarthy, S. R., Osher, S.: A new class of high accuracy TVD schemes for hyperbolic conservation laws. AIAA Paper 85-0363 (1985)

Chang, J. L. C., Kwak, D.: On the method of pseudo compressibility for numerically solving incompressible flows. AIAA Paper 84-0252 (1984)

Chang, J. L. C., Kwak, D., Dao, S. C.: A three dimensional incompressible flow simulation method and its application to the Space Shuttle main engine – Part I, Laminar Flow. AIAA Paper 85-0175 (1985a)

Chang, J. L. C., Kwak, D., Dao, S. C., Rosen, R.: A three dimensional incompressible flow simulation method and its application to the Space Shuttle main engine – Part II, Turbulent Flow. AIAA Paper 85-1670 (1985b)

Chang, J. L. C., Kwak, D.: Numerical study of turbulent internal shear layer flow in an axi-symmetric U-duct. AIAA Paper 88-0596 (1988a)

Chang, J. L. C., Kwak, D., Rogers, S. E., Yang, R.-J.: Numerical simulation methods of incompressible flows and an application to the Space Shuttle main engine. Int. J. Numer. Meth. Fluids, 8, 1241–1268 (1988b)

Chen, Y. S., Sandborn, V. A.: Computational and experimental study of turbulent flows in 180-degree bends. AIAA Paper 86-1516 (1986)

Chen, Y. S., Shang, H. M., Chen, C. P.: Unified CFD algorithm with a pressure based method. Proceedings of the 6th International Symposium on CFD, September 4–8, Lake Tahoe, NV (1995)

Chien, S.: Biophysical behavior of red blood cells in suspensions. In The Red Blood Cell, Vol. II, ed. by Surgenor, D. M., Academic Press, New York pp. 1031–1133 (1975)

Choi, D., Merkle, C. L.: Application of time-iterative schemes to incompressible flow. AIAA J., 23, No. 10, 1518–1524 (1985)

Chorin, A. J.: A numerical method for solving incompressible viscous flow problems. J. Comp. Phys., 2, 12–26 (1967)

Chorin, A. J.: Numerical solution of Navier-Stokes equations. Math. Comput., 22, No. 104, 745–762 (1968)

Chow, J. S., Zilliac, G., Bradshaw, P.: Initial roll-up of a wingtip vortex. Proceedings of the Aircraft Wake Vortex Conference, Vol. II, Washington, DC, October 29–131 (1991)

Chung, T. J.: Computational Fluid Dynamics, Cambridge University Press, Cambridge (2002)

Collins, W. M., Dennis, S. C. R.: Flow past an impulsively started circular cylinder. J. Fluid Mech., 60, 105–127 (1973)

Coutanceau, M., Bouard, R.: Experimental determination of the main features of the viscous flow in the wake of a circular cylinder in uniform translation – Part II. Unsteady flow. J. Fluid Mech., 79, 257–272 (1977)

Dacles-Mariani, J. S., Rogers, S., Kwak, D., Zilliac, G., Chow, J.: A computational study of a wingtip vortex flowfield. Proceedings of the 24th Conference in Fluid Dynamics, AIAA Paper 93-3010 (1993)

Dacles-Mariani, J. S., Zilliac, G. G., Chow, J., Bradshaw, P.: Numerical/experimental study of a wingtip vortex in the near-field. AIAA J., 33, No. 9, 1561–1568 (1995a)

Dacles-Mariani, J., Kwak, D., Zilliac, G.: Incompressible Navier-Stokes simulation procedure for a wingtip vortex flow analysis. Proceedings of the 6th International Symposium on Computational Fluid Dynamics, Lake Tahoe, NV, Sept. 4–8 (1995b)

Dacles-Mariani, J., Kwak, D., Zilliac, G.: Accuracy assessment of a wingtip vortex flowfield in the near-field region. Proceedings of the AIAA 34th Aerospace Sciences Meeting and Exhibit, Reno, NV, Jan. 15–18 (1995c)

Deardorff, J. W.: The use of subgrid scale transport equations in a three-dimensional model of atmospheric turbulence. J. Fluid Eng., 95, 429–438 (1973)

Dennis, S. C. R., Ingham, D. B., Cook, R. N.: Finite-difference methods for calculating/steady incompressible flows. J. Comp. Phys., 33, 325–339 (1979)

Dickinson, S. C.: An experimental investigation of appendage-flat plate junction flow, Vol. I and II, DTNSRDC Reports 86/051, 86/052, David Taylor Research Center, Bethesda, MD (Dec. 1986)

Drikakis, D., Rider, W.: High-Resolution Methods for Incompressible and Low-Speed Flows, Springer, Berlin (2004)

Dwyer, H. S., Soliman, M., Hafez, M.: Time accurate solutions of the Navier-Stokes equations for reacting flows. Proceeding of the 10th International Conference on Numerical Methods in Fluid Dynamics, Beijing, China, Springer, Berlin, pp. 247–251 (1986)

Eckerle, W. A., Langston, L. S.: Horseshoe vortex around a cylinder. Proceedings of the ASME International Gas Turbine Conference, Dusseldorf, West Germany, June 8–12 (1986)

Eskinazi, S., Yeh, H.: An investigation on fully developed turbulent flows in a curved channel. J. Aero Sci., 23, 23–34 (1956) (Also, The Johns Hopkins University, Mechanical Engineering Department, Internal Flow Research Report I-20, August, 1954)

Fasel, H.: Investigation of the stability of boundary layers by a finite-difference model of the Navier-Stokes equations, Part II. J. Fluid Mech. 78, 355–383 (1972)

Ferrandez, A., David, T.: Computational models of blood flow in the circle of Willis. Comp. Methods Biomech. Biomed. Eng., 4, No. 1, 1–26 (2000)

Ferrandez, A., David, T., Brown, M. D.: Numerical models of auto-regulation and blood flow in the cerebral circulation. Comp. Methods Biomech. Biomed. Eng., 5, No. 1, 7–20 (2002)

Ferziger, J. H.: Incompressible turbulent flows. J. Comp. Phys., 69, 1–48 (1987)

Ferziger, J. H., Peric, M.: Computational Methods for Fluid Dynamics, 3rd edn., Springer, Berlin (2002)

Figriola, R. S., Mueller, T. J.: On the hemolytic and thrombogenic potential occluder prosthetic heart valves from in-vitro measurements. ASME J. Biomech. Eng., 103, 83–90 (1981)

Flores, J.: Convergence acceleration for a three-dimensional Euler/Navier-Stokes zonal approach. AIAA Paper 85-1495 (1985)

Formaggia, L., Gerbeau, J. F., Nobile, F., Quarteroni, A.: Numerical treatment of defective boundary conditions for the Navier-Stokes equations. SIAM J. Numer. Anal., 40, No. 1, 376–401 (2002)

Fox, D. G., Lilly, D. K.: Numerical simulation of turbulent flows. Rev. Geophys. Space Phys., 10, No. 1, 51 (1972)

Garcia, R., Jackson, E. D., Schutzenhofer, L. A.: A summary of the activities of the NASA/MSFC pump stage technology team. Proceeding of the 4th International Symposium on Transport Phenomena and Dynamics of Rotating Machinery, Honolulu, Hawaii, April 5–8 (1992)

Garcia, R., McConnaughey, P., Eastland, A.: Activities of MSFC pump stage technology team. AIAA Paper 92-3232 (1992)

Garcia, R., McConnaughey, P., Eastland, A.: Computational fluid dynamics analysis for the reduction of impeller discharge flow distortion. AIAA Paper 94-0749 (1994)

Gatski, T. B., Grosch, C. E., Rose, M. E.: A numerical study of the two-dimensional Navier-Stokes equations in vorticity-velocity variables. J. Comp. Phys., 48, 1–22 (1982)

Ghia, K. N., Hankey, W. L., Hodge, J. K.: Study of incompressible Navier-Stokes equations in primitive variables using implicit numerical technique. AIAA Paper 77-648 (1977)

Gijsen, F. J. H., van de Vosse, F. N., Janssen, J. D.: The influence of non-Newtonian properties of blood on the flow in large arteries: steady flow in a carotid bifurcation model. J. Biomech., 32, 601–608 (1999a)

Gijsen, F. J. H., Allanic, E., van de Vosse, F. N., Janssen, J. D.: The influence of non-Newtonian properties of blood on the flow in large arteries: unsteady flow in a 90-degree curved tube. J. Biomech., 32, 705–713 (1999b)

Giller, C. A., Bowman, G., Dyer, H., Mootz, L., Krippner, W.: Cerebral arterial diameters during changes in blood pressure and carbon dioxide during craniotomy. Neurosurgery, 32, No. 5, 737–741 (1993)

Gillis, J. C., Johnston, J. P.: Turbulent boundary-layer flow and structure on a convex wall and its redevelopment on a flat wall. J. Fluid Mech., 135, 123–153 (1983)

Glowinski, R., Juarez, H., Pan, T.-W.: On the numerical simulation of incompressible viscous fluid flow around moving rigid bodies of elliptical shapes. In Numerical Simulations of Incompressible Flows, ed. by Hafez, M., World Scientific, Singapore (2002)

Grammeltvedt, A.: A survey of finite-difference schemes for the primitive equations for a barotropic fluid. Mon. Wea. Rev., 97, No. 5, 384–404, May (1969)

Gray, H.: Anatomy of the Human Body, Chapter VI The Arteries, 20th edn., ed. by Lewis, W. H., Lea & Febiger, Philadelphia (1918); Bartleby.com, New York (2000)

Gresho, M. P., Sani, R. L.: On pressure boundary conditions for the incompressible Navier-Stokes equations. Int. J. Num. Methods Fluids, 7, 1111–1145 (1987)

Gresho, M. P., Sani, R. L.: Incompressible Flow and Finite Element Method, Wiley, New York (1998)

Guerra, J., Gustafsson, B.: A numerical method for incompressible and compressible flow problems with smooth solutions. J. Comp. Phys., 63, 377 (1986)

Gunzburger, M. D., Nicolades, R. A. (Eds.).: Incompressible Computational Fluid Dynamics Trends and Advances, Cambridge University Press, Cambridge (1993)

Gustafsson, B., Lotstedt, P., Goran, A.: A fourth order difference method for the incompressible Navier-Stokes equations. In Numerical Simulations of Incompressible Flows, ed. by Hafez, M., World Scientific, Singapore (2002)

Hafez, M.: On the incompressible limit of compressible fluid flow. In Computational Fluid Dynamics for the 21st Century, ed. by Hafez, M. M., Morinishi, K., Periaux, J. and Satofuka, N., Springer, 255–272 (2001)

Hafez, M. (Ed.).: Numerical Simulations of Incompressible Flows, World Scientific, Singapore (2002)

Hafez, M., Dacles, J., Soliman, M.: A velocity vorticity method for viscous incompressible flow calculations. Proceedings of the 11th International Conference on Numerical Methods in Fluid Dynamics, Williamsburg, Virginia, June 27–July 1 (1988)

Hafez, M., Oshima, K.: Computational Fluid Dynamics Review 1995, Wiley, New York (1995)

Hafez, M., Oshima, K.: Computational Fluid Dynamics Review 1998, World Scientific, Singapore (1998)

Hafez, M., Soliman, M.: Numerical solution of the incompressible Navier-Stokes equations in primitive variables on unstaggered grids. In Incompressible Computational Fluid Dynamics, ed. by Gunzburger, M. D. and Nicolaides, R. A., Cambridge University Press, Cambridge (1993)

Hah, C., Loellbach, J. M., Lee, Y.-T.: Generation and transport of tip clearance vortices in a high Reynolds number axial flow rotor. In Computational Fluid Dynamics in Aeropropulsion. Proceeding of the ASME International Mechanical Engineering Congress and Exposition, San Francisco, CA, Nov. 12–17, pp. 31–44 (1995)

Harlow, F. H., Welch, J. E.: Numerical calculation of time-dependent viscous incompressible flow with free surface. Phys. Fluids, 8, No. 12, 2182–2189 (1965)

Harten, A., Lax, P. D., Van Leer, B.: On upstream differencing and Godunov-type schemes for hyperbolic conservation laws. SIAM Rev., 25, No. 1, 35 (1983)

Hirsch, C.: Numerical Computation of Internal and External Flows, Wiley, New York (1988)

Housman, J.: Time-derivative preconditioning method for multicomponent flow. PhD thesis, University of California, Davis (2007)

Housman, J., Kiris, C., Hafez, M. Preconditioned methods for simulations of low speed compressible flows. Comp. Fluids, 38, 7, 1411–1423 (2009)

Humphrey, J. A. C., Taylor, A. M. K., Whitelaw, J. H.: Laminar flow in a square duct of strong curvature, Part III. J. Fluid Mech., 83, 509–527 (1977)

Hunter, P. J., Smaill, B. H., Nielsen, P. M. F., Le Grice, I. J.: A mathematical model of cardiac anatomy. In Computational Biology of the Heart, ed. by Panfilov, A. V. and Holden, A. V., Wiley, New York (1997)

Hussain, A. K. M. F., Reynolds, W. C.: Measurements in fully developed turbulent channel flow. ASME Transactions, Series I., J. Fluid Eng., 97, 568–578 (1975)

Idelsohn, S. R., Costa, L. E., Ponso, R.: A comparative computational study of blood flow through prosthetic heart valves using the finite element method. J. Fluid Dyn., 18, No. 2, 97–115 (1985)

Issa, R. I.: Solution of the implicitly discretized fluid flow equations by operator-splitting. J. Comp. Phys., 62, 40–65 (1985)

Jameson, A., Schmidt, W., Turkel, E.: Numerical solution of the Euler equations by finite volume methods using runge-kutta stepping scheme. AIAA Paper 81-1259 (1981)

Jameson, A., Yoon, S.: Multigrid solution of the Euler equations using implicit schemes. AIAA J., 24, 1737–1743 (1986)

Kaul, U., Kwak, D., Wagner, C.: A computational study of saddle point separation and horseshoe vortex system. AIAA Paper 85-182 (1985)

Khosla, K., Rubin, S. G.: Direct primitive variable solution techniques for incompressible flows. In Numerical Simulations of Incompressible Flows, ed. by Hafez, M., World Scientific, Singapore (2002)

Kiehm, P., Mitra, N. K., Fiebig, M.: Numerical investigation of two- and three-dimensional confined wakes behind a circular cylinder in a channel. AIAA Paper 86-0035 (1986)

Kim, C. S., Kiris, C., Kwak, D., David, T.: Numerical models of human circulatory system under altered gravity: brain circulation. AIAA paper 2004-1092 (2004)

Kim, C. S., Kiris, C., Kwak, D., David, T.: Numerical simulation of local blood flow in the carotid and cerebral arteries under altered gravity. J. Biomech. Eng. Trans. ASME, 128, 194–202 (2006)

Kiris, C., Chang, I., Kwak, D., Rogers, S. E.: Numerical simulation of the incompressible internal flow through a tilting disk valve. AIAA Paper 90-0682 (1990)

Kiris, C., Chang, L., Kwak, D., Rogers, S. E.: Incompressible Navier-Stokes computations of rotating flows. AIAA Paper 93-0678 (1993)

Kiris, C., Rogers, S. E., Kwak, D., Chang, I. D.: Computation of incompressible viscous flows through artificial heart devices with moving boundaries. In Fluid Dynamics in Biology, Proceeding of the AMS-IMS-SIAM Joint Research Conference, ed. by Cheer, A. Y. and van Dam, C. P., American Mathematical Society, Providence, RI, pp. 237–247 (1991)

Kiris, C., Rogers, S. E., Kwak, D., Lee, Y.-T.: Time-accurate incompressible Navier-Stokes computations with overlapped moving grids. ASME Fluids Engineering Division Summer Meeting, Lake Tahoe, NV, June 19–23 (1994a)

Kiris, C., Kwak, D.: Progress in incompressible Navier-Stokes computations for the analysis of propulsion flows. NASA CP 3282, II, Advanced Earth-to-Orbit Propulsion Technology (1994b)

Kiris, C., Kwak, D., Benkowski, R.: Incompressible Navier-Stokes calculations for the development of a ventricular assist device. Comp. Fluids, 27, Nos. 5–6, 709–719 (1998)

Kiris, C., Kwak, D.: Parallel unsteady turbopump flow simulations for liquid rocket engines. Proceedings of the ACM/IEEE Conference, article No. 36 (2000)

Kiris, C., Kwak, D.: Numerical solution of incompressible Navier-Stokes equations using a fractional-step approach. Comp. Fluids, 30, 829–851 (2001) (Original version in AIAA Paper 96-2089)

Kiris, C., Kwak, D.: Aspects of unsteady incompressible flow simulations. Comp. Fluids, 31, 627–638 (2002)

Kiris, C., Kwak, D., Chan, W., Housman, J. A.: High-fidelity simulations of unsteady flow through turbopumps and flowliner. Comp. Fluids, 37, 536–546 (2008)

Kiris, C., Kwak, D., Rogers, S.: Incompressible Navier-Stokes solvers in primitive variables and their applications to steady and unsteady flow simulations. In Numerical Simulations of Incompressible Flows, ed. by Hafez, M., World Scientific, Singapore (2002)

Kiris, C., Kwak, D., Rogers, S., Chang, I-D.: Computational approach for probing the flow through artificial heart devices. ASME J. Biomech. Eng., 119, 452–460 (1997)

Kreiss, H. O.: On difference approximations of the dissipative type for hyperbolic differential equation. Comm. Pure Appl. Math., 17, 335–353 (1964)

Kuwahara, K., Komurasaki, S., Bethancourt, A.: Incompressible flow simulation by using multidirectional finite difference scheme. In Numerical Simulations of Incompressible Flows, ed. by Hafez, M., World Scientific, Singapore (2002)

Kwak, D.: Computation of viscous incompressible flows. von Karman Institute for Fluid Dynamics, Lecture Series 1989–04 (1989) (Also NASA TM 101090, March 1989)

Kwak, D., Chang, J. L. C., Shanks, S. P., Chakravarthy, S.: A three-dimensional incompressible Navier-Stokes flow solver using primitive variables. AIAA J., 24, No. 3, 390–396 (1986) (Original version: AIAA Paper 84-0253, AIAA 22nd Aerospace Sciences Meeting, Reno, Nevada, Jan. 9–12 (1984)

Kwak, D., Chang, J. L. C., Chang, Rogers, S. E., Rosenfeld, M.: Potential applications of computational fluid dynamics to biofluid analysis. International Symposium on Biofluid Mechanics, Palm Springs, CA, April 27–29 (1988)

Kwak, D., Reynolds, W. C., Ferziger, J. H.: Three-dimensional time dependent computation of turbulent flow. TF-5, Department of Mechanical Engineering, Stanford University (1975)

Leonard, A.: On the energy cascade in large-eddy simulation of turbulent fluid flows. TF-1, Department of Mechanical Engineering, Stanford University, or Adv. Geophys., 1, No. 18A, 237 (1973)

Leonard, A.: Computing three-dimensional incompressible flows with vortex elements. Ann. Rev. Fluid Mech., 17, 523–559 (1985)

Lin, S.-J., Yang, R.-J., Chang, J. L. C., Kwak, D.: Numerical simulation of flow path in the oxidizer side hot gas manifold of the Space Shuttle main engine. AIAA Paper 87-1800 (1987)

Lomax, H., Pulliam, T. H., Zingg, D. W.: Fundamentals of Computational Fluid Dynamics, Springer, Berlin (2001)

Loner, R., Yang, C., Cebral, J., Soto, O., Camelli, F.: On incompressible flow solvers. In Numerical Simulations of Incompressible Flows, ed. by Hafez, M., World Scientific, Singapore (2002)

MacCormack, R. W.: Current status of numerical solutions of the Navier-Stokes equations. AIAA Paper 85-0032 (1985)

Marchuk, G. M.: Methods of Numerical Mathematics, Springer, Berlin (1975)

McConnaughey, P., Cornelison, J., Barker, L.: The prediction of secondary flow in curved ducts of square cross-section. AIAA Paper 89-0276 (1989)

McCracken, M. F., Peskin, C. S.: A vortex method for blood flow through heart valve. J. Comp. Phys., 35, 183–205 (1980)

Merkle, C. L.: Preconditioning methods for viscous flow calculations. In Computational Fluid Dynamics Review 1995, ed. by Hafez, M. and Oshima, K., Wiley, New York (1995)

Merkle, C. L., Athavale, M.: Time-accurate unsteady incompressible flow algorithms based on artificial compressibility. AIAA Paper 87-1137 (1987)

Merkle, C. L., Tsai, P. Y. L.: Application of Runge-Kutta schemes to incompressible flows. AIAA Paper 86-0553 (1986)

Merrill, E. W.: Rheology of blood. Physiol. Rev., 49, 863–888 (1969)

Milnor, W. R.: Hemodynamics, 2nd edn., The Williams & Wilkins Co., Baltimore (1989)

Monson, D. J., Seegmiller, H. L., McConnaughey, P. K.: Comparison of LDV measurements and Navier-Stokes solutions in a two-dimensional 180-degree turn-around duct. AIAA Paper 89-0275 (1989)

Monson, D. J., Seegmiller, H. L.: An experimental investigation of subsonic flow in a two-dimensional U-duct. NASA TM 103931, July (1992)

Morgan, K., Harlan, D., Hassan, O., Sorensen, K., Weatherill, N.: Steady incompressible inviscid and viscous flow simulation using unstructured tetrahedral meshes. In Numerical Simulations of Incompressible Flows, ed. by Hafez, M., World Scientific, Singapore (2002)

Morkovin, M. V.: Flow around circular cylinder – a kaleidoscope of challenging fluid phenomena. In Symposium on Fully Separated Flows, ed. by Hansen, A. G., ASME, New York, pp. 102–118 (1964)

Moser, R. D., Moin, P.: Direct numerical simulation of curved turbulent channel flow. NASA TM 85974 (1984)

Nichols, W. W., O'Rourke, M. F.: McDonald's Blood Flow in Arteries: Theoretical, Experimental and Clinical Principles, 4th edn., Arnold, London (1998)

Nikfetrat, K., Hafez, M.: Numerical solution of the incompressible Navier-Stokes equations using Helmholz velocity decomposition. In Numerical Simulations of Incompressible Flows, ed. by Hafez, M., World Scientific, Singapore (2002)

Nikuradse, J.: Laws of turbulent flow in smooth pipes (1932) (English Translation) NASA TT F-10 (1966)

Olufsen, M. S., Peskin, C. S., Kim, W. Y., Pedersen, E. M.: Numerical simulation and experimental validation of blood flow in arteries with structure-tree outflow conditions. Ann. Biomed. Eng., 28, 1281–1299 (2000)

Orszag, S. A., Israeli, M.: Numerical simulation of viscous incompressible flows. Ann. Rev. Fluid Mech., 6, 281–318 (1974)

Orszag, S. A., Israeli, M., Deville, M. O.: Boundary conditions for incompressible flows. J. Sci. Comput., 1, 75–111 (1986)

Osswald, G., Ghia, K. N., Ghia, U.: Direct algorithm for solution of incompressible three-dimensional unsteady Navier-Stokes equations. AIAA Paper 87-1139 (1987)

Park, D. K.: The biofluidmechanics of arterial stenoses. M.Sc. Thesis, Lehigh University, Bethlehem, PA (1989)

Patankar, S. V.: Numerical Heat Transfer and Fluid Flow, Hemisphere Publishing Co., New York (1980)

Patankar, S. V., Spalding, D. B.: A calculation procedure for heat, mass and momentum transfer in three-dimensional parabolic flows. Int. J. Heat Mass Transf., 15, 1787–1806 (1972)

Patankar, S. V., Ivanovic, M., Sparrow, E. M.: Analysis of turbulent flow and heat transfer in internally finned tubes and annuli. Int. J. Heat Mass Transf., 101, 9929–9937 (1979)

Peake, D. J., Tobak, M.: Three-dimensional interactions and vortical flows with emphasis on high speeds. NASA TM 81169, March (1980)

Pedley, T. J., Stephanoff, K. D.: Flow along a channel with a time-dependent indentation in one wall: the generation of vorticity waves. J. Fluid Mech., 160, 337–367 (1985)

Perktold, K., Resch, M., Florian, H.: Pulsatile non-Newtonian flow characteristics in a three-dimensional human carotid bifurcation model. ASME J. Biomech. Eng., 113, 463–475 (1991)

Perktold, K., Rappitsch, G.: Computer simulation of local flow and vessel mechanics in a compliant carotid artery bifurcation model. J. Biomech., 28, No. 7, 845–856 (1995)

Peskin, C. S.: The fluid dynamics of heart valves: experimental, theoretical and computational methods. Annu. Rev. Fluid Mech., 14, 235–259 (1982)

Peskin, C. S., McQueen, D. M.: Modeling prosthetic heart valves for numerical analysis of blood flow in heart. J. Comp. Phys., 37, 113–132 (1980)

Peskin, C. S., McQueen, D. M.: A three-dimensional computational method for the blood flow in heart. J Comput. Phys., 81, 372–405 (1989)

Peyret, R., Taylor, T. D.: Computational Methods for Fluid Flow. Springer Series in Computational Physics, Springer, Berlin (1983)

Prandtl, L.: Effects of stabilizing forces on turbulence. NACA TM 625 (1931) (original version in 1929)

Pulliam, T. H.: Artificial dissipation models for the Euler equations. AIAA J., 24, 1931–1940 (1986)

Pulliam, T. H., Chaussee, D. S.: A diagonal form of an implicit approximate-factorization algorithm. J. Comput. Phys., 39, 347–363 (1981)

Quarteroni, A.: Modeling the cardiovascular system-a mathematical venture: Part I. SIAM News, 34, No. 5, June 10 (2001)

Quartapelle, L.: Numerical Solution of the Incompressible Navier-Stokes Equations, Birkhauser, Basel (1993)

Quarteroni, A., Tuveri, M., Veneziani, A.: Computational vascular fluid dynamics: problems, models and methods. Comput. Vis. Sci., 2, No. 4, 163–197 (2000)

Rai, M. M.: Navier-Stokes simulations of blade-vortex interaction using high-order accurate upwind schemes. AIAA Paper 87-0543 (1987)

Raithby, G. D., Schneider, G. E.: Numerical solution of problems in incompressible fluid flow: treatment of the velocity-pressure coupling. Numerical Heat Transfer, 2, 417–440 (1979)

Rapposelli, E., Cervone, A., Bramanti, C., d'Agostino, L.: Thermal cavitation experiments on a NACA 0015 hydrofoil. Proceeding of the FEDSM'03 4th ASME/JSME Joint Fluids Engineering Conference, Honolulu, Hawaii, July (2003)

Reneman, R. S., Merode, T., van Hick, K., Muijtjens, A. M. M., Hoeks, A. P. G.: Age-related changes in carotid artery wall properties in man. Ultrasound Med. Biol., 12, 465–471 (1986)

Reynolds, W. C.: Computation of turbulent flows. Ann. Rev. Fluid Mech., 8, 183–208 (1976)

Roach, P. J.: Fundamentals of Computational Fluid Dynamics, Hermosa Publishers, Albuquerque, New Mexico (1972) (revised in 1998)

Roe, P. L.: Approximate Riemann solvers, parameter vectors, and difference schemes. J. Comput. Phys., 43, 357 (1981)

Rogers, S. E., Chang, J. L. C., Kwak, D.: A diagonal algorithm for the method of pseudocompressibility. J. Comput. Phys., 73, No. 2, 364–379 (1987)

Rogers, S. E., Kwak, D., Kaul, U.: On the accuracy of the pseudocompressibility method in solving the incompressible Navier-Stokes equations. AIAA Paper 85-1689 (1985)

Rogers, S. E., Kwak, D.: Numerical solution of the incompressible Navier-Stokes equations for steady and time-dependent problems. AIAA Paper 89-0463 (1989)

Rogers, S. E., Kwak, D.: An upwind differencing scheme for the time-accurate incompressible Navier-Stokes equations. AIAA J., 28, No. 2, 253–262 (1990) (Also, AIAA Paper 88-2583, 1988)

Rogers, S. E., Kwak, D., Kiris, C.: Steady and unsteady solutions of the incompressible Navier-Stokes equations. AIAA J., 29, No. 4, 603–610 (1991a)

Rogers, S. E., Kwak, D.: An upwind differencing scheme for the steady-state incompressible Navier-Stokes equations. J. Appl. Numer. Math., 8, 43–64 (1991b)

Rosenfeld, M., Kwak, D.: Time-dependent solutions of viscous incompressible flows in moving coordinates. Int. J. Numer. Methods Fluids, 13, 1311–1328 (1991a) (Also, AIAA Paper89-0466, 1989)

Rosenfeld, M., Kwak, D., Vinokur, M.: A fractional-step method for unsteady incompressible Navier-Stokes equations in generalized coordinate systems. J. Comput. Phys., 94, No. 1, 102–137 (1991b) (Also, AIAA Paper 88-0718, 1988)

Rosenfeld, M., Kwak, D., Vinokur, M.: Development of a fractional-step method for unsteady incompressible Navier-Stokes equations in generalized coordinate systems. NASA TM 103912, November (1992)

Rosenfeld, M., Stephanoff, K. D., Park, D., Kwak, D.: A numerical and experimental simulation of pulsatile flow in a constricted channel. Proceedings of the 4th International Symposium on Computational Fluid Dynamics, UC Davis, CA, September 9–12 (1991b)

Rudy, D. M., Strikwerda, J. C.: A nonreflecting outflow boundary condition for subsonic Navier-Stokes calculations. J. Comput. Phys., 36, 55–70 (1980)

Salvetti, M.-V., Beux, F.: Liquid flow around non-cavitating and cavitating NACA 0015 hydrofoil. Mathematical and Numerical Aspects of Low Mach Number Flows, Porquerolles, France, Workshop Problem (2004)

Sandborn, V. A.: Measurement of turbulent flow quantities in a rectangular duct with 180-degree bend. NASA CP 3012, Advanced Earth-to-Orbit Propulsion Technology,

II, 292–304, Proceedings of the Conference at NASA Marshall Space Flight Center, May (1988)

Satofuka, N., Ishikura, M., Ishikawa, Y.: Use of lattice Boltzman method for computing incompressible flows. In Numerical Simulations of Incompressible Flows, ed. by Hafez, M., World Scientific, Singapore (2002)

Sharma, L., Ostermier, B., Nguyen, L., Dang, P., O'Connor, G.: Turbulence measurements in an axisymmetric turnaround duct air flow model. Rocketdyne Division, Rockwell International, Report RSS-8763, ATU-87-5237, October (1987)

Smagorinsky, J.: General circulation experiments with the primitive equations. Mon. Wea. Rev., 93, No. 3, 99 (1963)

Smith, N. P., Pullan, A. J., Hunter, P. J.: An anatomically based model of transient coronary blood flow, in the heart. SIAM J. Appl. Math., 62, No. 3, 990–1018 (2002)

Spalart, P. R.: Detached eddy simulation. Annu. Rev. Fluid Mech., 41, 181–202 (2009)

Spalart, P. R., Allmaras, S. R.: A one-equation turbulence model for aerodynamic flows. AIAA Paper 92-0439 (1992)

Spalart, P. R., Jou, W.-H., Strelets, M., Allmaras, S. R.: Comments on the feasibility of LES for wings, and on a hybrid RANS/LES approach. First AFOSR International Conference on DNS/LES, August 4–8 (1997)

Steger, J. L., Kutler, P.: Implicit finite-difference procedures for the computation of vortex wakes. AIAA J., 15, No. 4, 581–590 (1977)

Steinman, D. A., Ethier, C. R.: Effect of wall distensibility on flow in a two-dimensional end-to-side anastomosis. ASME J. Biomech. Eng., 116, 294–301 (1994)

Steinman, D. A., Thomas, J. B., Ladak, H. M., Milnor, J. S., Rutt, B. K., Spence, J. D.: Reconstruction of carotid bifurcation hemodynamics and wall thickness using computational fluid dynamics and MRI. MAGMA., 47, No. 1, 149–159 (2002)

Stephanoff, K. D., Pedley, T. J., Lawrence, C. J., Secomb, T. W.: Fluid flow along a channel with an asymmetric oscillating constriction. Nature, 305, 692–695 (1983)

Stevensen, D. M., Yoganathan, A. P., Williams, F. P.: Numerical simulation of steady turbulent flow through trileaflet aortic heart valves-II. Results on five models. J. Biomech., 16, No. 12, 909–926 (1985)

Taft, J. R.: Performance of the OVERFLOE-MLP and LAURA-MLP CFD codes and the NASA Ames 512 CPU Origin systems. HPCC/CAS 2000 Workshop, NASA Ames Research Center (2000)

Taneda, S., Honji, H.: Unsteady flow past a flat plate normal to the direction of motion. J. Phys. Soc. Japan, 30, 262–273 (1971)

Tannehill, J. C., Anderson, D. A., Pletcher, R. H.: Computational Fluid Mechanics and Heat Transfer, 2nd edn., Taylor & Francis, Washington, DC (1997)

Tarbell, J. M., Gunshinan, J. P., Geselowitz, D. B., Rosenburg, G., Shung, K. K., Pierce, W. S.: Pulse ultrasonic doppler velocity measurements inside a left ventricular assist device. J. Biomech. Eng. Trans. ASME, 108, 232–238 (1986)

Taylor, A. M. K. P., Whitelaw, J. H., Yianneskis, M.: Measurements of laminar and turbulent flow in a curved duct with thin inlet boundary layers. NASA CR 3367, January (1981)

Taylor, A. M. K. P., Whitelaw, J. H., Yianneskis, M.: Curved ducts with strong secondary motion: velocity measurements of developing of laminar and turbulent flow. J. Fluid Eng., 104, 350–359 (1982)

Taylor, C. A., Draney, M. T., Ku, J. P., Parker, D., Steele, B. N., Wang, K., Zarins, C. K.: Predictive medicine: computational techniques in therapeutic decision-making. Comput. Aided Surg., 4, No. 5, 231–247 (1999)

Temam, R.: Navier Stokes Equations, Revised edn., North Holland, Amsterdam (1979)

Tezduyar, T. E.: Stabilized finite element formulations and interface-tracking of interface-capturing techniques for incompressible flows. In Numerical Simulations of Incompressible Flows, ed. by Hafez, M., World Scientific Singapore (2002)

Thom, A.: The flow past circular cylinder at low speeds. Proc. R. Soc. Lond. B. Biol. Sci., Series A, 141, 651–666 (1933)

Thomas, A.: Laminar juncture flow-A visualization study. Phys. Fluids Lett., February (1987)

Thurston, G. B.: Rheological parameters for the viscosity, viscoelasticity and thixotropy of blood. Biorheology, 16, 149–162 (1979)

Turek, S.: Efficient Solvers for Incompressible Flow Problems, Springer, Berlin (1999)

Turkel, E.: Symmetrization of the fluid dynamic matrices with applications. Math. Comput., 27, 729–736 (1973)

Van Driest, E. R.: On turbulent flow near a wall. J. Aeronautical Sci., 23, No. 11, 1007–1011, 1036 (1956)

Van Dyke, M.: An Album of Fluid Motion, The Parabolic Press, Stanford, CA (1982)

Venkataswaran, S., Merkle, C. L.: Analysis of preconditioning methods for the Euler and Navier-Stokes equations. von Karman Institute Lecture Series, March (1999)

Venkataswaran, S., Merkle, C. L.: Evolution of artificial compressibility methods in CFD. In Numerical Simulations of Incompressible Flows, ed. by Hafez, M., World Scientific, Singapore (2002)

Vinokur, M.: An analysis of finite-difference and finite-volume formulations of conservation laws. NASA CR--177416, June (1986)

Warming, R. F., Beam, R. M., Hyett, B. J.: Diagonalization and simultaneous symmetrization of the gas-dynamic matrices. Math. Comput., 29, 1037–1045 (1975)

Wattendorf, F. L.: A study of the effect of curvature on fully developed turbulent flow. Proc. Roy. Soc., 148, 565–598 (1935)

Wendl, M. C., Agarawal, R. K.: Mass conservation and accuracy of non-staggered grid incompressible flow schemes. In Numerical Simulations of Incompressible Flows, ed. by Hafez, M., World Scientific, Singapore (2002)

Wesseling, P., Segal, A., van Kan, J. J. I. M., Costerlee, C. W., Kassels, C. G. M.: Finite volume discretization of the incompressible Navier-Stokes equations in general coordinates on staggered grids. Comp. Fluid Dyn. J., 1, No. 1 27–33 (1992)

White, F. M.: Viscous Fluid Flow, McGraw-Hill, New York, p. 123 (1974)

White, R. J., Averner, M.: Humans in space. Nature, 409, No. 6823, 1115–1118 (2001)

Yanenko, N. N.: The Method of Fractional Steps, Springer, Berlin (1971)

Yang, R.-J., Chang, J. L. C., Kwak, D.: A Navier-Stokes simulation of the Space Shuttle main engine hot gas manifold. AIAA Paper 87-0368 (1987) (Also J. Spacecraft Rockets, 29, No. 2, 253–259, 1992)

Yee, H. C.: Linearized form of implicit TVD schemes for the multidimensional Euler and Navier-Stokes equations. Comp. Math. Appl., 12A, Nos. 4/5, 413–432 (1986)

Yoganathan, A. P., Concoran, W. H., Harrison, E. C.: In vitro velocity measurements in the vicinity of aortic prostheses. J. Biomech, 12, 135–152 (1979a)

Yoganathan, A. P., Concoran, W. H., Harrison, E. C.: Pressure drops across prosthetic aortic heart valves under steady and pulsatile flow. J. Biomech, 12, 153–164 (1979b)

Yoon, S., Jameson, A.: An LU-SSOR scheme for the Euler and Navier-Stokes equations. AIAA Paper 87-0600 (1987)

Yoon, S., Kwak, D.: Artificial dissipation models for hypersonic external flow. AIAA Paper 88-3708 (1988)

Yoon, S., Kwak, D.: LU-SGS implicit algorithm for three-dimensional incompressible Navier-Stokes equations with source term. AIAA Paper 89-1964 (1989)

Yoshida, Y., Nomura, T.: A transient solution method for the finite element incompressible Navier-Stokes equations. Int. J. Numer. Meth. Fluids, 5, 873–890 (1985)

Zhao, S. Z., Ariff, B., Long, Q., Hughes, A. D., Thom, S. A., Stanton, A. V., Xu, X. Y.: Interindividual variations in wall shear stress and mechanical stress distributions at the carotid artery bifurcation of healthy humans. J. Biomech., 35, 1367–1377 (2002)

Zilliac, G. G., Chow, J. S., Dacles-Mariani, J., Bradshaw, P.: Turbulent structure of a wingtip vortex in the near field. AIAA Paper 93-3011 (1993)

Further Reading

Ahn, H. T., Shashkov, M.: Multi-material interface reconstruction on generalized polyhedral meshes. J. Comp. Phys. 226, 2096–2132 (2007)

Anderson, H. I., Billdal, J. T., Eliasson, P., Rizzi, A.: Staggered and non-staggered finite-volume methods for nonsteady viscous flows. Proceedings of the 12th International Conference on Numerical Methods in Fluid Dynamics, Oxford, England, 172–176 (1990)

Baker, A. J.: Finite Element Computational Fluid Mechanics, McGraw-Hill, New York (1983)

Bell, J. B., Colella, P., Howell, L.: An efficient second order projection method for viscous incompressible flow. In Proceeding of the AIAA 10th CFD Conference, ed. by Kwak, D., AIAA Paper 91-1560 (1991)

Belov, A., Martinelli, L., Jameson, A.: Three-dimensional computations of time-dependent incompressible flows with and implicit multi-grid driven algorithm on parallel computers. Proceedings of the 15th Interantional Conference on Numerical Methods in Fluid Dynamics, Monterey, CA, June 24–28, pp. 430–437 (1996)

Botella, O., Shariff, K.: Solving the Navier-Stokes equations using B-spline numerical methods. In Numerical Simulations of Incompressible Flows, ed. by Hafez, M., World Scientific, Singapore (2002)

Briley, W. R., McDonald, H., Shamroth, S. J.: A low mach number Euler formulation and application to time-iterative LBI schemes. AIAA J., 21, No. 24, 1467–1469 (1983)

Cebeci, T.: Convective Heat Transfer, Horizon Publishing, Long Beach, CA/Springer, Heidelberg, 1st edn., 1884, 2nd rev. edn., 2002

Chakravarthy, S. R.: Inviscid analysis of dual-throat nozzle flow. AIAA Paper 81-1201 (1981)

Chakravarthy, S. R.: Euler equations-Implicit schemes and implicit boundary conditions. AIAA Paper 82-0228 (1982)

Chakravarthy, S. R., Osher, S.: High resolution applications of the Osher upwind scheme for the Euler equations. AIAA Paper 83–1943 (1983)

Chang, K.-S., Sa, J.-Y.: The effect of buoyancy on vortex shedding in the near wake of a circular cylinder. J. Fluid Mech., 220, 253–266 (1990)

Choi, H., Hinze, M., Kunisch, K.: Instantaneous control of backward-facing step flows. App. Num. Math., 31, 133–158 (1999)

Choi, Y.-H., Merkle, C. L.: The application of preconditioning in viscous flows. J. Comp. Phys., 105, 207–223 (1993)

Crocco, L.: A suggestion for the numerical solution of the steady Navier-Stokes equation. AIAA J., 3, No. 10, 1824–1832 (1965)

Durbin, P. A., Pettersson Reif, B. A.: Statistical Theory and Modeling for Turbulent Flows, Wiley, Chischester (2001)

Edwards, J. R.: Toward unified CFD simulation of read fluid flows. Proceedings of the 15th AIAA Computational Fluid Dynamics Conference, AIAA Paper 2001-2524 (2001)

Fletcher, C. A. J.: Computational Techniques for Fluid Dynamics, Vols. I & II, Springer, Berlin (1988)

Gresho, P. M.: Incompressible fluid dynamics: some fundamental formulation issues. Annu. Rev. Fluid Mech., 23, 413–453 (1991)

Guillard, H., Viozat. C.: On the behavior of upwind schemes in the low Mach number limit. Comput. Fluids, 28, 63–86 (1999)

Guillard, H., Murrone, A.: A five equation reduced model for compressible two-phase flow problems. INRIA Technical Report 4778 (2003)

Guillard, H., Murrone, A.: Behavior of upwind schemes in the low Mach number limit: III. INRIA Technical Report 5342 (2004)

Guyton, A. C., Hall, J. E.: Medical Physiology, 10th edn., W. B. Saunders Company, Philadelphia, CA (2000)

Harlow, F. F., Amsden, A. A.: A simplified MAC technique for incompressible fluid flow calculations. J. Comp. Phys., 6, 322–325 (1970)

Henshaw, W. D., Petersson, N. A.: A split-step scheme for the incompressible Navier-Stokes equations. In Numerical Simulations of Incompressible Flows, ed. by Hafez, M., World Scientific, Singapore (2002)

Hirt, C. W., Amsden, A. A., Cook, J. L.: An arbitrary Lagrangian-Eulerian computing method for all flow speeds. J. Comp. Phys., 14, 227–253 (1974)

Hughes, T. J. R., Liu, W. K., Brooks, A.: Finite element analysis of incompressible viscous flows by the penalty function formulation. J. Comp. Phys., 30, 1–60 (1979)

Hughes, T. J. R., Mazzei, L., Jansen, K. E.: Scale-variand and turbulence models for large-eddy simulation. Comput. Vis. Sci., 3, 47–59 (2000)

Hussaini, M. Y., Kumas, A., Salas, M. D.: eds., Algorithm Trends in Computational Fluid Dynamics, Springer, Berlin (1993)

Jenkins, D. R.: Space Shuttle – The History of Developing the National Space Transportation System, The Beginning Through STS-75, Walsworth Publishing Company, Marceline, Missouri (1996)

Karni, S.: Multicomponent flow calculations by a consistent primitive algorithm. J. Comp. Phys., 112, 31–43 (1994)

Karni, S.: Hybrid multifluid algorithms. SIAM J. Sci. Comput., 17, No. 5, 1019–1039 (1996)

Kim, J., Moin, P.: Application of a fractional-step method to incompressible Navier-Stokes equations. J. Comp. Phys., 59, 308–323 (1985)

Kim, W.-W., Menon, S.: An unsteady incompressible Navier-Stokes solver for large eddy simulation of turbulent flows. Int. J. Num. Meth. Fluids, 31, 983–1017 (1999)

Klainerman, S., Majda, A.: Singular limits of quasilinear hyperbolic systems with large parameters and the incompressible limit of compressible fluids. Commun. Pure App. Math., 34, 481–524 (1981)

Klainerman, S., Majda, A.: Compressible and incompressible fluids. Commun. Pure App. Math., 35, 629–653 (1982)

Kolmogorov, A. N.: Local structure of turbulence in an incompressible fluid at very high Reynolds numbers. Doklady Akad Nauk SSSR, 30, 299–303 (1941)

Kwak, D., Kiris, C., Dacles-Mariani, J., Rogers, S., Yoon, S.: Incompressible Navier-Stokes solvers using primitive variables. In Computational Fluid Dynamics Review, ed. by Hafez, M. and Oshima, K., World Scientific, Singapore (1998)

Kwak, D., Rogers, S. E., Kaul, U. K., Chang, J. L. C.: A numerical study of incompressible juncture flows. Proceedings of the 10th Interantional Conference on Numerical Methods in Fluid Dynamics, Beijing, Peoples Republic of China, June 23–27 (1986)

Lambert, B. K., Taylor, L. K., Briley, W. R.: Evaluation of a preconditioned flow solver for a broad range of Mach number and temperature ratio. In Numerical Simulations of Incompressible Flows, ed. by Hafez, M., World Scientific, Singapore (2002)

Lee, D.: Data compression for incompressible flow simulation. In Numerical Simulations of Incompressible Flows, ed. by Hafez, M., World Scientific, Singapore (2002)

Lee, D.-H., Hwang, J.: Incompressible simulations on the flowfield around a square cylinder at ground proximity. In Numerical Simulations of Incompressible Flows, ed. by Hafez, M., World Scientific, Singapore (2002)

Liou, M.-S., Edwards, J. R.: AUSM-family schemes for multiphase flows at all speeds. In Numerical Simulations of Incompressible Flows, ed. by Hafez, M., World Scientific, Singapore (2002)

Lohner, R., Yang, C. Y., Cebral, J., Soto, O., Cmelli, F.: On incompressible flow solvers. In Numerical Simulations of Incompressible Flows, ed. by Hafez, M., World Scientific, Singapore (2002)

Mansour, N. N: Numerical simulation of the tip vortex of a low-aspect ratio wing at transonic speed. AIAA J., 23, 1143–1149 (1985)

Merkle, C. L., Choi, Y. H.: Computation of low speed compressible flows with time-marching methods. Int. J. Num. Meth. Eng., 25, 292–311 (1985)

Minion, M. L.: Higher-order semi-implicit projection methods. In Numerical Simulations of Incompressible Flows, ed. by Hafez, M., World Scientific, Singapore (2002)

Moretti, G., Abbett, M.: A time-dependent computational method for blunt body flows. AIAA J., 4, No. 12, 2136–2141 (1966)

Moretti, G.: The lambda-scheme. Comput. Fluids, 7, No. 3, 191–205 (1979)

Neel, R. E., Godfrey, A. G., McGrory, W. D.: Low-speed, time-accurate validation of GASP version 4. AIAA Paper 2005-686 (2005)

Newell, D. W., Aaslid, R., Lam, A., Mayberg, T. S., Winn, H. R.: Comparison of flow and velocity during dynamic autoregulation testing in humans. Stroke, 25, No. 4, 793–797 (1994)

Pan, D., Chakravarthy: unified formulation for incompresible flows. AIAA Paper 89–0122 (1989)

Phillips, W. R. C., Graham, J. A. H.: Reynolds-stress measurement in a turbulent trailing vortex. J. Fluid Mech., 147, 353–371 (1984)

Quarteroni, A., Formaggia, L.: Mathematical modeling and numerical simulation of the cardiovascular system. In Handbook of Numerical Analysis, ed. by Ciarlet, P. G. and Ayache, N., Guest Editor, 12 edn., Elsevier, 3–127 (2004)

Rehm, R. G., Baum, H. R.: The equations of motion for thermally driven buoyant flows. J. Res. Natl. Bur. Stand., 83, 297–308 (1978)

Rhie, C. M., Chow, W. L.: A numerical study of the turbulent flow past an isolated airfoil with trailing edge separation. AIAA J., 21, 2525–1532 (1983)

Rizzi, A., Viviand, H.: Numerical methods for the computation of inviscid transonic flows with shock waves. In Notes on Numerical Fluid Mechanics, 3, Vieweg, Braunschweig (1981)

Rosenfeld, M., Kwak, D.: Multigrid acceleration of a fractional-step solver in generalized curvilinear coordinate systems. AIAA J., 31, No.10, 1792–1800 (1993)

Roshko, A.: On the development of turbulent wakes from vortex streets. NACA TN 2913 (1953)

Rudy, D. H., Strikwerda, J. C.: A nonreflecting outflow boundary condition for subsonic Navier-Stokes calculations. J. Comp. Phys, 36, 55–70 (1980)

Ryzhov, O.: Advances in hydrodynamics stability theory: the wave/vortex eigenmodes' interaction. In Numerical Simulations of Incompressible Flows, ed. by Hafez, M., World Scientific, Singapore (2002)

Sa, J.-Y., Kwak, D.: A numerical method for incompressible flow with heat transfer. NASA TM-110444 (1997)

Sagaut, P.: Large Eddy Simulation for Incompressible Flows, Springer, Berlin (2001)

Scheuermann, G., Kollman, W., Tricoche, X., Wischgoll, T.: Evolution of topology in axi-symmetric and 3-D viscous flows. In Numerical Simulations of Incompressible Flows, ed. by Hafez, M., World Scientific, Singapore (2002)

Schnerr, G. H.: Modeling and computation of Unsteady cavitating flows. In Numerical Simulations of Incompressible Flows, ed. by Hafez, M., World Scientific, Singapore (2002)

Shin, B. R.: Stable numerical method applying a total variation diminishing scheme for incompressible flow. AIAA J., 41, No. 1, 49–55 (2003)

Shyy, W., Thakur, S. S., Ouyang, H., Liu, J., Blosch, E.: Computational Techniques for Complex Transport Phenomena, Cambridge University Press, New York (1997)

Shyy, W., Tong, S. S., Corea, S. M.: Numerical recirculation flow calculation using a body-fitted coordinate system. Num. Heat Transfer, 8, 99–113 (1985)

Steger, J. L., Warming, R. F.: Flux vector splitting of the inviscid gasdynamic equations with application to finite difference methods. J. Comp. Phys., 40, No. 2, 263–293 (1981)

Srinivasan, G. R., McCroskey, W. J., Baeder, J. D., Edwards, T. A.: Numerical simulation of tip vortices of wings in subsonic and transonic flows. AIAA J., 26, No. 10, 1153–1162 (1988)

Turkel, E.: Preconditioning methods for solving incompressible and low-speed compressible equations. J. Comp. Phys., 72, 277–298 (1987)

Turkel, E.: Preconditioning techniques in computational fluid dynamics. Ann. Rev. Fluid Mech., 31, 385–416 (1999)

Visbal, M.: Viscous flow simulation using compact differencing and filtering schemes. In Numerical Simulations of Incompressible Flows, ed. by Hafez, M., World Scientific, Singapore (2002)

Viviand, H.: Pseudo-unsteady systems for steady inviscid calculations. In Numerical Methods for
the Euler Equations of Fluid Dynamics, ed. by Angrand, F., Plenum, New York, p. 334 (1985)
Wesseling, P.: Principles of Computational Fluid Dynamics, Springer, Berlin, Heidelberg (2000)
Wilcox, D. C.: Turbulence Modeling for CFD, DCW Industries, La Cañada, CA (1993)
Wiltberger, N. L., Rogers, S. E., Kwak, D.: A comparison of two incompressible Navier-Stokes
algorithms for unsteady internal flow. NASA TM 108794, November (1993)
Yabe, T., Ogata, Y., Kawai, T.: Simulation of structure-fluid interaction by universal solver CIP for
solid, liquid, and gas in Cartesian grid. In Numerical Simulations of Incompressible Flows, ed.
by Hafez, M., World Scientific, Singapore (2002)
Zhang, J.-B., Kuang, Z.-B.: Study on blood constitutive parameters in different blood constitutive
equations. J. Biomech., 33, 355–360 (2000)

Index